C000050302

Livestock's Longer Shadow

Other books by Tim Bailey:

Miraculum Naturae: Venus's Flytrap
Dionaea: The Venus's Flytrap
Carnivorous Plants of Britain and Ireland

Livestock's Longer Shadow

Hope Lives in Kindness

TIM BAILEY

WITH DR. ALICE BROUGH

THE CHOIR PRESS

First published in the United Kingdom in 2021 by
The Choir Pres

ISBN 978-1-78963-215-6

Disclaimer

Although the author has made every effort to ensure that the information in this book is correct at the time of publication, and while this publication is designed to provide accurate information in regard to the subject matter covered, the author assumes no responsibility for errors, inaccuracies, omissions, or any other inconsistencies herein, and hereby disclaims any liability to any party for any loss, damage or disruption caused by errors or omissions, whether such errors or omissions result from negligence, accident or any other cause.

Author's Note

Throughout this book I have provided a lot of information and used numerous statistics, some of which will change over time. Other relevant information and statistics may also come to light which I have not been aware of and may make a material difference to topics I discuss. Where important and significant information of *substance* becomes available, I will endeavour to provide updates on my websites www.livestockslongershadow.com / www.veganoscene.com or through other avenues available to me.

Facts are facts, and associations are associations (to varying degrees of strength and *truths*), but all opinions I make throughout my book are *personal* and *must not* be taken as representing the views of any other body or person. As a subject expert on agricultural pollution, and from studying a considerable amount of evidence in areas where I'm not an expert, I've drawn the following conclusions:

- How good a diet is for health and longevity rests on what food is available at any time, its qualities, and ultimately what you choose to eat, or in some instances what you are able to choose, from what is available. Wherever available, the optimal human diet choice for good health and longevity points overwhelmingly towards a predominantly wholesome whole-food, non-processed, plant-based diet. Cake is nice, but an apple is better for you. An egg is a better choice than a beef steak, but neither are as healthy as a bowl of lentils or beans, and so forth.
- A diet with a predominance of animal and/or processed/ ultra-processed food is harmful to human health.
- The adverse impact on the environment and our planet's health and biological diversity from eating animal-based foods is considerable and unsustainable – and will never be sustainable until the industry stops trying to fit sustainability into livestock farming rather than vice versa.
- We want to eat, but do not need to eat, animal-based foods – it is not an essential part of the human machinery, we have a choice and as omnivores we can take it or leave it.
- Farmed animals are sentient beings who we exploit, make suffer, endure injury and experience a death conducive to profits and how much we are willing to pay – *humane* and *high welfare*

treatment of livestock is a misnomer, is not science-based and is simply used to conceal a whole host of inconvenient truths to maintain business as usual.

- As we do not need to eat farmed animals, the cruelty we subject them to is out of choice and not out of an unfortunate necessity – just because farming and eating animals is customary practice it does not make it right.
- Animals do not give their lives willingly to us, we take them.
- The human race parades speciesism, advocating rights from pain, injury, suffering and abuse to some which we call companion animals or pets; whilst advocating pain, injury, suffering and abuse to others which we call livestock and kill for food. We do this regardless of any material difference between the animal species and without any necessity or obligation to eat the ones we call livestock in order to live and thrive.
- Removing farmed livestock from our food chain offers the greatest opportunity to regenerate/rewild our landscape, remove the industries' impact on rainforests and climate change, and to sequester carbon.
- Factory farms are unnatural places where we do particularly unkind and unnatural things to billions of animals for our self-fulfilment; they promote antibiotic resistance and are a cauldron for future pandemics.
- Science must conform to fact, but facts that do not favour the livestock industry, and those associated with the industry, are all too often ignored or subverted either deliberately or in ignorance of fact.
- Hope for our health, environment, flora and fauna and our children's future lives in kindness.

If after reading this book you wish to consider making changes to your lifestyle it is extremely important to ensure you get all the nutrition you need for a healthy diet. This is helped by eating as diverse a range of plants and plant colours as you can to meet your calorific intake, rather than concentrating on the latest food fads. Where needed, you should always seek help from an *appropriate* professional health expert, particularly if you have an underlying health condition or are taking prescribed medicines for a health condition. Be sure to take a B12 supplement regularly, iodine, vitamin D and an omega-3 supplement made from microalgae farmed oil as necessary. Don't be fooled that a diet including animal-based food provides all your needs without supplements, including enough B12. If you do believe a diet including animal-based foods gives you all your nutrition, look in your medical cabinet for vitamins and mineral supplements and if you have any, ask yourself why. One of the easiest ways to track your nutrition and compare what I have to say, whatever your diet, is to use Cronometer: https://cronometer.com/.

Throughout, my focus is on human health, the environment and the unkind way livestock are treated, which are of a much higher order of importance than the feelings of those cemented to the livestock industry and to the ways animals are farmed. Notwithstanding that, I understand more than many how we have come to farm livestock and produce animal-based foods in the ways we do, and in particular since World War II.

I also fully understand, and accept, complexity; that unless global action is taken, stopping meat, fish, dairy and egg production in the UK (and the European Union) alone at this time would inevitably cause some shift elsewhere, leading to an increase in greenhouse gas emissions, cruelty to animals and other harmful aspects associated with the industry. However, a serious beacon of light and leadership is needed, and I would call on the UK to do what is only right and sincere and lead the world in earnest towards a truly kind and green economy. We would certainly be a healthier United Kingdom for doing it. Not changing just because someone else may do it instead or worse, would ultimately be a lose–lose situation for our planet – not least as they will probably do it any way without being led towards a better way. To support farmers, we need to ensure what we stop in the UK is not simply imported into the UK from elsewhere and to an even lower standard. Farmers are our future, not our past. It will not be cheap, change would be substantial, hugely disrupting and expensive, but not as costly as the consequences of business as usual.

Over the years working in agriculture, I have enjoyed many open and frank discussions with farmers and others for whom I have the utmost respect for feeding us and for providing food security. Many good people are also working to reduce the livestock industry's impact on the environment, and in particular greenhouse gas and ammonia emissions. As the saying 'Rome wasn't built in a day' reminds us, reducing emissions whilst we work towards ending livestock farming will need to go hand in hand. I wish it could end overnight, and whilst it can for individuals, it will take much longer for society as a whole. I hope to continue the good relationship I have towards making our planet a better place for people and all life that lives alongside us, despite the differences I have with others, and the difficulties change would cause us to face. The problems we face require honesty and cannot be dealt with in silence, or without kindness and hope. Loving animals is not about farming and eating them, but nurturing and caring for them in natural environments.

Ultimately mine is just a single voice and I would urge you to explore this subject in more depth and from angles than I can possibly cover in this book. To help you with that, I have provided a list of resources I found helpful in taking my journey. This is your journey, and I wish you well wherever it takes you.

Contents

Dedicated to all those who are working
to safeguard the future of animals, our health
and our planet

Acknowledgements

I began researching and writing *Livestock's Longer Shadow* towards the back end of 2016, about a year after I suffered a stroke. Whilst I had a good idea how long it would take me to produce a piece of work worthy of publishing, my constant additions of detail to evidence the content, working through the fatigue hanging over from my illness, working at weekends, and a whole host of other distractions, led me to miss my original 2019 deadline. I'm so glad I didn't publish in 2019, as over the last two years I have met and got to know some truly amazing and inspirational people who have helped me (knowingly or not) produce a much better book as a consequence. Not least Alice Brough, who kindly added content to my book. Alice, you are a truly wonderful, caring and knowledgeable person, and I thank you with all my heart.

Having been an obsessive author in previous times, my family sighed deeply and once more stood steadfastly behind me. To my wife Natalie, and my four children, Eloise, Connor, William and Georgiana, I love you and I thank you with my heart of hearts for putting up with my constant ramblings and for not being around as much as I should have been for you all – again! A special mention goes to you Eloise, the inspiration behind my kinder lifestyle, along with a big thank you for your wonderful contribution on speciesism.

Two other people to whom I'm particularly grateful are Ed Winters (aka Earthling Ed) and Tati von Rheinbaben. Thank you both so much for your support, for the work you do, and for the friendship you have extended to me. I extend equal admiration and thanks to Dr. Shireen Kassam, Founder and Director of Plant-based Health Professionals and the co-founder of Plant Based Health Online. When I needed help to truth my medical work the extraordinary, and workaholic, Shireen stepped in. I must also make a special mention to film Producer Alex Lockwood. Alex came to prominence as a BAFTA winner for his film *73 Cows*, telling the story of Jay Wilde, the first farmer in the UK to trade beef farming for sustainable organic farming. In 2020 I met Alex in person when he interviewed me for the film *The End of Medicine*, which he was working on alongside producer Keegan Kuhn and exec producers Mara and Joaquin Phoenix. When you are interviewed with challenging questions, it really makes you think and the experience has had a profound influence on me as a person. Thank you, Alex.

I also extend a special thanks and gratitude to Sivalingam Vasanthakumar, for his most appreciated contribution; to Andrew Wasley, of the Bureau of Investigative Journalism, for help with referencing; and to Christine Hugh-Jones and Margaret Tregear of the Campaign for the Protection of Rural Wales, for help with data on poultry numbers in Powys, Herefordshire and Shropshire.

The base silhouette artwork for the book's cover was drawn by freelance illustrator Vanessa Wells (http://www.vanessa.withbits.com/), who I have had the absolute pleasure to work with several times over the years. Thank you, Vanessa, for all your fabulous artwork. I'm also indebted to my great friend Graham Jones, Inside Outside Design (https://ioutsidedesign.wixsite.com/e-learning), who also lent his considerable talent to the cover of *Livestock's Longer Shadow* and for the book's website: www.livestockslongershadow.com.

Finally, I could not have written my book without the invaluable words of the many accomplished authors and YouTubers whose work I have had the pleasure to read, listen to, learn and take my reference from. Thank you!

Introduction

Whether maliciously or, as is more often the case, due to ignorance, the mainstream of Western culture is hell-bent on ignoring, disbelieving, and, in some cases, actively twisting the truth about what we should be eating – so much so that it can be hard for us to believe that we've been lied to all these years. It's often easier to simply accept what we've been told, rather than consider the possibility of a conspiracy of control, silence, and misinformation.

T. Colin Campbell PhD, *Whole: Rethinking the Science of Nutrition* (2014)

What my book is about

Most people have little idea how eating animal-based foods harms our health, our planet and animal welfare. We accept without question, as a *fait accompli*, that the illnesses we suffer in later life are a consequence of getting old, which we can do little about. We are only too eager to listen and jeopardise our health on the opinions of unqualified people, provided they tell us what we prefer to hear. We want to believe the animals we eat do not suffer and are happy to give up their lives for the modicum of care we may show them, despite no animal choosing to be exploited, suffer and die. We are told farmers are the guardians of our land and the environment, yet the naturalness of our landscape is over-cultured and dysfunctional and our environment polluted by the livestock we choose to put in our mouths.

In the UK, as elsewhere in the world, mixed messages are rampant and spread confusion, with animal rights, environmental and healthcare lobbyists firmly pitted against their livestock industry counterparts over these and other issues.

This book is intended to cut through at least some of the noise and is suitable for anyone wanting to know more about how we treat our health, planet and animals through the way we farm and consume livestock products. The book looks in detail at the problems of livestock farming from a UK perspective, addressing the topic as a *whole*, as it should be. The UK is not unique and the many issues I describe about livestock farming are common around the world, and where they are often no better and sometimes a lot worse. In this respect my book is for all and not just those who live and eat animals in the UK.

My book will not be comfortable reading for those who want to, or have to, believe plant-based foods are less sustainable than meat, and that there are many, if not more, examples of unsustainable plant-based alternatives; and that saving the planet with livestock alternative foods is corruptive messaging. My book will also not be comfortable for those who similarly need to believe that the soggy UK is the world's best place to farm livestock *sustainably* without causing pollution; that osteoporosis is a dairy deficiency disease, or a lack of biodiversity a livestock deficiency disease; that animal saturated fats are a health food; that rotation of grass and livestock grazing, and the muck that they put back into the soil, is what sustains all biodiversity; and that the suffering and injury we subject livestock to is necessary and something we can be proud about.

Livestock farming undoubtedly contributes greatly to the ensuing climate crisis; to disastrous ecological impacts, including deforestation and biodiversity loss; to serious environmental damage, including the pollution of the water, soil and air; to severe violations of human and animal rights; and to antibiotic resistance and diseases, including acting as a potential Petri dish for zoonotic diseases. Furthermore, livestock farming can in no way be considered an improvement on nature or nature-based solutions. When we turn our back on Mother Nature, we lose.

If you have a natural, species-rich habitat, then remove all the vegetation, trees, shrubs and herbs, and the wildlife that existed with it, plough the land, plant it with a single species of grass and put a cow or a sheep on it, it does not improve the land's biodiversity. Quite the opposite. Equally, continually grazing cows and sheep on the land does not make up for the carbon lost and contribute to carbon-neutrality. Planting hedgerows and the odd tree, and dollops of manure, adds very little to what is sacrificed for livestock farming. Equate this to images of burning and land clearance in the Cerrado, Brazil, to make way for livestock grazing, or crops to feed livestock, against any claim that the livestock which follow are the cornerstone species for a functioning ecosystem. Yet, in the UK, as elsewhere in the world, we are told to believe livestock farming is vital to biodiversity and that without livestock farming what else would we do with this otherwise economically deficient, ineffectual, *unimproved* wasteland?

Since World War II agriculture has relied heavily on industrialisation to provide food security and to help keep farmers in profit, including the extensive use of chemicals and medicines. Vast amounts of public money have also been spent creating markets to use unnecessary animal-based foods. However well intended or misguided, this has resulted in increased environmental degradation and exploitation of animals. Today we are on a precipice and without unprecedented change our natural assets will continue to decline, perhaps beyond reach, and plunge farming into a financial crisis. We need nature-based farming solutions today, not tomorrow, to resolve our situation, ideally putting livestock farming into the history books, along with all animal unkindness that goes with it.

It is a fact that agriculture, dominated by a livestock industry which uses most of the UK's agricultural land, has contributed greatly to the UK being one of the most nature-depleted countries in the world. Not just what we no longer see on the surface, but within the equally biologically dysfunctional soil that sits beneath, which so much of life depends upon.

Unfortunately, the livestock industry's solution to UK people taking a responsible attitude to their health, the environment and animal welfare by reducing or stopping their consumption of animal-based foods is to fight against it, grow the business and to export it to other parts of the world instead. A cluster-muck of a model that will propel us closer towards planetary Armageddon, which unfortunately is being equally embraced by all the other developed countries, in a collective race to the bottom.

A better solution would be to embrace change and for the UK and farmers to invest in healthy plant-based foods instead and make the UK a world leader in food sustainability and security. For those farmers who are unable to make that change, to admirably reward them for rewilding grasslands and looking after them as a public good for biodiversity, pollution and flood prevention and genuine carbon sequestration. Change is not about leaving farmers behind; it is about investment in farming which makes the improvements needed for the sake of us all. It will take time to unpick our current traditions and habits but must be done while maintaining the livelihoods of farmers, rural communities and other businesses reliant on livestock farming. This includes consumers abstaining from even more abysmally treated, environmentally damaging and health-harmful animal-based products imported from countries in any future free-trade deals, like Australia and the USA, which would otherwise damage the substantial changes in land use and farming practices we need to see.

Plant-based food production is not without issues; there is no room to be disproportionately righteous, complacent and otherwise intelligent in that respect, but nevertheless it is not on the same scale as livestock farming, particularly when the livestock problem is looked at as a whole and not in silos.

Our future depends on dealing with these issues head-on, through changing the ways we produce and consume food. Each single danger is a concern in its own right but collectively as a whole is a crisis, and one which will only get worse if we do not challenge and transform our ways without delay and in unprecedented ways. The UK likes to believe we are somehow better than most other countries when it comes to livestock farming; while that has some merit, hitting something three times rather than four is not much of an achievement to shout about.

I started my own journey learning the truth about livestock farming at 7 a.m. on 2 February 2016, when I had a stroke. In reality I began that journey many years before, but up until that time, and for a few months after, chose silence over the inconvenient truths. I was fortunate the stroke was not severe enough to leave me with any obvious after-effects, and while it was a bad experience at the time it ultimately proved to be a stroke of luck.

I begin this book by recounting the story of my stroke of luck, and how it encouraged me to look at the potential causes which contributed to it. What I didn't know at the time was where that journey would take me, discovering not only what contributed to my stroke but how a lifetime of eating animal-based foods is associated with many other poor health outcomes and a lower average lifespan, and all the silence and misinformation which surrounds it.

It became particularly evident to me that overeating animal-based food essentially limits the amount of vital nutrients and fibre that the body needs to be healthy throughout life that are only available from plants. The bigger the dietary gap between the two, the bigger the concern. When it comes to animal-based foods we are not simply talking about moderate consumption but the need for very low consumption – perhaps as little as 5 per cent of our daily calorie intake according to some. This could include adopting a 'traditional' Mediterranean Blue Zone, or a complete whole-food plant-based diet. It is also no good replacing meat, dairy, fish and eggs with calories from sugar and other harmful saturated fats either, as you simply end up with a similarly poor health outcome. Furthermore, it is not helpful eating significant amounts of animal-based foods loaded with salt, such as butter and cheese. This equally applies to eating anything other than a small amount of processed plant-based foods, which can be full of refined sugar, saturated fat and salt – vegan *junk* food, if you like. Achieving the best health outcomes from food depends on what you replace your existing diet with. In my own journey I quickly learned that we are better off eating a completely, or at least predominantly, whole-food plant-based diet (WFPBD). For me, having had a stroke, there is no room in my health for meat, dairy, fish and eggs, or anything else the livestock industry serves up.

The best part of a WFPBD is that by default it addresses animal welfare issues and helps the environment too, two outcomes that are equally important. It is a sustainable WFPBD I compare animal-based foods against in *Livestock's Longer Shadow*, not a vegan diet per se. The Vegan Society's definition of veganism begins: 'Veganism is a philosophy and a way of living which seeks to exclude – as far as possible and practicable – all forms of exploitation of, and cruelty to, animals for food, clothing or any other purpose.' I consider myself a philosophical vegan, but one that eats a predominantly WFPBD (I don't say no to the odd cake, etc) and who limits my wider impact on the environment too and not just by the food I eat.

I quickly found from my studies that we cannot *moderately* eat meat, dairy, fish and eggs out of our health crises, any more than we can the unkindness we show to animals and the harm such errant consumption does to our environment. If every citizen of the world moderately ate animal-based foods we would need several Earths to feed us. Being *moderate* may help make us feel better about our eating habits, but it doesn't help our health, environment or the animals.

Despite this, people will cling to, as evidence, the odd centenarian who defies logic when it comes to eating animals all their lives, in the same way as those centenarians who have smoked all of their lives. In doing so, neglecting the many thousands more who have suffered and died long before their time, or even checking how else these centenarians lived their lives as a whole.

My experience also reignited my interest in climate change, which I studied as part of an environmental science degree at the University of Plymouth. My time at Plymouth was life-changing in many respects. To that I owe a great deal of gratitude to my former tutor, John Bull. In my first tutorial group meeting with John, we discussed a wide range of subjects but concentrated on climate change in particular, and John asked us to come up with potential solutions to several problems he raised. Each time we did, John challenged us on the consequences of our solutions, and we kept finding the problems much more complicated to solve than we first thought as we repeatedly swapped one trouble for another. The next three years of my studies were going to be a lot more complicated than I had thought.

The tutorial lasted about an hour and by the end we all came away feeling thoroughly depressed about the state of the world, where we were heading without unprecedented change and the size of the task at hand. Fortunately, the feeling of depression only lasted a night and by the next morning I was totally invigorated to help save the world, my fellow students similarly. John had worked some magic and made us come back into the university the next day highly motivated to make a *real* difference and with an appetite to look at the problems we faced as a whole. I took that attitude into my work life and have always aimed to ensure I have a broad overview of the environment and not to get myself pigeonholed on single issues. Furthermore, to ensure my work did not simply look to contain problems but to address them at their source.

My first job after graduation was a Waste Regulation Officer at Devon County Council, before taking a senior role a couple of years later at Somerset County Council. This was in the early 1990s, when businesses were being woken to the fact that the waste they generated and its safe management were of importance. Often companies would expand with little or no thought about what would happen to the extra waste they would generate. Waste then wasn't a profitable business, but selling commodities was and their environmental footprint was something they neglected or looked to externalise.

With an interest in agriculture, I began to specialise in the *recovery* of industrial wastes to land for agricultural benefit. Over time I became highly judicious in agronomy and safe landspreading management to avoid pollution, which I have enthusiastically built on throughout my career and still do today. There is always more to learn and some brilliant people to learn from, including those I disagree with. After the first few years cutting my teeth in waste, I have spent the majority of my career knee-deep in more waste - livestock slurry.

I've written *recovery* in italics because waste disposal was, and is still too often, the order of the day. Good management and timing, to ensure organic manure wastes spread to farmland met a soil and crop need or was without a significant risk of pollution, wasn't common. Farmers were similarly complacent with the animal slurry they produced and unfortunately too many still are today. Regrettably doing the right thing does not earn money and with little risk of being visited by a regulator, and with it an incentive to do the right thing, and with agricultural lobbyists and retailers keen to protect profits over the environment, farm pollution was, and continues to be, commonplace.

So begins the next chapter of my book, in which I concentrate on livestock's longer pollution shadow, covering most forms of pollution, climate change and ecological harm caused by the livestock industry both home and abroad. This is a subject area I'm well placed to talk about after a career spent largely in farm waste pollution management, where I have reached the top of my profession. My experience of the livestock industry, borrowing words from animal welfare activist Ruth Harrison, can be summed up as follows:

If pollution occurs, especially in the name of commerce, and the risk of being caught is low, or the consequence of being caught is less than can be gained, it will more likely happen, and where the price of doing the right thing is too high, will cause ignorance of fact and be defended to the last by otherwise intelligent people.

I do not express this view lightly and I have no financial interest in being believed. What I say within this chapter, as in the other chapters, will undoubtedly be defended against to the last by otherwise intelligent people who *do* have a vested interest. I have no interest in gaslighting people with myths or unevidenced claims and ambitions based on words rather than actions, that is for others, and throughout I have been careful to be conservative in my accounts and to avoid exaggeration. To do otherwise would only serve to undermine my book and credibility.

Getting the changes needed in livestock farming to make *real* differences can be akin to turning a supertanker around – very slow going. If the evidence is against the industry, they will argue the law. If the law is against them, they will argue the evidence. If the law and the evidence are against them, they will claim it is out of touch with current on-farm practice and/or lobby like hell to get their way. If they don't get what they want, they will get as many concessions as they can, regardless of the evidence and the law. In the case of the latter, this may be less about them and more about those who listen to them.

Livestock farming is not just unkind to our environment and unsustainable as a *whole*, it is unkind to the animals we farm too (and other wildlife), which is an unescapable truth. Factory farms in particular are unnatural places, where unnatural and unkind things are done to a lot of animals in the name of commerce. Places where animals live in and around their own excrement for all of their shortened lives. Places where vets are used, and are sometimes fully employed, to lower the incidence of the stress, injuries, diseases and mortality caused to livestock by the very systems and conditions they are held in. Interventions of *animal welfare*, which would not be necessary if livestock were kept and treated in a natural way.

It is not unknown for the livestock industry to claim that such production systems do not dictate animal welfare and that it's the treatment of individual animals that is important. This claim cannot be taken with any degree of seriousness, with the reality limited to keeping animals as healthy as possible to optimise production and profit despite the system. We need to see ourselves in the animal's eyes.

The unkind way we treat farm animals completes my book, covering the life of livestock from conception to death. Farm animals are born and placed into a life of human servitude, where, like all other countries, UK law allows farmers to cause *necessary suffering* or *injury* in the name of commerce. What constitutes allowable cruelty to livestock is principally decided by tradition, common practice and cost, largely decided by the industry itself, tempered somewhat by consumer noise. This in an industry which routinely looks towards new and unnatural ways of animal husbandry which can extract more value from already mistreated and exhausted animals in the race to the bottom to keep the industry in profit and cheap foods on our plates. We do not create and use the animals we want to farm for their benefit.

Farmers love farming and the industry claim farmers love the animals they farm, but the latter is a strange love where kindness is substituted for blindness. Being unkind can only be justified if such treatment is completely unavoidable for food which is absolutely essential for human life. It isn't, and the ultimate aim of this book is to show you just how needless livestock farming is, and how we cannot be proud of any aspect of livestock farming which involves abusing animals and causing them injury. Even when that cruelty is just limited to their death. For instance, can we be proud of separating dairy calves from their mothers, confining sows in crates in which they can barely move, gassing pigs, causing animal illness and performing sexual acts on animals in the name of commerce, etc? I would hope not.

Once realised it is not necessary to eat animals, the animal abuse and injury involved in livestock farming, including their transport and final slaughter, becomes a choice, so too the damage livestock farming does to our health, to our planet and to humankind as a whole. The choice of where we go from here is ours, and I hope my book inspires you to disturb the silence. There is no time left to wait for a failing livestock industry and governments weighed down by the millstones of their own creations to face up to the facts and truths. It is up to us through what we put into our mouths. Our future depends on it.

If all that was not enough of a reason, animal agriculture, and in particular intensive poultry and pig farming, is a tinderbox for the next serious zoonotic disease. 2020 has seen a spike in bird flu outbreaks on UK and European poultry farms, including the strains H5N2 and the more concerning H5N8 influenza. It is also only a decade since the UK had a visit from the H1N1 (and now seasonal) swine flu pandemic, originating in Mexico in 2009, which although mild for most who contract the virus is serious and even deadly for some. The people most at risk are familiarly those with underlying health problems, of which eating animal-based foods is a contributing lifestyle factor. If we keep on eating animals and destroying wildlife habitats to feed them, we risk spending more of our future lives locked up and waiting for cures. We must stop eating animals, we can't afford not to, and the Covid-19 pandemic is teaching us a harsh lesson. No country is impervious to being the next Wuhan, and lest we forget, I must mention the UK's own bovine spongiform encephalopathy outbreak (BSE/'mad cow disease'), human variant Creutzfeldt-Jakob disease (vCJD), which the state and livestock industry served up on our plates during the 1980s and 1990s.

What my book is not about

My book is not about blame or shame. Farming livestock is deeply embedded in the blood and souls of farmers, our traditions and culture. Most farmers are good people and the things we do to farmed animals have become customary practices which do not start and stop on livestock farms. It is a shared responsibility, with society. Farmers, and others in the animal-based food industry, pollute the environment and harm animals for us. Please keep that in mind throughout reading this book.

Livestock farming, and the suffering and painful injuries the animals experience, is justified as an unfortunate necessity to feed people and make the best use of land for commerce. For instance, in terms of animal welfare, farmers and vets performing sexual acts on farm animals to industrialise their reproduction would once have been viewed as abnormal but is done to such an extent today, in the name of commerce, that the practice has become normalised and accepted within the industry, and even celebrated. It even creates new jobs and innovations which further its disturbing cause. Society accepts, or is blind to, this abuse and other abnormal/unnatural practices, particularly on factory farms, as the price for plentiful and cheap meat, dairy, farmed fish and eggs. Often bluntly put, my book is about challenging customary practices, awakening and informing to promote change from the unkind, unsustainable, polluting, ecologically harmful, unhealthy and totally unnecessary farming models we have evolved to ones instead based on hope and kindness.

Change takes time; we do not have a lot of that and we need to work and support each other for the benefit of all, without leaving anyone behind, at the earliest possible time. Whilst Flash Gordon only had fourteen hours to save the Earth, we have perhaps ten to fifteen years before irreversible change. The clock is ticking and getting louder with every minute of inaction. I hold no moral high ground, I'm just sharing what I know and have learnt in one book from a career spent largely knee-deep in livestock pollution and from everything I have read, seen, suffered and otherwise experienced. I ask that you objectively read what I have to say, whatever your persuasion, and walk in my shoes for the moment. If you think what I have to say about livestock farming isn't true, that it is necessary to abuse and injure billions of animals, that we do not harm our health by eating animal-based foods and that we are not farming and polluting our way towards extinction, then I welcome that conversation.

If we are to survive on this planet, we need to uncover our eyes, unplug our ears, ungag our mouths and see, hear and verbally question everything about animal-based food production. We need to stop pretending we are the victims, because the only casualties here are the animals we harm, our planet and our health. We need to change from the cultures and traditions that feverishly defend livestock farming and which look the other way to the abuse and injury we subject vulnerable animals to. We need to completely reshape our relationship with food production and the natural world. It would be a failure not to act – we simply have no time left to fail.

Hope lives in kindness.

Tim S. Bailey

CHAPTER I

In Silence a Cancer Grows

'Fools,' said I, 'You do not know
Silence like a cancer grows
Hear my words that I might teach you
Take my arm that I might reach you'
But my words, like silent raindrops fell
And echoed in the wells of silence

Paul Simon, 'The Sound of Silence' (1964)

Each year billions of animals are farmed unkindly for their meat, milk and eggs, and over two trillion fish caught and harmed for their flesh, fins and oils. Their calls for love, compassion and mercy at death, like silent raindrops, fall and echo in the wells of silence. Like fools we do not hear or shed a caring tear. Our planet's future, our health and the rights of the animals we eat or harm in nature depend on how we produce and consume food. If we do not take responsibility for managing these things and more, our silence like a cancer will continue to grow. We must break the shackles that bind us and take the right actions now to save ourselves and future generations by putting an end to the silence. Hope for humankind depends on our kindness.

CHAPTER II

My Journey

What counts in life is not the mere fact that we have lived.
It is what difference we have made to the lives of others
that will determine the significance of the life we lead.

Nelson Mandela (2011)

1. A Stroke of Luck

a. My Journey Part I

I was born in 1966 and grew up in Worcester, England, in a somewhat left-field family even for the time, at least in comparison to the lives of my friends. Money was tight, many of my clothes were second (or sometimes third) hand-me-downs, my mother and father (often out of necessity) were great makers and recyclers, and a large part of my food came from the family allotment down the road. A traditional meal was largely five vegetables, meat and a fruit pudding; what some would call today an optimal British diet.

My father grew many vegetables and fruits on two (felt like three!) large plots, which my brothers and I spent many hours weeding, digging, planting and harvesting, often when most of my friends were out playing. So much so, that the allotment became demonstratively known as the slave yard. My mother was equally at home growing fruit and vegetables and proliferations of flowers, a pollinator's delight, and with a knowledge of plant names I envy.

Whilst I often cursed my time spent on the allotment, not least in the dark, cold winter months, that, and with the nature of my upbringing, gave me a great deal of love and respect for the environment. It was no surprise that in 1989 I found myself at the University of Plymouth studying for an Environmental Science degree. A time when the United Nations Framework Convention on Climate Change (UNFCCC) was in the making, which in 1992 committed signatories to reduce greenhouse gas emissions. The Kyoto Protocol was adopted in Kyoto, Japan, on 11 December 1997, implementing the UNFCCC objectives to negate global warming. It is of great sadness to find ourselves in such a predicament today, all these years on.

On leaving university I began a career in waste regulation with Devon County Council. During my time I specialised in the regulation of industrial wastes spread to land for agricultural and ecological benefit, and without risk of causing pollution. Over the years my work became increasingly centred on agricultural pollution in general, working upwards from a local to a national level, and sometimes European. Today I still retain a very close interest in our ensuing climate emergency. Most of my work life though involves livestock farming, slurry management and pollution prevention. More often than not as successfully as King Canute holding back the sea, but not because I'm bad at my job.

Since 2002 I have largely eaten a vegetarian diet. With a vegetarian wife, and my eldest daughter from age eight (later a vegan), it made things easier for me rather than for any moral, ethical or health reason. Whilst I ate meat from time to time, along with my eldest son, my animal-based diet consisted mostly of eggs and dairy and occasionally fish. If meat-eating visitors came, I'd usually turn to chicken.

Life carried on pretty much like that until the 2 February 2016 at 7 a.m., the morning after a trip to the European Commission in Brussels. That morning I was struck by a life-changing event. Standing in the doorway next to my bathroom, talking to my wife, I experienced a sudden 'click' sensation in the right side of my brain. A wave of weakness flooded over the left side of my body, causing me to slump into the doorframe. I had little doubt what had happened, and slurred: 'Call an ambulance, I think I've just had a stroke.' With my bed one stride in front of me I took a right-footed lurch of faith and just about managed to reach and roll myself onto the bed. My wife's first thought was that I'd had a sudden migraine which, given a longstanding history of migraines since a road rage incident, was an expected reaction.

As I lay on the bed, I became less coherent. My wife called 111, the National Health Service (NHS) helpline for urgent medical concerns, and explained my symptoms. Within minutes an ambulance was arranged and a very short time later two heroes arrived. By then I had recovered some movement on my left side and, propped up on either side by the ambulance crew, was able to part-support myself as I was ushered downstairs and into the ambulance. I do not remember much of the journey to the hospital or the computerised tomography (CT) brain scan I had rapidly on my arrival.

At first the physician in attendance was unsure whether I'd had a stroke, but it was confirmed by an MRI scan. I'd suffered an ischaemic stroke in my right thalamus. My first *stroke of luck* was that the clot had not caused enough damage to permanently impair me. My own good fortune bleakly confronted by seeing several severely damaged stroke victims come and go. By the time I left hospital, I was able to walk, albeit very wobbly, with the aid of my two youngest children. After a few weeks of the most amazing care and support from my wife and children, I was able to walk safely enough without support.

For the first few months I struggled with heavy fatigue, visual problems and headaches, which often laid me up in bed for hours at a time. Nine months later I managed to start back at work, though it took me a further year to get more or less back to my normal self and working hours. My second *stroke of luck* was the great support my employer and work colleagues provided during my recovery, and in particular my line manager.

On the day of my stroke I was forty-nine, muscular, lean and extremely fit. My elder brother Eddie and I have a long-standing tradition that on the eve of each decade of our lives we would be as fit as the ten years before, give or take a little unavoidable attrition! In fact, I was practically eight months ahead of my target, with a measured biological age of less than forty years. My blood pressure was 120 over 80 and my total cholesterol was 4.6 mmol/L. On both counts my levels were within NHS guidance levels. My stroke risk was low, but someone has to make up the statistics. It was just one of those things, I thought, and for the rest of my life I would need to take statins and a blood thinner.

Before leaving hospital, I asked the physician signing my discharge papers whether I could do anything else other than take statins. I received a firm 'No'.

b. Cholesterol & Atherosclerotic Plaques

When you have an illness like a stroke it encourages you to look into potential causes, including diet. My stroke was caused by a dislodged blood clot, most probably originating from an artery in my head or neck. The blood clot did not appear out of the blue, but would have stemmed from an atherosclerotic plaque built up over time.

An atherosclerotic plaque is a deposit of fat and other substances that accumulate in the endothelial lining of artery walls.

As the plaque builds up it causes the arteries to harden and narrow, restricting blood flow and oxygen supply to vital organs.[1] Eventually, it can cause a complete blockage. Or, as in my case, the plaque may soften, rupture, a piece break away, travel distally and cause a blockage elsewhere, cutting off the blood/oxygen supply and triggering a stroke.

There are several lifestyle factors that contribute to the formation of atherosclerotic plaques, including eating unhealthy foods, smoking, drinking excess alcohol and being overweight. Medical conditions, such as high blood pressure, diabetes, and high blood cholesterol, increase your risk of having a stroke. Your risk of having a stroke also rises as you get older due to poor lifestyle choices. Blood cholesterol combined with the company it keeps (e.g. animal saturated fat and protein), and the company it doesn't keep (e.g. whole plant-based foods, substituted for animal saturated fat and protein) is a primary risk indicator, but not all our cholesterol is bad for us.

Cholesterol is present in two forms: low-density lipoprotein (LDL cholesterol), often called LDL or 'bad' cholesterol; and high-density lipoprotein (HDL cholesterol), often called HDL or 'good' cholesterol.

LDL cholesterol is considered 'bad' because in certain circumstances it is involved in plaque formation, raising your chance of having a stroke, heart attack or another vascular disease. HDL cholesterol is considered 'good' because it lowers this risk by grabbing hold of excess LDL and carrying it away to your liver for disposal. In fact, both forms of cholesterol are made in your liver from precursors and are necessary for our existence. For instance, cholesterol is used to protect our nerves, construct our cells, produce sex hormones and make bile.

Our liver makes all the cholesterol our body needs. Extra cholesterol comes from the food we eat, namely animal-based foods, which are not needed provided your liver is functioning properly. It is the cholesterol you don't need, in conjunction with the company it keeps within a poor diet, which gives LDL its 'bad' name.

The NHS does not state *safe* cholesterol levels, they just provide a general guide. In general, the levels you should aim for are below 5 mmol/L for total cholesterol, 3 mmol/L or below for LDL and 1 mmol/L or above for HDL. However, an individual's level will vary and you are urged to consult with a medical professional.[2] Whenever I had my cholesterol tested my total was always just below 5 mmol/L. I can't ever remember being told my specific LDL and HDL, though they would have been tested too.

The formation of atherosclerotic plaques (inflammatory lesions) begins with damage to the endothelial cells lining our artery walls, as already mentioned. LDL contributes to this damage when the LDL particles become oxidised in our blood stream.

Oxidation of LDL cholesterol in our bloodstream is a normal process, caused by chemical interactions with free radicals. Free radicals are unstable molecules that can have both positive and negative consequences. In the right amount they are used positively to fight infections, amongst other things. Too much and they can damage our cells and destabilise molecules like LDL, making them harmful. The problem is not that this happens, but the extent to which it happens. If you have too many free radicals and too much LDL cholesterol, you can end up with an excessive amount of oxidised LDL cholesterol, resulting in inflammatory responses which can lead to atherosclerosis and other health issues.

Our bodies neutralise excess free radicals with antioxidants. If we do not have enough antioxidants, free radicals run riot and cause oxidative stress. Prolonged exposure to oxidative stress not only harms our cells but can cause DNA damage too, increasing the risk of developing cancer. To keep free radicals in check it is vital to our health to eat foods rich in antioxidants. Tiny amounts are found in animal-based foods, with the majority found in fruits, vegetables and other plant-based whole foods. Antioxidant foods are best eaten regularly throughout the day, in particular if you eat a lot of inflammatory foods like animal-based products, refined carbohydrates and oils. Our bodies are always working to bring us into a neutral state, and as such it is important to keep our antioxidants regularly topped up.

Oxidised LDL particles cause damage when they enter and irritate the artery walls. In response, monocytes, a type of circulating white blood cell, are sent to immerse and destroy them. Unfortunately, in destroying the oxidised LDL particles, the monocytes become what are known as macrophage foam cells; cauldrons of free radical activity that further oxidise and damage the surrounding proteins and other structural molecules of the arterial lining, which makes things worse.

LDL in animal foods is not solely to blame for atherosclerosis. Other foods and the way they are cooked can also contribute, such as fried food, processed foods, simple sugars and acidic drinks. To make matters worse, these foods are often eaten in combination with animal-based foods, to create the perfect storm.

The fats in whole-food plant-based diets (WFPBDs), such as nuts and seeds, are not associated, probably because they come together in a complete package and with plenty of fibre. A WFPBD, and the variety of fibre it contains, feeds and promotes the good bacteria in our guts. Meat, other animal-based foods and processed foods encourage bad bacteria, which I discuss later.

Consuming animal-based foods day after day, particularly in excess, causes repeated injury, and our body's response to that injury over time grows these atherosclerotic plaques to a point when they eventually become injurious to our health. Atherosclerotic plaques are not a predestined fact of ageing, they are caused by our lifestyle choices and start forming in childhood, or even before. Saturated fat has also been shown to raise LDL, a subject I return to next.

Whilst knowing your total LDL and HDL cholesterol is helpful, according to Dr. Michael Klaper, M.D., an American physician for over forty years, your arterial health risk is better understood by measuring your inflammatory markers too. Markers which can detect the progression from early plaque formation to the risk of imminent plaque rupture. I am not aware, despite loads of investigations in hospital, of having been tested for any of these markers.

Table 1: Plaque inflammatory markers[3]

Plaque formation	Oxidation	Disease initiation/ plaque growth	Plaque maturation/ vulnerable plaque	Acute coronary syndrome
Marker	F_2-IsoPs / OxLDL	ADMA Microalbumin hsCRP	MPO Lp-PLA$_2$ Activity	Troponin T CK-MB

Source: Cleveland Heart Lab (2018).

What surprised me most from my research was what I discovered about statins, the drug type prescribed by doctors to prevent strokes and heart attacks. What I didn't realise was that statins are taken to suppress the liver's own cholesterol production. They do not remove the LDL we consume in the animal foods we eat. Statins simply slow down the progression of atherosclerotic plaques and lower your future risk.

At no time during my recovery was that explained. I wanted a cure. I wanted to reverse my disease. I wanted a healthy life without a future risk. But I had to learn all this for myself. Furthermore, I learned that statins can have many negative side effects, and could cause harm to your liver and other organs over time.

Over the years I had literally been eating my way to a stroke, without realising. All of a sudden, the cholesterol- and saturated-fat-laden dairy, eggs and more occasional chicken, sausages and bacon I ate became less appealing (with saturated fat consumption a bigger contributor to blood cholesterol levels than cholesterol, per se). My health and family were much more important than my taste for animal-based foods.

Fortunately, if you have atherosclerotic plaques the condition is reversable. The solution is what you put in your mouth, a well-balanced WFPBD, and leaving your liver to produce all the cholesterol you need.

A word of caution though: it is not advisable to come off medication like statins without seeking medical advice. I continued to take statins, and a blood thinner, for a long time after my stroke and changing to a WFPBD. Statins may be required in some people, even with a healthy diet.

After I converted to a WFPBD diet I was eager to see the benefits. So, in 2018 I had my cholesterol privately tested. Again in 2019.

In 2018 my LDL cholesterol was 0.7 mmol/L, my HDL cholesterol 1.8 mmol/L and my total cholesterol 2.8 mmol/L. My LDL was very low, and I was concerned that combining a WFPBD with statins over the longer term may be injurious to my health in other ways. Ideally, I wanted my liver to create all the cholesterol I needed naturally. It was around that time I stopped taking statins, which I talked to my doctor about. Tests a year later saw a rise in my LDL cholesterol to 2.0 mmol/L, my HDL dropped to 1.5 mmol/L and my total cholesterol rose to 3.9 mmol/L.

Atherosclerotic plaques are not just associated with strokes and heart attacks. They are linked to the accrual of plaques in the brain, causing reduced blood flow, which in turn can lead to a series of damaging, silent, mini-strokes. Accumulative damage can also lead on to vascular cognitive impairment and subsequently vascular dementia.[4]

Even the risk of Alzheimer's disease is increased by the same risk factors that lead to heart attacks and strokes.

Erectile dysfunction is another potential sign of high LDL and atherosclerosis, as plaque formation clogs the vessels to the penis, reducing blood flow. Whilst there are non-physical, psychological causes of erectile dysfunction too, if you are having this problem it may be a good idea to get it checked out. If caused by narrowing and clogging of your blood vessels, it could very much be an early warning sign of a risk of a stroke and/or a heart attack. Or even a future risk of vascular dementia. A risk factor which can be reduced or reversed over time with a WFPBD.

Cholesterol and saturated fat in animal-based foods is not a good news story for the livestock industry. Great effort is made to dismiss their significance in atherosclerotic plaque formation and other diseases. To do this, the industry uses a plethora of industry-based and industry-funded research to sow doubt. *Sympathetic* research which is often at odds with independent research.

This is hardly a surprise given the multi-billion value of the livestock industry and the many businesses and jobs directly and indirectly associated with it. Sowing doubt is a useful weapon as consumers of meat, fish, eggs and dairy are only too willing to be reassured about their poor dietary habits.

The livestock industry uses a number of methods to get favourable results from research they fund. One way is to compare eggs with eggs. For instance, take a typical animal-laden Western diet high in saturated fat and compare it to a Mediterranean and a vegetarian diet made up to contain the same amount of saturated fat. A parity of saturated fat across the three diets is achieved by *unnaturally* loading the comparison diets with things like palm oil and cheese, which are known to act in a similar way.

The same methodology can be used to compare individual animal-based foods, like eggs. Here, comparison diets are manipulated to give a similar cholesterol to a diet rich in eggs. The conclusion reached, and publicised, is that eggs do not raise your LDL compared to other diets. Simple but effective, so long as journalists pick up on the headline fed to them rather than the detail hidden within the study.

Researchers may also perform tests on people already suffering from a condition, like high cholesterol. Eating animal-based foods have been shown to raise your LDL, but eventually if you keep adding cholesterol to your diet it reaches a plateau. Feed people more cholesterol at this point and their level changes very little. The research conclusion would be that the further addition of eggs, dairy or meat to the diet did not significantly increase LDL. That becomes the headline, not that the subjects tested had a high LDL in the first place, caused by eating too much animal-based food.

Another method the industry uses is *favourable* comparisons. For instance, between meat alternatives that are high saturated fat vs. the leanest equivalent meat products they can find.

Whilst writing this I thought it would be useful to do my own fat–meat comparison using a well-known UK pork sausage brand, Richmond. Richmond produces an award-winning meat-free alternative, which can be compared against their nearest equivalent sausage. I added in a few other sausages for good measure, including the Beyond Meat plant-based sausage and the leanest meat sausage I could find.

Table 2: Comparison of meat-free and pork meat sausages

Sausages (336g)	Richmond Meat-free	Beyond Meat sausage	Richmond Pork sausages, thin	Pork sausage fresh	Breakfast sausage	Debbie & Andrew Clean and Lean sausage
Calories (kcal)	489.11	840	853.44	1,092	1,471.68	632.3
Protein (g)	29.8	70.7	40.3	62.3	50.8	66.9
Fat (g)	18.3	53.1	53.8	91.6	140	38.6
Saturated fat (g)	13.6	22.1	27.2	29.7	45.6	13.8
Sodium (mg)	2,381	2,210.5	2,956.8	2,735	3,326	2,434
Fibre (g)	12.3	13.3	-	-	-	1.6
Carbohy-drates (g)	59.1	22.1	53.8	4.8	2.1	5.0

Source: Calculated using Cronometer (https://cronometer.com/).[5]

If I were looking to write an industry-favourable article based on this study, I would only compare the saturated fat in the Debbie and Andrew Clean and Lean sausage against the Beyond Meat sausage. That would be the headline. Not that the Richmond Meat-free sausage had a similar saturated fat content, less total fat, fewer calories and 10.7 g more fibre.

Industry-funded researchers do not like to compare animal-based diets against well-balanced WFPBDs. That would be industrial-scale commercial suicide.

Citations:

1.   NHS, *Atherosclerosis*. https://www.nhs.uk/conditions/atherosclerosis
2.   NHS, *Cholesterol Levels*. https://www.nhs.uk/conditions/high-cholesterol/cholesterol-levels/
3.   Cleveland Heart Lab (2018), *Inflammatory Biomarkers and their Association with Atherosclerosis*. http://www.clevelandheartlab.com/wp-content/uploads/2018/11/CHL-D0 03-JUL2018-Artery-Wall-OnePager.pdf
4.   Stroke Association (2018), *A complete guide to vascular dementia*: https://www.stroke.org.uk/sites/default/files/user_profile/a_complete_gui de_to_vascular_dementia.pdf

5. Cronometer. *Track your nutrition, fitness & health data.*
https://cronometer.com/

c. Saturated Fat & Cardiovascular Disease

Diversionary tactics are rampant when it comes to the potential impact on our health of eating any animal-based foods, and no less the role of saturated fat. However, not all saturated fats found in these foods are equal in terms of their impact on our health. Here I predominantly refer to palmitic and myristic saturated fatty acids, which are strongly associated with poor health outcomes, rather than fats like stearic acid and potentially lauric acid which may be innocuous[1] in a balanced diet. Innocuous doesn't mean beneficial, and as such are simply a waste of calories when you could be eating fats which actually promote good health.

Saturated fat is associated with raising your level of LDL cholesterol, which in turn raises your risk of cardiovascular diseases. Whilst animal-based sources are the prime source, certain vegetable oils like coconut and palm are also linked.

Up until the end of the twentieth century most research pointed to this link, and after that time independent research started to dry up. It was job done. However, such a negative message, and intense pressure to reduce saturated fatty acids linked to animal-based foods by governments and non-governmental organisations, was a barrier to increasing demand.

After 2000 livestock industry research stepped in to fill the *gap*, and the research results began to point *otherwise*. The divergence of evidence came to a head in a well-reported *good news* research story of a study on saturated fat which appeared in the 12 June 2014 edition of the *Time* magazine, 'Eat Butter: Scientists labelled fat the enemy: Why they were wrong'. The study concluded:

> *Current evidence does not clearly support cardiovascular guidelines that encourage high consumption of poly-unsaturated fatty acids and low consumption of total saturated fats*[2]

Dr. Michael Greger M.D., of Nutritionfacts.org, describes how these kinds of studies can be done, and reach the conclusions they do, by using: 'the observational studies that mathematically would be unable to show any correlation ... industry friendly researchers [then] combine all the observational studies that don't have the power to provide significant evidence.'[3]

This 2014 study, a repackaged version of an earlier 2010 meta-analysis, elicited a terse response from the Chair of Harvard University's nutritional department, as described by Dr. Greger:[4]

> *As the Chair of Harvard's nutrition department put it, their conclusions regarding the type of fat being unimportant are seriously misleading and should be disregarded, going as far as suggesting the paper should be retracted, even after the authors corrected a half dozen different errors.*

Another study, published in 2015, led to similar headlines that saturated fats found in meat and dairy were not as bad for our health as previously thought, after the researchers failed to find an association between saturated fats and all-cause mortality, cardiovascular disease, ischaemic stroke and type 2 diabetes.[5]

However, in response to challenges about these headlines, the first author of the study, Dr. de Souza, had the following to say:

> *We could not confidently rule out an increased risk of death from heart disease with higher amounts of saturated fat, and we should not ignore stronger and consistent evidence from better designed studies that eating less saturated fat and more polyunsaturated fat from vegetable oils reduces 'bad' cholesterol levels and that diets that replace saturated fat with these fats, as well as whole grains, reduces the chance of developing or dying from heart disease.*[6]

This was added to by Professor Tom Sanders, a retired nutritionist at King's College London, who said the study should come with its own health warning as it largely relied on people's memory of what they ate some time ago, which is notoriously unreliable:[7]

> *Memory-based dietary recall is subject to substantial bias particularly for food items seen to be good or bad with under-reporting becoming more prevalent among those who are obese ... It would be foolish to interpret these findings to suggest that it is OK to eat lots of fatty meat, lashings of cream and oodles of butter.*

The NHS waded in too, on the 12 August 2015, commenting that the study was reported accurately in *The Independent*, which included the researcher's warnings about the study's limitations and noting 'the authors say they can only have "very low" confidence in their findings, because of the methodological limitations of the individual studies that contributed data.'[8]

More recently, in 2020, physicist-cardiologist Richard M Fleming, M.D. commented on the issue:[9]

This claim – that saturated fat and cholesterol are not responsible for the development of inflammatory coronary artery disease – cannot be taken seriously, and completely ignores the fact that the Inflammation and Heart Disease Theory itself, includes and explains the impact cholesterol and saturated fat have on this inflammatory process.

A WFPBD has been used to reverse cardiovascular disease and other chronic illnesses in real-world situations, with renowned heart disease expert Caldwell Esselstyn, M.D., paving the way. The Dean Ornish, M.D., *Program for Reversing Heart Disease* is another leading example. Take the animal out, put the plants in and the body can recover.

Other authors of studies include, but are not limited to: Neal Barnard M.D., founder of the not-for-profit Physicians Committee for Responsible Medicine (PCRM)) with his work on saturated fat and cholesterol and their association with heart disease and vascular dementia; John McDougall, M.D.; Michael Greger, M.D., founder of the not-for-profit NutritionFacts.org; Arjun and Shobha Rayapudi, M.D.s, co-founders of Gift for Life; Milton Mills, M.D.; Gemma Newman, M.D.; David Katz, M.D.; Brenda Davis, R.D., dietitian and nutritionist; Michael Klaper, M.D.; bariatric surgeon turned plant-based doctor Garth Davis, M.D.; cardiologist Danielle Belardo, M.D.; cardiologist Robert Ostfeld, M.D. M.S.; cardiologist Joel Kahn, M.D.; physician and heart surgeon Ellsworth Wareham, M.D.; cardiologist Nicole Harkin, M.D.; Laura Freeman, M.D., co-founder of Plant Based Health Online (PBHO); Shireen Kassam, M.D., co-founder of PBHO and founder and director of Plant-based Health Professionals; cardiologist Lamprini Risos, M.D.; public health nutritionist Tracye McQuirter; Kim Williams, M.D., former president of the American College of Cardiology; and by no means least, and there are many others, the inspirational T. Colin Campbell, PhD, a lead scientist of the China–Cornell–Oxford Project and author of The *China Study*, along with his son Thomas M. Campbell II, M.D.

Between them, these medical doctors and health professionals have hundreds of years of experience, many on the operating table. Each of them provides concise, evidenced-based and professional medical advice on the causes and interventions we need to take to prevent and reduce the burden of chronic diseases.

Specifically, in the UK, there is the Plant-based Health Professionals UK, a group of medical doctors and healthcare professionals from within the NHS, dedicated to providing education and advocacy on whole-food plant-based nutrition to prevent and treat chronic disease.[10]

These medical professionals have stepped forward to break the silence on the causes of chronic diseases, specifically from processed foods, sugary drinks and animal-based products.

On 28 May 2020, Plant-based Health Professionals UK sent a letter, with over 200 signatories, urging NHS leaders and the government to 'pass bold post Covid-19 legislation to allow for rapid, nationwide changes to the obesogenic and unsustainable food environment in which we currently live. This environment has added to the UK's Covid-19 pandemic death toll.'

They proposed six changes:[11]

- Further taxation and cessation of subsidies for junk food/fast foods/soft drinks producers as well as industrial animal farming.
- Ban on junk food advertising and any market manipulations which aim to increase consumption of unhealthy, processed foods or drinks.
- Subsidies to support the UK population in adopting a predominantly whole-food plant-based diet for both human and planetary health.
- Promotion of a plant-based food system through the adoption of conservation agriculture (no-till) systems to lower carbon emissions and reduce the risk of entering a 'post-antibiotic era'.
- Protection of the oceans by discouraging over-fishing and reducing the reliance on fish consumption, finding alternative sources of long-chain omega-3 fats.
- Showcase healthy, sustainable eating through public-sector catering in schools, hospitals and prisons.

The livestock industry, with their burden of vested interests, continues to defend animal-based foods despite the overwhelming evidence against the diet it promotes. Its institutes, jobs and livelihoods depend on it.

When it comes to my health, and that of my family, I choose to listen to the likes of the Plant-based Health Professionals UK, who tell it as it is, urging NHS leaders and government to 'act in the interests of the people and the planet rather than the interests of the corporate sector, who have contributed to the current climate and healthcare crisis through aggressive lobbying and marketing techniques'.

I do not turn to the livestock industry for nutritional health advice, nor to anyone else like them, who are not remotely qualified and who are burdened with ulterior motives.

Citations:

1. Katz, David L. (2018). *The Truth About Food: Why Pandas Eat Bamboo and People Get Bamboozled*. Dystel & Goderich, New York, NY

2. Chowdhury, R. et al. (2014). 'Association of Dietary, Circulating, and Supplement Fatty Acids with Coronary Risk.' *Ann. Intern Med*. March 2014, 160(6): pp. 398-406

3. Dr. Michael Greger M.D. (4 October 2016), Nutritionfacts.org: *How to Design Saturated Fat Studies to Hide the Truth*

4. Ibid.

5. Russell J de Souza et al. (2015). 'Intake of saturated and trans unsaturated fatty acids and risk of all cause mortality, cardiovascular disease, and type 2 diabetes: systematic review and meta-analysis of observational studies.' *BMJ*, DOI 215;351:h3978, https://www.bmj.com/content/351/bmj.h3978

6. Connor, Steve, 'Saturated fats in meat and dairy not as bad for health as previously thought, study finds', *The Independent*, 12 August 2015. https://www.independent.co.uk/life-style/health-and-families/health-news/saturated-fats-in-meat-and-dairy-produce-not-as-bad-for-health-as-previously-thought-study-finds-10450663.html

7. Ibid.

8. NHS (12 August 2015), *Claims that 'butter is safe' and 'margarine deadly' are simplistic*. https://www.nhs.uk/news/heart-and-lungs/claims-that-butter-is-safe-and-margarine-deadly-are-simplistic/

9. Fleming, R.M. (16 February 2020). *Rapid Response: Saturated is a major issue and is part of the Inflammatory process we call Coronary Artery Disease. Re: Saturated fat is not the major issue. BMJ* 2013;347:f6340. https://www.bmj.com/content/347/bmj.f6340/rr-1

10. Plant-based Health Professional UK. https://plantbasedhealthprofessionals.com

11. Plant-based Health Professionals UK (28 May 2020), *Open Letter to NHS Leaders and the UK Government*. https://plantbasedhealthprofessionals.com/wp-content/uploads/Open-lett er-FINAL-with-signatures-and-references-04.pdf

d. Trans Fatty Acids

Another unhelpful category of fat is trans fatty acids (trans fats); these are commonly associated with processed vegetable oils (e.g. spreads) and junk food. They also occur naturally in animal meat (including chicken), eggs and cow's milk.[1]

Trans fats are a type of unsaturated fatty acids, which are known to raise LDL cholesterol, decrease HDL, and therefore increase the risk of developing heart disease, strokes, liver dysfunction, type 2 diabetes, Alzheimer's disease and other chronic illnesses.

Trans fats are not necessary and can be mostly avoided by eating a predominantly WFPBD.

According to advice from the European Commission[2] the risk from dying from heart disease is between 20 to 32 per cent higher when 2 per cent of the energy we eat daily comes from trans fats.

Trans fats are not banned in the UK; instead the government has a position which relies on voluntary reduction in the food industry.

It is not uncommon for meat and dairy proponents to raise the issue of vegetable margarines being the problem, but butter also contains trans fats, as do other animal-based products. Around 0.47 g per tablespoon of butter, according to the USDA. Butter also contains a high level of saturated fats, which has been linked to raising blood LDL cholesterol.

Most brands of margarines sold in the UK today contain very little trans fats. Whilst food manufacturers and retailers have made efforts to reduce and eliminate trans fats in these products, you can't do the same for meat and dairy products.

Table 3, from 2013 data published by the National Diet and Nutrition Survey (NDNS), shows the percentage contribution to total trans fatty acids intake from foods in the UK (adults), from back in 2000/1.[3]

Table 3: Percentage contribution of the main sources of trans fatty acids in the UK diet (adults, 2000/1)[4]

Food group	Examples of foods contributing to trans fatty acid intake (2000/1)	Percentage contribution to total trans fatty acids intake in the UK (adults)
Cereal products	Biscuits (9%), buns, cakes and pastries (retail) (8%), made using partially hydrogenated fats and oils.	26
Meat and meat products	Burgers, kebabs, meat pies and pastries. Naturally present in beef and lamb at low levels.	21
Fat spread	Butter, margarines and spreads made using partially hydrogenated oils.	18
Milk and milk products	Naturally present in milk and milk products, at low levels.	16
Potatoes and savoury snacks	Chip retail (4%), some savoury snacks (1%).	6

Source: Data from the National Diet and Nutrition Survey (NDNS), from Henderson et al. (2003).

Things have moved on since this NDNS review and the statistics will have changed in light of the reformulation of foods we eat today.

For example, according to the British Nutrition Foundation, frying and coating oils used for manufacturing the vast amount of packaged savoury snacks in the UK no longer contain partially hydrogenated oils, and this is also the case for bakery products.[5]

With the removal of partially hydrogenated fats and oils in cereal products, plant-based spreads and potato and savoury snacks, today animal-based trans fats have jumped to the top of the table.

Whilst these changes have helped lower overall UK consumption of trans fats to below 2 per cent, it does not suddenly make animal trans fats healthy and of no concern. Not least given the amount of animal trans fats found in processed junk foods. Many high consumers of animal-based foods in the UK will fall above the 2 per cent threshold and will have a higher risk of disease.[6] Having less than 2 per cent trans fats in your diet only means you have a reduced risk.

Citations:

1. European Commission, *Trans Fats in Foods: A New Regulation For EU
    Consumers*. https://ec.europa.eu/food/sites/food/files/safety/
    docs/fs_labelling-nutrition_transfats_factsheet-2019.pdf
2. Ibid.
3. British Nutrition Foundation, *Fats about Trans Fats*.
    https://www.nutrition.org.uk/attachments/045_Facts%20about%20trans%
    20fats.pdf
4. Henderson, L. et al. (2003c), 'Energy, protein, carbohydrate, fat and alcohol
    intake', in *The National Diet and Nutrition Survey: Adults Aged 19-64
    Years,* Volume 2. HMSO, London.
5. British Nutrition Foundation, *Facts about Trans Fats*.
    https://www.nutrition.org.uk/attachments/045_Facts%20about%20trans%
    20fats.pdf
6. Henderson, L. et al. (2003c), 'Energy, protein, carbohydrate, fat and alcohol
    intake', in *The National Diet and Nutrition Survey: Adults Aged 19-64
    Years,* Volume 2. HMSO, London.

e. Type 2 Diabetes

There are two types of diabetes:

- Type 1 diabetes, where the body's immune system attacks and
  destroys the cells that produce insulin in the pancreas.
- Type 2 diabetes is a result of insulin resistance, whereby the body
  is producing enough insulin but the cells are unable to respond to
  it, resulting in elevated blood glucose.[1]

Insulin resistance normally develops slowly over time. People close to,
and at risk of type 2 diabetes, are known as pre-diabetics. Both types of
diabetes are dangerous without treatment.

It is widely accepted that type 2 diabetes is caused by a build-up of
saturated fatty acids (SFAs) in our muscles (intramyocellular lipid). This
is directly linked to the consumption of saturated fat and fructose. SFAs
clog up muscle cell doorways, preventing the hormone insulin our body
makes from opening them up to let the glucose in. The more SFAs in our
muscles, the greater the insulin resistance. However, while SFAs in
animal foods is a major player it is important to reduce consumption of
refined grains and sugars too as they are also responsible for type 2
diabetes.

To keep blood glucose within a safe range, extra insulin must be injected to force sugar into our muscle cells. Stop eating SFAs, refined sugars and in particular animal SFAs and the condition can be reversed.

Even a single meal containing SFAs can cause a degree of insulin resistance, although it would go unnoticed.

Progressive high consumption of SFAs does not just block glucose getting into our muscles but can kill off insulin-producing beta cells in our pancreas too. Oxidised LDL can also kill beta cells. Extra insulin is then needed to be injected to compensate for insulin resistance and any impaired beta cell function.

Once a beta cell is killed it can't be replaced, but the good news is that type 2 diabetes can be reversed by removing the source of the problem – animal-based foods containing SFAs. Food for thought!

Refined (processed) vegetable fats, and in particular coconut oil and palm oils, are also best avoided. These oils are safer to eat within their whole-food source, as nature intended, from foods like nuts and seeds.

Type 2 diabetes may not be an outcome for everyone eating a hypercaloric diet rich in SFAs and refined sugars, as there will be a genetic component in up to 2 to 3 per cent of the population. The same way that not everyone who smokes will get lung cancer, and with some living to a ripe old age. Nevertheless, not progressing towards a serious underlying health problem does not mean your health is not being impaired.

The condition can be reversed by dropping animal and refined vegetable oil SFAs from your diet and adopting a WFPBD. Regular exercise is helpful too. Being overweight is also a risk factor, which in most people melts away on a WFPBD.

People do *not* need to live with type 2 diabetes.

Whilst being overweight and obesity is associated with type 2 diabetes, and a loss of weight helps improve the condition, the condition is highly dietary related. We need to unblock our muscles and let the glucose in.

Unfortunately, there is presently no cure for type 1 diabetes, a condition I'm very familiar with, having a daughter with the condition. However, a well-balanced diet can help manage sugar levels, with emerging data suggesting that a WFPBD can significantly reduce insulin requirements. A diet full of refined sugar, oils and unhealthy fats is not helpful.

Citations:

1   NHS, *Diabetes*. https://www.nhs.uk/conditions/Diabetes/

f. Cancer

In October 2015, the International Agency for Research on Cancer (IARC), part of the World Health Organisation (WHO), concluded that there was sufficient evidence to classify processed meat as carcinogenic to humans (Group 1) and red meat as probably carcinogenic to humans (Group 2A).[1]

Processed meat refers to any meat that has been treated or transformed by salting, curing, smoking, fermentation or other processes to improve meat preservation or to enhance the flavour of meat. Pork and beef feature very highly in this category, but it also includes other red meats, poultry, offal and meat by-products such as blood.[2] Red meat refers to all types of mammalian muscle meat such as beef, pork, veal, lamb, mutton, horse and goat.[3]

Cancer from processed meat is primarily related to colorectal cancer, based on evidence from many epidemiological studies which show the development of cancer in exposed humans.[4] A non-conclusive association was also found for stomach cancer.

One of the most popular processed meats associated with cancer is bacon, largely linked to nitrites commonly used as a preservative. During digestion, nitrites in bacon are converted into potential cancer-causing N-nitroso compounds (NOCs). The way bacon is usually cooked (e.g. in a hot frying pan) also adds to this potential by introducing cancer-causing heterocyclic amines (HCAs) and polycyclic aromatic hydrocarbons (PAHs). When it comes to HCAs, PAHs and cancer, the more bacon you eat and the browner you cook it, the better. A further cancer risk, and other health concerns, are associated with the company bacon often keeps (eggs, sausages and fried processed bread).

Red meat consumption is primarily linked to colorectal cancer, based on epidemiological studies showing an association between eating red meat and developing colorectal cancer, along with strong mechanistic evidence.[5] Links to pancreatic cancer and prostate cancer have also been evidenced.

In 2020 the National Farmers Union (NFU) published a document highlighting nine myths made against the British red meat and dairy industry: *The facts about British red meat and milk*. Pertinent here is Myth 6, 'Red meat and dairy products are bad for your health', which also includes studies which suggest an association with:

- a protection against weight gain and obesity, significantly reducing the risk of type-2 diabetes and associated cardiovascular disease.

- a possible reduction in the risk of some cancers, including colorectal cancer.

I cover most of the first bullet point in other parts of this chapter. I will not add much here other than to say that you can always find studies (most probably industry-funded) that tell you what you want to believe/report or are misunderstood/misreported, either deliberately or in ignorance.

The second bullet point relates to a research review by Givens (2018), *Dairy foods, red meat and processed meat in the diet: implications for health at key life stages.*[6]

Drawing on reports from the World Cancer Research Fund International/American Institute for Cancer Research (WCRF/AICR),[7] Givens makes a case for an association between the consumption of dairy products and milk and a reduced risk of colorectal cancer, and that high intakes of milk/dairy are not associated with increased risk of breast cancer.

A core piece of evidence is a 2007 WCRF/AICR report,[8] which states that milk consumption probably protects against colorectal cancer. A subsequent WCRF/AICR report gave the same association, with overall results confirming the earlier study's message that total dairy products and milk, but *not* cheese *or* other dairy products (mainly butter, yoghurt, ice cream and fermented milk), are associated with a reduced risk of colorectal cancer.

The WCRF/AICR also looked into the association between dairy food consumption and the risk of breast cancer, which they found limited and without conclusion. In 2017[9] they reported again on an association with a reduced risk for pre-menopausal breast cancer from the consumption of total dairy, but not for milk. However, there was insufficient data for postmenopausal breast cancer to reach a firm conclusion.

In contrast, the 2007 WCRF/AICR study[10] reported that total dairy was associated with a possible increase in prostate cancer, whilst milk was associated with a substantial increased risk of advanced prostate cancer (see IGF-1). They downgraded this conclusion in 2014,[11] with total dairy and milk showing no significant association with *three* of the prostate cancer types. However, they found an association with increased risk for low-fat milk and cheese. The conclusion made was that: 'for a higher consumption of dairy products, the evidence suggesting an increased risk of prostate cancer is limited', in conflict with prospective cohort studies, which show a consistent link, especially with aggressive or fatal forms of prostate cancer.[12]

WCRF/AICR also reported that there is evidence to suggest that a higher consumption of dairy products (and diets high in calcium) increases the risk of prostate cancer. No evidence points to a decreased risk.[13]

The relationship with higher consumption is important, as no industry wants to sell less, only more. We can also get easily side-lined on single health issues like prostate cancer. Dairy products are associated with a host of other diseases too, including a risk of strokes, cardiovascular disease and other vascular diseases such as dementia. This all needs to be treated as one when discussing whether red meat and dairy products are bad or good for our health.

Notwithstanding that, some evidence for some dairy products points positively in respect of colorectal cancer. In regard to colorectal cancer, the mechanism which appears to be protective is calcium. Calcium can be consumed safely, and in sufficient quantity, from a WFPBD, and without the other connected health risks associated with dairy products.

Givens (2018)[14] also looked into the role of processed meat and red meat in cancer, citing the WHO (2015) findings.[15] In relation to WHO (2015), Givens states:

Earlier evidence of a significant increase in risk of colorectal cancer from the consumption of red meat and particularly processed meat has been reinforced by the inclusion of more recent studies.

There is no mention of this conclusion in NFU Myth 6, which only reports on the potential, and limited, positives mentioned by Givens and none of the negatives.

Givens (2018)[16] declaration of interests includes recent/current funding from UK Biotechnology and Biological Sciences Research Council, the UK Medical Research Council, The Dairy Council, The Agriculture and Horticulture Development Board, Dairy, The Barham Foundation and various companies.

Like Givens (2018)[17] the WCRF/AICR similarly concluded the evidence was convincing that red meat and processed meat were a cause of colorectal cancer, reporting in 2010[18] that the risk of colorectal cancer associated with processed meat is approximately twice that of red meat. Their report also stated: 'Despite the relative consistent outcomes from metanalysis of prospective studies, the causative mechanisms whereby red meat and processed meat increase the risk of [colorectal cancer] remains unclear.'

However, in his report Givens (2018)[19] does refer to rodent studies done by Alexander et al. (2015) which suggest a role for dietary haemoglobin as it, and red meat, promotes the development of aberrant crypt foci, a generally agreed pre-cancer feature. Furthermore, that haem may catalyse the endogenous production *of N*-nitroso compounds and certain aldehydes, both of which are carcinogenic.[20]

Despite a very thorough search I was unable to find these messages in any NFU or Agriculture and Horticulture Development Board (AHDB) publication. Or any other livestock publications for that matter.

The research studies form only a tiny part of Myth 6, which prefers to concentrate on the benefits of individual nutrients. These are largely non-arguable, except for excessive haem iron consumption from eating red meat. The problem with all the good nutrients mentioned is that they are packaged with all the associated unhealthy components found in red meat and dairy products. The beneficial nutrients are also found readily in a WFPBD, without the associated baggage. B12 may be missing, and iodine sometimes lacking, in a WFPBD, but these are easily and cheaply supplemented.

It is very easy to quote positively from research that suggests that certain foods, including some dairy foods, can reduce problems like colorectal cancer. If such a benefit is true, it is hardly a good news story if the same food is associated with causing ill health in another way. The whole truth is what is most important, and we must ensure we do not become sidestepped by reductionist associations.

Similarly, we must be aware of health associations between related livestock sectors. For instance, dairy-beef cows, whose red meat is linked to colorectal cancer. Eat certain dairy products and potentially reduce your colorectal cancer risk. Eat a dairy-beef cow with it and probably increase your risk of colorectal cancer and other cancers. The dairy industry supplies over 50 per cent of the UK beef herd. Further food for thought!

The amount of 'probable' cancer we consume has a bearing on our chance of getting an illness like colorectal cancer. For processed and red meat, claims have been made, based on research, that if you eat less than 70 g[21] you will be at low risk as it will be *within the safe range*.[22] This limit, though, also assumes you are eating lots of healthy foods to go with it from the WFPBD range.

According to the Meat Advisory Panel (2017)[23] around 44 per cent of British men and 25 per cent of women consume between 71 and 141 g red and processed meat daily. So, plenty of improvement is needed.

A limit is a recommendation; it does not remove total risk from getting an illness like colorectal cancer. You simply have a lower risk if you eat less than 70 g per day of processed and red meat. Your risk profile will depend on other factors too, which can't be accounted for in a recommended limit. For example, the qualities of the red/processed meat eaten, the way we choose to cook and consume them, what we choose not to consume with them (e.g. healthy, whole plant-based foods) and whether or not an individual has an underlying health condition and the seriousness of that condition.

Other health risk factors are also ignored in recommended limits. For example, animal-based foods are associated with raising your LDL 'bad' cholesterol level. All health outcomes need to be considered together, to enable consumers of meat, fish, egg and dairy to make an informed decision. For example, in terms of all-cause mortality dairy foods fair better than other animal proteins, including fish, but are associated with a significantly higher mortality than the consumption of plant-based proteins.[24,25] As with all diets, what you eat has a strong bearing on whether a particular food source, including animal foods, is beneficial or harmful. If the quality of food you eat is poor, substituting an ultra-processed food with dairy would be beneficial, but if you are eating a wholesome WFPBD, adding dairy, or meat, fish and eggs, would lower your health outcome. What is or isn't better for you is a matter of context.

Eating red and processed meat above Public Health England's (PHE) 70 g per day limit has been linked to increasing cancer risk by around 20 per cent. In a UK population of 66.5 million, that would amount to an awful lot of people and would be on top of those who have an increased risk even if they eat less than 70 g per day (compared to those who eat a wholesome WFPBD). Why eat food that probably will (red meat) and does (processed meat) increase the risk of cancer over your lifetime, when you don't have to, and when the alternatives, whole plant-based foods, are protective against cancer and other diseases?

Reading around this subject, I chanced upon the AHDB's *Share the facts about red meat and health*.[26] The information provided includes a number of red meat health images and promotional facts. Nothing too much of a surprise, given AHDB's role is supporting livestock farmers. One image in particular caught my attention, originally produced by the Meat Advisory Panel with the following claim:

NHS Live Well & The Scientific Advisory Committee on Nutrition (SACN) Recommend We Eat Up To 500 g Cooked Red & Processed Meat Per Week.

I could not find any evidence that the NHS *recommends* we eat *up to* 500 g of cooked red and processed meat a week (70 g per day – cooked weight). The NHS simply advises that we cut consumption to below 70 g per day. The NHS also states that 'Some meats are high in fat, especially saturated fat', and 'eating a lot of saturated fat can raise cholesterol levels in the blood, and having cholesterol raises your risk of heart disease'. The NHS also provides guidance on making healthier choices when buying and cooking meat. It would be folly for the NHS to recommend people eat up to 500 g of meat a week, not least as an individual's health status would be variable.

As a guide, an average portion of steak is around 180 g (70 g is equivalent to only a 2.5-ounce steak!), beef steak mince 100–125 g, a portion of roast beef 225–375 g, a burger 200 g, two sausages 115 g and a slice of ham or bacon 12–13 g. It soon adds up, along with all the animal saturated fat and cholesterol!

Ultimately, if you have to limit the intake of red and processed meat because of the increased risk of cancer they pose, and other diseases associated with them, why are we eating them at all, when, as omnivores, we have a choice rather than an obligation to eat them? Assuming a good variety of plant-based foods are consumed along with red and processed meat, then the plant foods will likely provide most of your protein need too. Add more replacement WPBFs into your diet and you would have no need to worry about being deficient in protein, or iron. The more plants the better.

If all that's not enough to convince you, meat and other animal foods do not contain fibre either and along with other bad dietary habits result in the average UK adult eating less than 18 g of fibre per day.[27] The UK minimum daily fibre recommendation for adults is 30 g, but more is better, to lower your risk of heart disease, stroke, type 2 diabetes and bowel cancer.[28] In contrast, our hunter-gatherer ancestors ate up to 100 g of fibre a day.

For cancer, refer also to: Insulin-like Growth Factor-1 (IGF-1) [page 37], Choline & *L*-carnitine [page 38] and Haem Iron [page 40]. For more on protein see How Much Protein Do We Need? [page 60]

Citations:

1. World Health Organisation (2015), *Q&A on the carcinogenicity of red and processed meat*.
https://www.who.int/news-room/q-a-detail/q-a-on-the-carcinogenicity-of-the-consumption-of-red-meat-and-processed-meat

2. Ibid.

3. Ibid.

4. Ibid.

5. Ibid.

6. Givens, D. I. (2018), 'Dairy Foods, Red Meat and Processed Meat in the Diet: Implications for Health at Key Life Stages'. *Animal*, 12(8), pp. 1709-1721

7. World Cancer Research Fund International/American Institute for Cancer Research WCRF/AICR, *Prostate cancer: How diet, nutrition and physical activity affect prostate cancer risk*.
https://www.wcrf.org/dietandcancer/prostate-cancer

8. World Cancer Research Fund/American Institute for Cancer Research (WCRF/AICR) (2007), *Food, nutrition, physical activity and the prevention of cancer: a global perspective*. American Institute for Cancer Research, Washington, D.C.

9. World Cancer Research Fund/American Institute for Cancer Research (WCRF/AICR) (2017), *Continuous update project report: diet, nutrition, physical activity and breast cancer*. https://www.wcrf.org/breast-cancer-2017

10. World Cancer Research Fund/American Institute for Cancer Research (WCRF/AICR) (2007), *Food, nutrition, physical activity and the prevention of cancer: a global perspective*. American Institute for Cancer Research, Washington, D.C.

11. World Cancer Research Fund/American Institute for Cancer Research (2014), *Continuous update project report: diet, nutrition, physical activity, and prostate cancer*.
http://www.wcrf.org/sites/default/files/Prostate-Cancer-2014-Report.pdf

12. Willett, W. C., and Ludwig, D. S. (2020), 'Milk and Health'. *N Engl. J Med* 382, pp. 644-54

13. World Cancer Research Fund International/American Institute for Cancer Research WCRF/AICR, *Prostate cancer: How diet, nutrition and physical activity affect prostate cancer risk*.
https://www.wcrf.org/dietandcancer/prostate-cancer

14. Givens, D. I. (2018), 'Dairy Foods, Red Meat and Processed Meat in the Diet: Implications for Health at Key Life Stages'. *Animal*, 12(8), pp. 1709-1721

15. World Health Organisation (2015), *Q&A on the carcinogenicity of red and processed meat*.
https://www.who.int/news-room/q-a-detail/q-a-on-the-carcinogenicity-of-the-consumption-of-red-meat-and-processed-meat

16. Givens, D. I. (2018), 'Dairy Foods, Red Meat and Processed Meat in the Diet: Implications for Health at Key Life Stages'. *Animal* 12(8), pp. 1709-1721

17. Ibid.

18. World Cancer Research Fund/American Institute for Cancer Research (2014), *Continuous update project report: diet, nutrition, physical activity, and prostate cancer.*
http://www.wcrf.org/sites/default/files/Prostate-Cancer-2014-Report.pdf

19. Givens, D. I. (2018), 'Dairy Foods, Red Meat and Processed Meat in the Diet: Implications for Health at Key Life Stages'. *Animal*, 12(8), pp. 1709-1721

20. Alexander, D., Weed, D., Miller, P. and Mohamed, M.A. (2015), 'Red Meat and Colorectal Cancer:A Quantitative Update on the State of the Epidemiologic Science'. *Journal of the American College of Nutrition* 34, pp. 521–543

21. NHS, *Meat in your diet: Eat well.*
https://www.nhs.uk/live-well/eat-well/meat-nutrition/.

22. Meat Advisory Panel (2017), *Red Meat: Cutting through the confusion.*

23. Ibid.

24. Willett, W. C., and Ludwig, D. S. (2020), 'Milk and Health'. *N Engl. J Med* 382, pp. 644-54

25. Song, M., Fung, T.T., Hu, F. B. et al. (2016), Association of animal and plant protein intake with all-cause and cause-specific mortality. *JAMA Intern Med* 176, pp. 1453-63.

26. AHDB, *Share the facts about red meat and health.*
https://ahdb.org.uk/redmeatandhealth.

27. NHS, *How to get more fibre into your diet.* https://www.nhs.uk/live-well/eat-well/how-to-get-more-fibre-into-your-diet/

28. Ibid.

g. Insulin-Like Growth Factor-1

Insulin-like Growth Factor-1 (IGF-1) is a growth hormone we naturally produce, mainly in our liver, which has a similar molecular structure to insulin. We need IGF-1 to grow from birth into adulthood, but too much IGF-1 in adulthood can cause health problems.

Feeding the body too much animal protein causes the liver to pump out more IGF-1 than we need, sending out a message to our cells to grow. An unwanted accelerated growth promotion, which can transform normal cells into cancer cells, help them survive and then help them spread elsewhere in the body. Elevated IGF-1 is also associated with atherosclerosis and type 2 diabetes.

A prime source of IGF-1 is cow's milk (to achieve high milk yields cows are bred to produce higher levels of IGF-1),[1] which, from Willet and Ludwig (2020), is strongly correlated with breast cancer, prostate cancer, and other cancers, but most consistently with the increased risk of prostate cancer [2]

Plant-based proteins are not associated with having the same affect. In fact, according to a 2002 study by Allen et al., a plant-based diet is actually associated with lower circulating levels of total IGF-1.[3]

Although IGF-1 levels will eventually fall after you stop eating animal-based foods, most people eat these foods regularly throughout the day and keep their IGF-1 level harmfully topped up.

Citations:

1. Echternkamp, S. E., Aad, P.Y., Eborn, D. R. and Spicer, L. J. (2012), *Increased abundance of aroma- tase and follicle stimulating hormone receptor mRNA and decreased insulin-like growth factor-2 receptor mRNA in small ovarian follicles of cattle selected for twin births. J Anim Sci* 90, pp. 2193-200

2. Willett, W. C., and Ludwig, D. S. (2020), 'Milk and Health'. *N Engl. J Med* 382, pp. 644-54

3. Allen, N. E. et al. (2002), 'The Association of Diet with Serum Insulin-like Growth Factor I and its Main Binding Proteins in 292 Woman Meat-Eaters, Vegetarians, and Vegans'. *Cancer, Epidemiology, Biomarkers & Prevention* 11(11), pp. 1441-8.

h. Choline & L-carnitine

The NFU's sixth myth in cancer also misses other health concerns associated with the intake of choline and *L*-carnitine found in animal-based foods.

Choline, found in red meat, egg yolks, dairy products, other animal-based foods, and medical supplements, has been linked with cancer, stroke, diabetes, heart attacks and colon cancer. The role of choline is described by the Cleveland HeartLab:

Choline interacts with gut bacteria to make the blood more prone to clotting by helping to produce a compound called, trimethylamine N-oxide (TMAO).

Elevated blood levels of TMAO have been linked to increased risk of heart attack, stroke and death, and research shows that TMAO may directly contribute to the narrowing of artery walls through plaque build-up.[1]

Toxic TMAO is an organic compound produced from trimethylamine (TMA). Choline, *L*-carnitine and lecithin are converted to TMA by bacteria in the gut. TMA then passes into the blood stream, where it is taken and oxidised into TMAO by the liver, and recirculated around in the blood.[2]

Recent research showing the risk of TMAO, by Weifei Zhu et al. (2017), follows other research linking cardiovascular and other diseases to the conversion of dietary nutrients like choline, lecithin and *L*-carnitine into toxic TMAO.[3]

The health risk associated with ingesting carnitine in red meat and other animal-based foods is unnecessary, as our body naturally produces all it needs through the synthesis of carnitine in the liver and kidneys. Our bodies do not need choline from animal-based foods, as it is sufficiently provided through a balanced WFPBD (plant foods also contain a small amount of carnitine).[4]

Research has shown that plant sources of choline and carnitine eaten within a healthy WFPBD diet does not result in TMAO, as a WFPBD diet does not produce TMA-promoting gut bacteria. When animal foods are eaten, different *bad* gut bacteria develop, which in turn metabolise the carnitine and choline into TMA.[5]

This is unfortunate for the egg industry, which promotes choline in eggs as one of its nutritional benefits, and in knowledge of the research available.[6]

TMAO is removed from our bodies by our kidneys, and within eight hours or so is back to baseline. This natural process is beneficial to the livestock industry in rigging their own studies, which is exactly what the egg industry has done. In one industry-funded study,[7] the researchers take measurements from test subjects twelve hours after eating eggs, and thereby producing results which would not show an increase of plasma TMAO.[8] Ho hum!

Citations:

1. Cleveland Heart Lab (2017). *Choline, TMAO and Heart Health.* http://www.clevelandheartlab.com/blog/choline-tmao-heart-health/.
2. Dr. Michael Greger M.D. (26 April 2013), NutritionFacts.org. *Carnitine, Choline, Cancer, & Cholesterol: The TMAO Connection,* Volume 13
3. Zhu, W., Wang, Z., Wilson Tang, W. H. and Hazen, S. L. (2017), 'Gut Microbe-Generated Trimethylamine N-Oxide from Dietary Choline is Prothrombotic in Subjects'. *Circulation* 135(17), pp. 1671-73.

4. Dr. Michael Greger M.D. (26 April 2013), NutritionFacts.org. *Carnitine, Choline, Cancer, & Cholesterol: The TMAO Connection,* Volume 13
5. Ibid.
6. Ibid.
7. West, A.A. (2014), 'Egg n-3 fatty acid composition modulates biomarkers of choline metabolism in free-living lacto-ovo-vegetarian woman of reproductive age'. *J Acad Nutr Diet* 114(10), pp. 1594-600.
8. Dr. Michael Greger M. D. (31 December 2019), NutritionFacts.org. *How the Egg Industry Tried to Bury the TMAO Risk*

i. Haem Iron

Iron is an essential nutrient, which performs many vital functions in the body, including the transport of oxygen. We can't live without it, but we can live without animal-based iron (haem iron). Our bodies absorb only a fraction of the iron present in foods. Haem iron, found in large amounts in red meat, is more easily absorbed than plant-based non-haem iron, but that is where any advantage stops.

Although haem iron is better absorbed, it is associated as a potential catalyst for the endogenous production of *N*-nitroso compounds and certain aldehydes, both of which are carcinogenic.[1]

Haem iron is also associated with a higher risk for heart disease. It is thought haem iron acts as a potential pro-oxidant, which has been shown to contribute to the development of atherosclerosis by 'catalysing the production of hydroxyl-free radicals and promoting LDL oxidation.'[2]

In a study by Wei Yang et al. (2014),[3] in a dose–response analysis, an increase in haem iron intake of 1 mg per day appeared to be significantly associated with a 27 per cent increase in risk of coronary heart disease.

Research by Joanna Kaluza et al. (2013)[4] found an association between haem iron and strokes, where in a population-based, prospective study of men, intake of haem iron, but not non-haem iron, was associated with an increased risk of total stroke.

Furthermore, higher haem iron intake and increased body iron stores have been found to be significantly associated with a greater risk of type 2 diabetes; whilst dietary total iron, non-haem iron, or supplemental iron intakes were not significantly associated with type 2 diabetes risk – 16 per cent for a 1 mg per day increment of haem iron intake.[5]

Yet another study showed up to a 12 per cent increase in risk of cancer for every 1 mg per day of haem iron.[6]

Turning to gene expression signatures, Tram Kim Am et al. (2014) looked to 'characterise the mechanisms underlying meat-related lung carcinogenesis, and to test the haem-iron hypothesis' by examining 'the gene expression patterns in a subset of the same subjects with available dietary meat information and gene expression data derived from fresh lung adenocarcinoma tissues'.[7] Using the lung tumour samples, they identified a significant haem-related gene expression, which may also extend beyond lung cancer.

Most of us do not need to worry about iron as a WFPBD contains a sufficient source of non-haem iron in foods like: whole-grain cereals and breads; dried beans and legumes (e.g. chickpeas and lentils); dark green leafy vegetables; dried fruits, nuts and seeds. All the iron you need from plants without all the associated risks of haem iron in red meat. If you need more iron then it can be simply, and cheaply, supplemented.

Citations:

1. Alexander, D., Weed, D., Miller, P. and Mohamed, M.A. (2015), 'Red Meat and Colorectal Cancer: A Quantitative Update on the State of the Epidemiologic Science'. *Journal of the American College of Nutrition* 34, pp. 521–543, cited in Givens, D. I. (2018), 'Dairy Foods, Red Meat and Processed Meat in the Diet: Implications for Health at Key Life Stages'. *Animal* 12(8), pp. 1709-1721

2. Hunnicutt, J. et al. (2014), 'Dietary Iron Intake and Body Iron Stores Are Associated with Risk of Coronary Heart Disease in a Meta-Analysis of Prospective Cohort Studies.' *J. Nutr.* 144(3), pp. 359-366

3. Yang, W. et al. (2014), 'Is heme iron intake associated with the risk of coronary heart disease? A meta-analysis of prospective studies.' *Eur J Nutr.* 53, pp. 395-400

4. Kaluza, J. et al. (2013), 'Heme Iron Intake and Risk of Stroke: A Prospective Study of Men.' *Stroke* 2013(44), pp. 334-339

5. Bao, W. et al. (2012), 'Dietary iron intake, body iron stores, and the risk of type 2 diabetes: a systematic review and meta-analysis.' *BMC Medicine* 10, article 119

6. Fonseca-Nunes, A. et al. (2014), 'Iron and Cancer risk – A Systematic Review and Meta-analysis of the Epidemiological Evidence.' *Cancer, Epidemiology, Biomarkers & Prevention* 23(1), pp. 12-31

7. Lam, T. K. et al. (2014), 'Heme-related gene expression signatures of meat intakes in lung cancer tissues'. *Mol Carcinog.* 53(7), pp. 548-56

j. Calcium, Osteoporosis & Female Cow Sex Hormones

Calcium is the most abundant mineral in our bodies, 99 per cent of which is stored in our bones and teeth. The rest is distributed in our blood, muscles and other tissue.

Calcium is necessary to keep our bones strong and healthy, and for muscular, nerve, hormone, enzyme and other functions too. Calcium's most commonly appreciated value is its importance in preventing osteoporosis.

Calcium is also claimed, amongst other things, to be beneficial towards reducing cardiovascular disease, regulating blood pressure and hypertension and reducing the risk of certain cancers, such as colon, rectum and prostate cancers. In its carbonate form it is also frequently used as an antacid to treat heartburn.

The dairy industry keenly promotes the value of calcium in milk and other dairy products for strong teeth and bones, to prevent osteoporosis.

A daily supply of calcium, along with vitamin D and other nutrients, is necessary to support our bones' continuous remodelling process. We use calcium elsewhere too, for example in our nails and hair and to replace what we lose through our sweat, urine and faeces.

The disorder osteoporosis, a degenerative weakening of bone density and quality, results in fragile bones which break more easily. The condition develops slowly over time, often going unnoticed until a fracture occurs. Fractures in older people are most common at the wrist, spine and hip and in children in the lower forearm.[1] There are about 500,000 bone fractures caused by osteoporosis each year in the UK, or 1,400 per day.[2]

This is very high for a country that is one of the world's biggest consumers of calcium-rich dairy products, and this disparity is not restricted to the UK. The incidence of osteoporosis is much higher in countries which consume more dairy per capita. America and Switzerland are the leading consumers of dairy and have the highest occurrences of osteoporosis in the world. In contrast, countries that eat very little dairy-based foods have the lowest occurrence:

The countries with the highest rates of osteoporosis are the ones where people drink the most milk and have the most calcium in their diets. The connection between calcium consumption and bone health is actually very weak, and the connection between dairy consumption and bone health is almost non-existent.[3]

In Europe, women are four times more likely to suffer fractures than men (EU6, 2015)[4] at a rate of 16 million women to four million men. Together this puts a huge strain on our health services and is a growing problem.[5]

The reason for increased rates of fracture is not satisfactorily known. There are likely to be several lifestyle factors, which, more or less, influence the onset of osteoporosis.

Calcium deficiency, reduced sunlight exposure, body mass index, as well as a lowering of physical activity are in the mix. Other lifestyle factors, including what we put into our mouths day in and day out, will be important too, as it can change the make-up of our gut bacteria and influence which nutrients we take in, and how.

To combat osteoporosis, it is very easy to reach for another 200 ml of milk or a few more grams of cheese, which the dairy industry keenly promotes. Or you could take supplements the pharmaceutical industry would like you to take. However, not all of the 20 million men and women in the EU with osteoporosis are deficient in dairy products.

A failure to work your bones, whatever your source of calcium, is an important factor. You can't eat protein, sit back and hope to build strong muscles while doing nothing. The same is true for calcium and your bones. Exercise is essential in adults to maintain bone health. Children build bones as they grow, but exercise is also an important component.

It is important to recognise that osteoporosis is *not* a dairy deficiency disease. If it was, we would all would have shrunk and crumbled away long before we started drinking milk around 7,500 years ago.

There are studies afoot, including in the UK, which show guideline levels for calcium in children are not being reached as they get older. This has been linked to a significant drop in milk consumption, with avoidance linked to stunted growth and poorer bone health.[6]

Other studies on milk and meat in developed countries show the inclusion of these foods reduces the risk of stunted growth. If children are not getting the right nutrition then stunting and other health problems will occur. Growth problems in children can just as easily be reversed with a WFPBD. Stunting is *not* an animal-food deficiency disease either, and we mustn't fall into this trap. It is no good solving these problems with meat, eggs, fish and dairy in children whilst storing up other diseases for later in life associated with eating the same foods. Animal foods may be necessary if a good source of appropriate plant foods is not available, yes, and I'm not suggesting otherwise, just pointing out the truth.

In our ancestors' history, the majority of people were unable to drink and digest milk without getting sick. Even now around two-thirds of the world's population have a degree of lactose intolerance (closer to 10–15 per cent for some Northern Hemisphere countries, including the UK, whose people have selectively adapted to drinking milk). Lactose intolerance can cause symptoms such as flatulence, irritable bowel syndrome and diarrhoea. Intolerance is higher within people of Asian or African-Caribbean descent.

Milk as a baby is vital, and as such we are born with the lactase enzyme which allows us to drink our mother's milk. Once weaned, we no longer produce the enzyme and we start to develop an intolerance. However, along the line, since consuming cow's milk, a degree of lactose tolerance persistence has developed within populations.

Evidence suggests that the amount of animal-based calcium in our diet is likely to be part of the growing osteoporosis crisis. We need to start weaning ourselves off it and instead eat a good variety of WFPBD sources of calcium. We are not baby calves, and milk lacks many of the more essential nutrients found in plants. Some nutrients we know about, and others we are yet to discover, which may play an important role in making the best use of the calcium we ingest.

At this point I'm not suggesting people suddenly stop consuming dairy products. After decades of the industry and government telling us dairy is essential, the onset of the junk food diet, a lack of plant-based eating and a lack of alternatives/palatability to alternatives has created a current calcium-milk dependency we need to overcome. We need a transformative change in what we grow and eat, along with a programme of education to help people make better health choices. We need to wean ourselves off dairy.

Unless you know what to do, or have got the right medical advice, baby steps are best, replacing your calcium need bit by bit, as you introduce more healthy plant-based sources. At first, if you have been eating a poor diet, you may not have enough of the right *good* bacteria to cope with a dramatic shift of diet. This can lead to gut issues, until your gut microbiome adjusts. Whenever in doubt, seek the *right* medical advice.

Regardless of what many of us want to concede, for most of our evolution, which stretches back around 20 million years, we ate plants. Our ancestors only started eating meat around two million years ago and milk around 7,500 years ago.

Neither meat nor dairy are essential. A balanced WFPBD consisting of fruit, green vegetables like broccoli and kale, legumes, seeds and nuts give us all the calcium, *and protein,* we need. We are the only species that drinks the mammary fluids of another.

Table 4 shows the calcium content of a variety of plant-based foods.[7]

Table 4: Calcium packed foods

Food	Calcium (mg/100 g)	Food	Calcium (mg/100 g)
Peanuts	92	Soft tofu	111
Pistachio nuts	105	Mung beans	27
Pumpkin seeds	133	Green peas	27
Sunflower seeds	78	Quinoa	17
Cashew nuts	43	Spinach	136
Walnuts	98	Collard greens	232
Wholewheat bread	161	Adzuki beans	28
Soybeans (green)	145	Kale	254
Extra firm tofu	176	Brussels sprouts	36
Lentils	19	Broccoli	40
Chickpeas	49	Sweet potatoes	30
Black beans	27	Chinese cabbage	105
Kidney beans	35	Butternut squash	48
Green/yellow split peas	14		
Broad bean (Fava)	36		

Source: My Food Data.

Milk is also designed to grow a calf at an extraordinary rate in a very short space of time to a weight of several hundred kilos. Milk contains very high levels of naturally occurring growth hormones, like estrogen and progesterone, which could impact on our health.

The level of estrogen in cow's milk remains relatively low for the first 120 days (below about 150 pg/ml) before rising rapidly in later stages of pregnancy to about 950 pg/ml.[8] For perspective, this is 10–20 times less estrogen than you would find in a birth control pill.[9]

Whilst apparently low, when cow's milk is ingested, tests on men have shown a considerable rise in levels of urine estrone (525 to 810 ng/h), estradiol (155 to 270 ng/h), estriol (2 to 88 ng/h) and pregnanediol (12 to 16 ng/h) after drinking milk.[10]

For children, average age eight, after drinking around two cups of milk, about 600 ml, within hours it causes the level of these sex hormones to shoot up (Table 5).

Table 5: Comparison between basal excretion volumes (basal) and maximum excretion volumes (peak) of urine estrone, estradiol, estriol and pregnanediol before and after intake of cow milk in prepubertal children

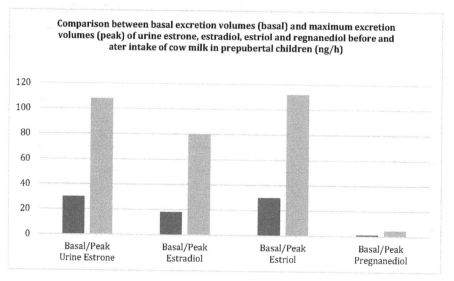

From this study the research authors, Kazumi Maruyama et al. (2010),[11] concluded that:

> *The present data on men and children indicate that estrogens in milk were absorbed, and gonadotropin secretion was suppressed, followed by a decrease in testosterone secretion. Sexual maturation of prepubertal children could be affected by the ordinary intake of cow milk.*

In modern dairy farming the amount of sex hormones present in milk has increased over the years. The most sex hormones are found in concentrated dairy products like cheese and butter. It has been hypothesised that the consumption of these sex hormones has contributed to lowering the age girls get their first period. Whether dairy food plays a role in this is undecided, with obesity most closely associated, but the ingestion of female cow sex hormones is a potential factor which can be avoided on a WFPBD.

Estrogens are also known to have a core role in the cause of endometrial cancer. The potential risk was studied by the Harvard School of Public Health, a prospective cohort study with 68,019 female participants in the Nurses' Health Study aged 34-59.[12] In conclusion, the researchers observed a marginally significant overall association between dairy intake and endometrial cancer and a strong association among postmenopausal woman who were not using estrogen-containing hormones. Total dairy intake was not significantly associated with a risk of preinvasive endometrial cancer.

Plant estrogens (phytoestrogens), including those found in soya beans, do not behave in the same way as animal estrogens. In fact, they have been shown to have anti-estrogenic effects.

Table 6 provides a comparison between human and cow's milk. Human milk contains around 1.3 g per 100 g of protein, whilst whole cow's milk contains around 3.3 g per 100 g of protein. The high protein level is another reason calves grow quickly, around forty times faster than human babies.

Table 6: Comparison of the mineral and vitamin components of cow's milk and human milk[13]

Component	Cow's milk (semi-skimmed, pasteurised) per 100 g	Human milk (mature) per 100 g
Sodium	43	15
Potassium	156	58
Calcium	120	34
Magnesium	11	3
Iron	0.02	0.07
Iodine	30	7

Source: Food Standards Agency (2002).

Table 7: Composition of some of the key nutrients found in human milk[14]

Component	Mean value for mature human milk (per 100 ml)
Energy (kj)	280
Energy (kcal)	67
Protein (g)	1.3
Fat (g)	4.2
Carbohydrate (g)	7.0
Sodium (mg)	15
Calcium (mg)	35
Phosphorus (mg)	15
Iron (mcg)	76
Vitamin A (mcg)	60
Vitamin C (mg)	3.8
Vitamin D (mcg)	0.01

Source: NHMRC Dietary Guidelines for Children and Adolescents in Australia, 2003.

A growing number of people are turning away from cow's milk and drinking plant-based alternatives like oat and pea milks, which also have a much smaller environmental footprint. The dairy industry's favourite target, almond milk, has its own issues, and if non-European almonds are used in its manufacture it's a plant milk I avoid. The main issue with almond milk is the amount of water used to grow the almonds in places like California. However, naturally irrigated almonds, grown in places like Spain, can be sourced. I will return to this subject later.

When I converted to oat milk and then also to pea milk, before changing to a WFPBD, and in disgust of the dairy pollution I dealt with on a daily basis, I quickly developed a taste. I enjoy these milks more than I did dairy. As a bonus from the switch, it cured IBS issues I had suffered from the age of six – within a few weeks!

A concern when replacing milk is a reduction of iodine and B12 in your diet. Both are essential nutrients which are present in milk but which can be easily and cheaply supplemented; vitamin D too. Some plant milks, like Oatly oat milk, have iodine and B12 added. Iodine is also found in iodised salt.

Vitamin B12 (cobalamin) is another frequent target of the livestock industry against plant-based diets in general but can be readily and cheaply taken as a supplement and is found in many fortified foods too. In the UK many dairy cows, beef cattle, sheep and goats are supplemented with cobalt to enable the synthesis of B12 by the animal's rumen microorganisms. Chickens and pigs may also receive a vitamin and mineral supplement containing B12.

Farmers regularly use supplements in livestock diets, so what's the difference? You can safely take a B12 supplement, whilst missing out the middle cow, sheep, etc. And why take a B12 supplement via a dead animal, along with cholesterol, saturated fat, animal protein and excessive haem iron which are associated, amongst other ills, with well-studied and documented health concerns, when there is an alternative? Furthermore, it is not an unfortunate necessity to cause pain and suffering to animals to get your B12. B12 is an extremely poor livestock industry target against a WFPBD, but it is essentially all they've got (but haven't really). Add all the environmental impacts embedded in livestock farming and its case closed for a WFPBD.

Citations:

1. Viva Health (2012). *Break Free: How to build healthy bones and what really matters in the prevention of osteoporosis.* https://www.vivahealth.org.uk/bones/resources/break-free
2. National Osteoporosis Society, 2018
3. Lanou, A. (2018), quoted in *Scientists link drinking milk with Osteoporosis.* https://iphysio.io/osteoporosis/
4. International Osteoporosis Foundation, *Broken Bones, Broken Lives: A roadmap to solve the fragility fracture crisis in Europe.* http://share.iofbonehealth.org/EU-6-Material/Reports/IOF%20Report_EU.pdf
5. Ibid.
6. Bates, B. et al. (2014), *National Diet and Nutrition Survey, results from Years 1-4 (combined) of the rolling programme (2008/2009-2011/12).* Public Health England and the Food Standards Agency, London, UK
7. My Food Data. https://tools.myfooddata.com/nutrition-facts.php?food=172421&serv=wt1
8. Maruyama, K. et al. (2009), 'Exposure to exogenous estrogen through intake of commercial milk produced from pregnant cows.' *Pediatrics International,* 52(1), pp. 33-38
9. Dr. Michael Greger M. D. (25 Sept 2019), NutritionFacts.org. *The Effects of Hormones in Dairy Milk on Cancer.* https://nutritionfacts.org/video/the-effects-of-hormones-in-dairy-milk-on-cancer/

10. Maruyama, K. et al. (2009) 'Exposure to exogenous estrogen through intake of commercial milk produced from pregnant cows.' *Pediatrics International* 52(1), 33-38

11. Ibid.

12. Ganmaa, D. et al. (2012), 'Milk, dairy intake and risk of endometrial cancer: a 26-year follow-up.' *International Journal of Cancer* 130(11), pp. 2664-2671

13. Food Standards Agency (2002), 'Table 2.0, A Comparison the mineral and vitamin components of cow's milk and human milk', cited in Viva, *A Comparison Between Human Milk and Cow's Milk.* https://www.viva.org.uk/white-lies/comparison-between-human-milk-and-cows-milk.

14. National Health and Medical Research Council (2003), *Dietary Guidelines for Children and Adolescents in Australia, Incorporating the Infant Feeding Guidelines for Health Workers*

k. The Blue Zones & The China Study

Dr. Colin Campbell's *The China Study*[1] led to the publication of over 8,000 statistically significant associations between diet and chronic disease, many of which were associated with people who ate the most animal-based foods. Conversely, it showed positive health associations and a tendency to avoid chronic disease within people eating a predominantly WFPBD.

The renowned Drs Campbell, Esselstyn, Greger and many others have shown that various diseases we suffer in the West can be a consequence of our diet and lifestyle. They are not a consequence of growing old. They are diseases which can be reversed, or at the very least mediated, by eating a well-balanced WFPBD.

We are still waiting for an animal-based diet which can reverse cardiovascular disease, diabetes, etc.

A WFPBD features heavily in populations with the longest life expectancy, dubbed Blue Zones and popularised by Dan Buettner.[2] In Blue Zone diets, meat-eating is very low and occasional, with vegetarians consuming only low amounts of dairy.

Among the longest-living people in the world are the Californian Adventists in Loma Linda, vegetarians and vegans. Similarly, the Blue Zone in Okinawa, Japan's 'land of immortals', where less than 1 per cent of fish and meat, and less than 1 per cent dairy, feature in their diet. The Okinawans mainly eat vegetables, beans and other legumes, rice and purple and orange sweet potatoes. Sweet potatoes make up about 70 per cent of their calories.

The traditional Okinawan diet is one of the world's most anti-inflammatory diets, being rich in antioxidants. It is high in carbohydrates (85 per cent), low in protein (< 10 per cent) and low in fat (6 per cent), with around 2 per cent saturated fat. They combine their diet with daily, but general, physical activity and also live a more relaxed lifestyle than most.

Another is the Sardinia Blue Zone. Largely plant-based, except for occasional meat; featuring whole-grain bread, beans, garden vegetables and fruit.

Replace Blue Zone diets with animal-based foods, junk food, refined carbohydrates and oils, and health outcomes, quality of life and longevity diminish.

This effect was also shown in *The China Study*, which compared the health status of China's relatively genetically homogenous population across different geographical locations.[3]

A catalyst for the study was the incredible difference in cancer rates found across China, discovered in the early 1970s following a major survey instigated by the then premier of China, Zhou Enlai, who at the time was dying of cancer.[4] The study became known as *The Cancer Atlas*.

The study showed a clear geographical variation across China in rates of cancer, up to 100 times (10,000 per cent) more.[5] This begged the question, why? Was it influenced by diet, environment, poverty vs. affluence, etc? *The Cancer Atlas* also gave the researchers access to disease mortality rates for a host of other different disease kinds, including specific cancers, heart disease and infectious diseases.[6]

Grouping diseases, the researchers found diseases of affluence (nutritional extravagance) consisted of cancer (colon, lung, breast, leukaemia, childhood brain, stomach and liver), diabetes and coronary heart disease. In contrast, diseases of poverty (poor nutrition and poor sanitation) consisted of pneumonia, intestinal obstruction, peptic ulcer, digestive disease, pulmonary tuberculosis, parasitic disease, rheumatic heart disease, metabolic and endocrine disease other than diabetes, diseases of pregnancy, and many others.[7]

The researchers extended their own study to sixty-five counties located in rural to semi-rural parts of China, involving 6,500 adults. By the end of the study, they had over 8,000 statistically significant associations between lifestyle, diet and disease variables.[8]

The China Study found that 'plant-based diets are linked to lower blood cholesterol, whilst animal-based foods were allied to higher blood cholesterol; animal-based foods are linked to higher breast cancer rates and plant-based foods to lower rates; fibre and antioxidants from plants are linked to lower risks of cancers of the digestive tract; and plant-based diets and an active lifestyle result in a healthy weight, yet permit people to become big and strong.'[9]

In America 15–16 per cent of total calories come from proteins, with upwards of 81 per cent of that from animal-based foods. In rural China, 9 to 10 per cent of total calories come from protein, with around 10 per cent from animal-based foods.[10] A marked difference between the countries, which also tallies with a marked difference in health outcomes between them.

Take a person largely eating a plant-based diet, change it to a Western diet rich in animal-based protein, and disease follows. It doesn't matter if a person moves to a more affluent part of their country or to another country; adopting a Western diet results in the same outcome. Genetic differences between populations make little difference.

Citations:

1. Campbell, T. Colin and Campbell, Thomas M. (2016), *The China Study: Revised and Expanded Edition: The Most Comprehensive Study of Nutrition Ever Conducted and the Startling Implications for Diet, Weight Loss, and Long-Term Health.* BenBella Books, Dallas, TX
2. Dan Buettner (2010). *Blue Zones: Lessons for Living Longer from the People Who've Lived the Longest.* National Geographic., Washington, D. C.
3. Campbell, T. Colin and Campbell, Thomas M. (2016), *The China Study: Revised and Expanded Edition: The Most Comprehensive Study of Nutrition Ever Conducted and the Startling Implications for Diet, Weight Loss, and Long-Term Health.* BenBella Books, Dallas, TX
4. Ibid.
5. Ibid.
6. Ibid.
7. Ibid.
8. Ibid.
9. Ibid.
10. Ibid.

I. My Journey Part II

Since having a stroke, I have learned a lot about myself, food and how food impacts your health.

During my early research, I also took time to look into the current status of climate change and South American deforestation. I was attracted to a number of papers on the production of soya beans, a large amount of which is imported into Europe as animal feed. Of particular interest was Pablo Lapegana's excellent book, based on his PhD thesis, *Soybeans and Power: Genetically Modified Crops, Environmental Politics and Social Movements in Argentina*, which gave me an important glimpse into the murky world of soya bean cultivation for livestock production.

I also learned the UK imported 2,255,426 t of soya bean meal in a year, 2015–2016,[1] primarily from South America and in particular Argentina. Around one million hectares of former rainforest is necessary to grow that amount of soya bean meal, largely from genetically modified soya beans.

The amount was disturbing. Whether I was eating meat, eggs, dairy (at least non-organic) or farmed fish, with it I was eating a rainforest and contributing to climate change. That did not sit comfortably with me, on top of already being thoroughly depressed by the tide of pollution associated with UK and global livestock farming in general.

It was on that awakening that I declared to my family I was going to adopt a whole-food plant-based diet (WFPBD), much to the delight of my eldest daughter. It was a bit of a shock to my wife, vegetarian since eleven, who followed my lead a few months later.

To ensure I wasn't missing out on any nutrition I made use of the Cronometer nutrition tracker app.[2] Pleasantly, I discovered my new diet was more nutritionally complete.

Within a few weeks bloating and stomach cramps I'd suffered since the age of six disappeared.

Migraines I regularly suffered disappeared too and I was able to stop taking the medication I had been on since 2001. I have not suffered a migraine since – touch wood. I rarely even get a headache.

I also found that a WFPBD would help lower my risk of having kidney stones, provided I didn't overeat on spinach. Good job too, as I have suffered kidney stones five times over the years, four of them in excruciating pain, with three taking up valuable bed space in hospital.

If that wasn't enough, the severe hay fever I suffered from as a child, whilst it's not gone away completely, is now mild. I can go days without a tablet, when before I had to take a tablet and a nasal spray every day from May right through until the end of summer.

At the start I mentioned I was very fit and well-muscled before my stroke. The first two years after my stroke I lost a lot of muscle mass. Steadily I regained my original strength and then some, powered by extra raw hemp protein and complex carbohydrates from my WFPBD. Personal to me, though experienced by many others, the gains I made exceeded anything I have done using animal-based proteins. All the benefits, without any of the bad stuff.

I have even been able to go head-to-head with lads thirty years my junior in fitness challenges at my local gym. In a press-up challenge, only one person was able to beat me, and only then after he made three attempts. I left mine at the first attempt, with nothing else to prove – at least to myself.

When I had my stroke, I was lucky. It wasn't damaging enough to cause me any permanent health issues. But the real biggest *stroke of luck* is where that experience has taken me.

Citations:

1. HM Customs, 2015-2016
2. Cronometer. *Track your nutrition, fitness & health data*.
    https://cronometer.com/

2. Whole-food Plant-based Diet

The wonder of a whole-food plant-based diet (WFPBD) is the sheer variety and taste experiences many people unknowingly miss out on. The types of food available far outweigh animal-based foods, and can be cooked and enjoyed in many ways.

Many salad vegetables can be eaten raw, though it is sometimes advisable not to eat particular raw vegetables, like sprouted mung beans, as they may harbour unwanted bacteria which can't all be washed away. Source is all important here, as well-prepared salad sprouts you can buy in supermarkets are fine.

Contamination of plant foods usually results from exposure to animal manures. Where there is a risk, and you have underlying health concerns, are pregnant, etc., foods like sprouted mung beans can be fully steamed, cooked and eaten or chilled before adding to sandwiches and salads. No need to miss out.

Dried beans, like kidney beans, need to be soaked and properly cooked before eating. Cooking can be avoided using tinned beans, and other tinned foods such as lentils. Beans and lentils are an excellent source of protein and other essential nutrients, and are inexpensive compared to most animal protein. Tins can also be recycled over and over again.

Some foods like spinach should not be eaten in large volumes. Spinach is high in oxalate, which is rarely associated with an increased risk of oxalate kidney stones. No need to avoid spinach; it is very good for you, and can be enjoyed with a wide variety of other plant foods as part of a well-balanced, antioxidant-rich meal.

The addition of WFPBD-friendly curry pastes and other sauces and ingredients can be added to cooked vegetables. Your choice is endless, with no need to compromise on the tastes and textures you enjoy. Simply select the right vegetables and combinations. There has never been a better time to stop eating meat, or making meat the minor, rather than the moderate to major, part of your diet.

There are numerous talented WFPBD cooks and nutritionists and a wide range of excellent cookbooks to meet any taste. In the vegan community there is an incredible amount of support.

Do not be put off by a few failed YouTube vegans who hit the headlines. They are scarce and make headlines for switching back to animal-based foods, not because a WFPBD is a health issue but from the restricted *vegan* diets and lifestyle they choose to follow and promote. If you are having issues, then seek professional health advice from a qualified nutritionist.

The range of foods available is huge and goes well beyond the following:

Whole grains:	Wheat, oat, brown rice, rye, barley, buckwheat, quinoa, wholewheat couscous, corn, amaranth, spelt, teff, bulgur, millet, etc.
Fruit:	Apple, pears, bananas, peaches, nectarines, citrus fruits, pineapple, berries (raspberry, blackberry, blueberry, strawberry, goji, bilberry, acai, cranberry, gooseberry, boysenberry, loganberry, mulberry, tayberry, grapes).
Vegetables:	Leafy greens, brassicas and other (broccoli, kale, spinach, brussels sprouts, cabbage, cauliflower,

kohl rabi, pak choi, leek, celery, Swiss chard, parsley, asparagus, lettuce), pod and seed (bean, peas, sweetcorn, okra), roots and tubers (beetroot, parsnip, carrot, celeriac, turnip, potato, sweet potato, yam, radish, ginger), squashes, stalks and shoots, bulbs (onion, garlic), mushrooms (button, chanterelle, cremini, shiitake oyster, enoki, portabella, porcini, morel), vegetable fruits (bell pepper, chilli pepper, tomato, eggplant) and squashes (zucchini, butternut, spaghetti, pumpkin).

Fats: Avocado, nuts and seeds.

Legumes: Beans (mung, lima, butter, soya, fava, azuki, kidney, pinto, cannellini, black, black-eyed, runner, green), peas (incl. green and yellow split peas), chickpeas, lentils, peanuts.

Nuts and seeds: Cashews, pecan, walnut, pistachio, chestnut, Brazil, hazelnut, macadamia, almond, sunflower, pumpkin, flaxseed, chia, cacao, pine, cedar (type of pine nut), acorn, coconut, ginkgo nuts, olives.

Plant-based milks: Oat, pea, soya, coconut, almond, cashew, hazelnut, hemp, walnut, rice.

Herbs and spices: Basil, coriander, dill, mint, parsley, rosemary, chives, fennel, oregano, sage, thyme, turmeric, curry, black and red pepper.

Other proteins: Tofu, tempeh, plant-based powders (pea, soya, hemp, rice, pumpkin, Lentein).

No food is without its own environmental footprint, and the UK livestock sector is right to question a few vegetable foods. However, the majority of plant-based foods have much lower carbon, water, land and other environmental footprints.

I choose not to drink almond milk made from *Californian* almonds, a common target of the livestock industry, due to the nut's water footprint. I prefer oat and pea milks, but this doesn't mean avoiding plant-based milk made from rain-fed *European* almonds, which *do not* have the water footprint that concerns the UK dairy industry so very much. Another plant-based food the UK livestock industry likes to erroneously target is soya beans associated with South American deforestation, which are nearly all used to feed their livestock. The vast majority of plant-based UK and European Union soya bean foods are not associated with this deforestation.

Avocados also get cited by the livestock sector, and quinoa from Peru is yet another target, but you can buy organic and British-grown quinoa. Really, you can! The vast majority of avocados and quinoa are also eaten by non-vegans. Other plant foods may have an issue, but not on the scale of animal-based foods. Often, it is simply a case of being *source aware* when selecting plant foods, and if you are still concerned, you have thousands of other choices.

Genetically modified (GM) soya bean meal used as an animal feed, which most of it is, has been linked to reduced animal health outcomes, particularly in pigs.[1] In pigs, symptoms include, chronic diarrhoea, birth defects, reproductive problems, reduced appetite, bloating, stomach ulcers, weaker and smaller piglets, and reduced litter size.[2]

This may not be the soya beans, or even because most soya meal feed comes from genetically modified crops, but the level of RoundUp glyphosate weed killer present in the beans (RoundUp perhaps as a whole rather than glyphosate in isolation). A Danish farmer, who is not alone and is backed by scientific studies and other farmers, after experiencing health problems in his pigs, switched to non-GM soya bean meal and started to see health improvements within a few days.

The very few nutrients and vitamins you may struggle to get, like B12 and iodine, can be supplemented. A lot of cereals, non-organic plant milks and nooch (nutritional yeast) are supplemented with B12, and some plant milks have added iodine. The equivalent amount of calcium found in cow's milk is added to most non-organic plant milks. Eat enough foods like ground flaxseeds, walnuts, chia seeds, hemp seeds and even Brussel sprouts and you will get sufficient omega-3 fatty acids. Try not to concentrate on one source but a range, as each contain their own assortment of nutrients. Variety is truly the spice of life when it comes to plant-based nutrition.

Plant sources typically contain the essential form of omega-3 in the form of alpha-linolenic acid (ALA), which your body must first convert into the usable forms, eicosapentaenoic acid (EPA) and docosahexaenoic acid (DHA). The conversion rate of ALA to EPA and DHA varies, but eat the right amounts of these foods and you will get enough. A vegan algal oil supplement can also be taken, giving a direct source of EPA and DHA whilst cutting out the middle fish. Fish get their omega-3 from eating the algae too, or from eating the fish that eat the algae. You don't have to eat fish. Vegetable ALA sources are also rich in lots of other nutrients. Flaxseeds can even be used as an egg substitute in baked goods.

Whichever way you look at it, there is an alternative to animal-based foods.

Whatever our side of the plants vs. animals debate, what most of us should all happily agree on is that we do not eat enough plants or enough variety of plants, eat too much animal-based foods, and should avoid, or at least greatly limit, processed carbohydrates like white breads, pasta and refined sugars.

A WFPBD is certainly up there when it comes to achieving the best health outcomes, with no substantive evidence, at least that I have seen, to suggest small quantities of the best reared animal-based foods (e.g. up to 5 per cent) eaten alongside a predominantly health-promoting, plant-based diet will cause undue harm. Add in animal welfare concerns, the environment, or taken as the power of three, then a WFPBD is an absolute no-brainer with no evidence of harm, provided you do not choose a junk-food or restricted version of the diet.

A book which is really useful when it comes to untangling the *noise* in nutrition, and to keep you grounded in the face of sensational media headlines, Joe Bloggs down the pub, or a celebrity chef selling their latest fad-diet book before piling the weight back on, is *The Truth About Food: Why Pandas Eat Bamboo and People Get Bamboozled* by David Katz (2018).[3] This is a book I came across late in the day in my research, but which I would put towards the top of my list for must-reads – if only to help keep any of my own unintended bias in check.

David Katz has since joined up with Mark Bittman, author and Special Advisor on Food Policy at Columbia's Mailman School of Public Health, to publish the book *How to Eat: All Your Food and Diet Questions Answered* (2020).[4] This book is much shorter than *The Truth About Food*, and directly answers many questions on food and diet that have puzzled me over the last four years on my personal food journey to better health. I wish it was the first book I had read, and I thoroughly recommend it as a starting point. As with *The Truth About Food*, David Katz, along with Mark Bittman, address the subject objectively through the lens of science.

Citations:

1. Dr. Eva Sirinathsinghji (2012), *GM Soy Linked to Illnesses in Farm Pigs*. http://www.i-sis.org.uk/GM_Soy_Linked_to_Illnesses_in_Farm_Pigs.php.
2. Ibid.
3. Katz, David L. (2018). *The Truth About Food: Why Pandas Eat Bamboo and People Get Bamboozled*. Dystel & Goderich, New York, NY
4. Bittman, D. and Katz, D. L. (2020), *How to Eat: All Your Food and Diet Questions Answered*. Houghton Mifflin Harcourt, Boston, MA

3. Animal-based Foods

Removing or limiting meat, fish, eggs and dairy products and increasing whole, plant-based foods in your diet has been shown to improve health and longevity.

The transition to a WFPBD can be challenging or daunting to many at first, and it is important during the adjustment that you get all the nutrition you need while on your health journey.

Regardless of the health benefits, some people may not be ready to change completely. However, you can still reduce your exposure to animal unkindness in the interim by making better choices.

Whether you eat animal-based foods during transition, or you just want to reduce consumption, limit them to less than 10 per cent of your calorie intake and choose free-range and organic *genuine* pasture-raised eggs, beef, pork, poultry and dairy whenever possible. Buy local, and direct from the farm if workable (which saves transport emissions, but also question where they get their animal feed from as it may not be quite so local!). Many livestock farmers would prefer to sell direct, but many are not able due to current consumer/retail patterns. Consumers can help change that pattern through their choices, and help free farmers from unwanted shackles. Unfortunately, *local* does not translate into animal kindness and the avoidance of pollution – don't be conned.

The animal-based foods you buy will be more expensive, but you would be eating less of them, and instead making cheaper vegetables your plate's centre stage. Your plate does not need to be more expensive.

Be aware, *high animal welfare* is a subjective marketing label or gimmick, which at best means in the UK we are a less *unkind* than many other countries. Animal-based foods are portrayed as essential by the livestock industry, and so a level of unkindness is seen as necessary to feed us. The reality, though, is that animal products are not an essential food for our health and well-being, and so there is absolutely no need for any level of unkindness towards animals. Whilst we have developed a food system that has placed a reliance on livestock at this time, it does not need to be our future.

Eating less animal-based foods is a quick way to reduce our environmental and animal-welfare footprint. Switching to organic animal-based foods is not necessarily a guarantee you will be causing less pollution than conventional or even factory-farmed livestock. The only way to better guarantee your footprint, and to be kind to all animals, is not to eat them.

4. How Much Protein Do We Need?

In general, research suggests that adults need around 0.65 g of protein per kg body weight per day[1] based on a lean body weight (your inner *lean* self and not your total body weight). However, people vary in nature and to guarantee no one goes short, the actual protein requirement is adjusted by a two-times standard deviation to 0.83 g/kg *lean* body weight. This is similar to the US government recommended daily allowance (RDA) bar of 0.8 g/kg. Calculated for myself, based on 0.83 g/kg, gives me a protein need of 61.5 g/kg.

These figures are not too far away from the US suggested average daily protein requirements of 56 g per day for men and 46 g per day for women. In the UK the estimated Reference Nutrition Intake is set at 0.75 g per kilogram of body weight. This is comparable to the US target, which for adult men and women (19–50 years) equates to 56 g per day and 45 g per day respectively.[2] Growing infants and children and pregnant and breastfeeding women require a bit more protein. If you are an athlete and/or into your bodybuilding, you will need extra protein.

When training I routinely add another 25–50 g to my daily protein requirement. I also ensure I eat sufficient complex carbohydrates to keep myself fully fuelled. Most importantly, whatever exercise I do, I match what my body needs on a daily basis to ensure I do not go short on calories. Having trained seriously since the age of fifteen, I do that instinctively. In fact, since turning to a WFPBD, and eating very little processed food, not that I ate much junk food before, I find that straightforward. My body tells me when I genuinely need more, rather than screams at me to eat more sugars and fats which I don't need and would ultimately help put on the wrong kind of pounds. The net result is I comfortably build muscle and stay lean.

As would be expected, the meat, fish, egg and dairy industry want us to eat lots of their protein. They want to sell you their protein and it's not their concern whether you are eating too much. What you eat is your responsibility. They do not pick up the healthcare costs should you become sick, or sicker, from eating their products, a situation equally well-suited to the pharmaceutical industry, who benefit in kind from treating our illnesses.

For instance, as covered above, animal protein is associated with elevated levels of TMAO, IGF-1 and inflammation, which contribute to the development of heart disease, cancer and a host of other health-limiting diseases, and so it is better to use plant-based proteins to meet your needs. Plant-based proteins are not inferior to animal-based protein, although animal protein may be more bioavailable at the time of ingestion. However, provided you eat a good range of plant-based foods you will get all the amino acids you need, lots of antioxidants and a host of other health-preserving and health-promoting benefits besides. As omnivores we have a choice when it comes to protein, we just need to make the better choice.

Plants contain all nine essential amino acids you need, with hemp, soya beans, chia seeds and quinoa nearest to the amounts found in meat. A diverse, healthy WFPBD will give you the level of amino acids you need to thrive. If in any doubt, think about how herbivores like gorillas, whose strength is much greater than ours, get their muscle. Think of the phrase 'that man is as strong as an ox', and ask yourself how the ox became so strong. Do not be fooled by animal industry marketing that you need meat to get all your amino acids. Be assured, it can just as easily come from plants. Furthermore, you do not have to eat the *perfect* amounts of each amino acid each day, as the body banks amino acids for up to seventy-two hours[3] – making it extremely hard not meet your body's needs.

However, all that said, it could be a good idea to eat plant-based proteins above the recommended limit to counteract the differences in bioavailability with animal-based proteins. My personal concern is that limits are based largely on our existing diet, and until compelling evidence is available I add at least an extra 10 g per day of plant-based protein to my diet.

Citations:

1. Dr. Rangan Chatterjee (2018), *The 4 Pillar Plan: How to Relax, Eat, Move and Sleep Your Way to a Longer, Healthier Life* Penguin Life, London, UK
2. British Nutrition Foundation, *Protein.* https://www.nutrition.org.uk/nutritionscience/nutrients-food-and-ingredients/protein.html?start=2
3. Bittman, D. and Katz, D. L. (2020), *How to Eat: All Your Food and Diet Questions Answered.* Houghton Mifflin Harcourt, Boston, MA

5. UK Meat, Milk & Egg Consumption

According to the Agriculture Horticulture and Advisory Board (AHDB), total meat consumption per person in the UK in 2016 was 79.3 kg.

Table 8: Total meat consumption (UK, per person in 2016)

Meat Type	Weight
Beef & Veal	18.2 kg
Lamb	4.8 kg
Pig meat	26.2 kg
Poultry	30.0 kg
Total	79.3 kg

To reflect actual consumption, this needs to be adjusted to retail weight: 0.7 for beef and veal, 0.78 for pig meat and 0.88 for sheep and poultry.

Table 9: Total meat consumption based on retail weight (UK, per person in 2016)

Meat Type	Weight
Beef & Veal	12.74 kg
Lamb	4.2 kg
Pig meat	15.94 kg
Poultry	26 kg
Total	59.3 kg

This quantity does not include fish and shellfish, which when weight adjusted (0.43) is around a further 8.5 kg per person per year (2014).[1]

Dairy consumption per person per year is around 220 litres, as a whole milk equivalent, and we eat around 197 eggs.[2] Both figures sound high but reflect quantities consumed in concentrated forms like cheese, cream and butter and in processed pastries and cakes.

So how does all of this add up in terms of protein?

One litre of milk contains about 35 g of protein and a medium egg about 5.7 g.[3] For raw meat (uncooked) per kg, beef and veal has 222 g, lamb 200 g, pig meat 200 g and poultry 240 g. Fish has around 200 g (based on most UK citizens eating cod). Now we can start doing the maths.

Table 10: Average annual meat, fish, milk and egg protein consumption (UK, per person)

Meat, Milk & Eggs	Weight	Protein
Beef & Veal	12.74 kg	2.8 kg
Lamb	4.22 kg	0.844 kg
Pig meat	15.94 kg	3.19 kg
Poultry	26.4 kg	6.34 kg
Fish (cod)	8.5 kg	1.7 kg
Milk	220 kg	7.7 kg
Eggs	197 kg	1.12 kg
Total		23.69 kg

In a year the UK average meat, fish, milk and egg protein consumption is 23.69 kg, which equates to around 65 g a day.

This protein consumption compares well to the European Environment Agency (EEA) calculation[4] of 60 g per person per day, which excludes eggs. If nothing else, it provides comfort in the accuracy of my own accounting. The same EEA data suggests EU citizens eat around 43 g a day of plant-based proteins. Taken together, the average UK (and EU) citizen eats around 103 g of protein a day, well above the UK daily recommendation for women (45 g) and men (56 g).

Clearly, we are far from a protein-deficient nation. Where we are eating more protein than we need, logically, the protein we need to cut back on is animal protein – and in particular processed meat. This is an average, so you always need to check out your own consumption.

A diverse WFPBD ensures we get all our essential amino acids, vitamins, minerals and dietary fibre – the latter being notably and harmfully absent from animal-based foods. It comes with the added bonus of not harming animals or our health, nor inflicting so much ethical and environmental damage. We do not need to eat refined sugar, highly processed oils, saturated fats and trans fats either.

Citations:

[1.] Seafish (2015), *Seafood industry Factsheet, Seafood Consumption.*
[2.] Egg Info, *Industry data.*
    https://www.egginfo.co.uk/egg-facts-and-figures/industry-information/data
[3.] Dairy UK, *Nutritious Dairy.* https://www.milk.co.uk/nutritious-dairy/
[4.] European Environment Agency: FAOSTAT, 2017a, b

6. If We Give Up Meat, Can We Grow Enough Plant Protein?

The UK population is around 66.5 million (2018). Based on an average protein need of 50 g per day, 50:50 men and women, the UK's total annual protein need is 1,213,625,000 kg (1.214 million tonnes).

The usable agriculture area in the UK is 17.5 million hectares, covering 72 per cent of land in the UK. The total croppable area is 6.1 million hectares, with a cereal crop area of 3.2 million hectares.[1] For wheat, the UK harvest was 14.8 million tonnes and oilseed rape 2.2 million tonnes from 562,000 hectares.[2] It must be noted that, like financial investments, yields can go up and down. Table 11 below provides a breakdown of the majority of the croppable land, along with my average estimates of protein in the product's raw state.

Table 11: Croppable land and average raw protein

Crop	Thousand hectares	Tonnes	Protein tonnes
Arable crops	4.577		
Wheat	1,792	14,837,000	1,780,000 (12%)
Barley	1,177	7,169,000	716,000 (10%)
Oats	161	875,000	105,000 (12%)
Rye, mixed corn & triticale	52		
Oilseed rape	562	2,167,000	282,000 (13%)
Linseed	26	46,000	6,000 (13%)
Potatoes	145	6,218,000	367,000 (5.9%)
Sugar beet (not for stock feeding)	111	8,918,000	
Peas for harvesting dry (animal/human)	40		
Peas for harvesting dry (animal)		85,000	19,500 (23%)
Peas for harvesting dry (human)		75,000	17,250 (23%)
Field beans	193	771,000	208,000 (27%)
Maize	197		
Horticultural crops	168		
Vegetables (a) grown outdoors (b) & protected	118		
Orchard fruit	11		
Soft fruit & wine grapes			
Outdoor plants and flowers	13		
Glasshouse crops	3		
Temporary grass under 5 years	1,148		
Total protein			3,501,000

(a) For example, cabbages, carrots, cauliflowers, calabrese, lettuces, mushrooms, onions, tomatoes.

(b) Of which grown outdoors, 117 hectares.

Ignoring horticultural crops and maize land, for arable alone we produce 3.501 million tonnes of protein. With a UK population of 66.5 million that gives each person a 52.6 kg share of the protein, or 0.144 kg a day (144 g). That's nearly three times our daily protein requirement.

UK soft fruit production (strawberries, raspberries, blackcurrants and other) was 142,100 tonnes in 2016.[3] This is up from 61,400 tonnes in 1996.[4]

According to British Growers, the UK is over 40 per cent self-sufficient in indigenous fruits and less than 60 per cent self-sufficient in vegetables and salad.[5] Of this, vining peas (frozen & canning) are grown on about 33–34,000 hectares (producing around 132,000 tonnes); carrots around 9,000 hectares (producing 731,000 tonnes); asparagus, 5,000 tonnes; brassicas, 440,000 tonnes; onions, 433,460 tonnes; lettuce, 122,000 tonnes; watercress, 2,000 tonnes; leeks, 34,000 tonnes; rhubarb, 25,000 tonnes; tomatoes, 75,000 tonnes; top fruit (desert apples and pears), 276,000 tonnes; cherries, 5,000 tonnes; and plums, 2,500 tonnes.[6]

Protein content of all this produce ranges from around 2.6 g per kg of apples to the heady heights of 23 g per kg of watercress. Strawberries and raspberries come out at 6.7 g and 12 g respectively per kg. Carrots, the largest tonnage and one of the most commonly eating vegetables, contain around 9.3 g of protein per kg.

Table 12: Protein and calcium packed foods[7]

Food	Protein (g/100 g)	Food	Protein (g/100 g)
Peanuts	25.8	Adzuki beans	7.5
Pistachio nuts	20.2	Soft tofu	7.2
Pumpkin seeds	20.0	Mung beans	7.0
Sunflower seeds	20.8	Green peas	5.4
Cashew nuts	17.6	Quinoa	4.4
Walnuts	15.2	Spinach	3.0
Wholewheat bread	12.5	Collard greens	3.0
Soybeans (green)	12.4	Kale	2.9
Extra firm tofu	10.1	Brussels sprouts	2.6
Lentils	9.0	Broccoli	2.4
Chickpeas	8.9	Sweet potatoes	1.6
Black beans	8.9	Chinese cabbage	1.5
Kidney beans	8.7	Butternut squash	1.0
Green/yellow split peas	8.3		
Broad bean (Fava)	7.6		

Source: My Food Data.

Compared to arable crops, the extra protein per UK citizen from horticultural crops is extremely small, only raising protein from 144 g to 145 g. The significance is much greater for those who eat their five fruit and vegetables a day, and a lot more again for those choosing to eat a balanced WFPBD. However, these foods also pack you full of other nutrients, including phytonutrients, antioxidants and the fibre you need for a healthy gut and reduced risk of disease – many of which are not available in animal-based foods.

On top of the croppable land mentioned is an additional 1.148 million hectares of temporary grass that could be brought into direct human food production. However, the land offers a break from year-on-year arable cropping and cultivation, which can deplete soil organic matter and undermine fertility. A grass break is particularly useful in the absence of organic manures to help rebuild soil fertility, along with a diverse arable crop rotation. I have therefore excluded the land from my calculations, and the small amount of permanent grassland that could also be reasonably brought into arable cropping.

An assortment of crops in rotation and other measures (conservation tillage, reduced use of plant protection products, green manures etc), can be used towards tackling decades of damage caused by monocropping and continuous ploughing by dirt (rather than soil) farmers. The inclusion of other crops, like hemp, quinoa, lentils and more peas/beans, would be beneficial and support a UK-wide WFPBD.

Animal-based proteins are not a need, they are a want. A want that has been manifested on our lives. As a nation we became healthier during World War II, when our consumption of animal-based products declined due to rationing. Following the introduction of the 1947 Agriculture Act, designed to increase home-grown food production and food security, including meat, dairy and eggs, we began to become less healthy again. Along with our environment.

We carried on in the same vein when we joined the EU Common Market in 1973, and as we marched on through the 1980s, 1990s and the 2000s, increasingly eating more meat in a processed form, which has further harmed our health. We have never been so ill as a nation, with our growing life expectancy a measure of lower child mortality and the effectiveness of antibiotics and other drugs rather than a measure of a healthy diet. The damage we are doing to our health starts at an early age and expresses itself when we get older; more so as we continue to indulge our bad eating habits.

Animal-based agriculture is a business-supported sector. Farmers are in business to make a living. Any *bad news* is not *good news* for money, investments and livelihoods and must be gaslit away to preserve business (and pollution) as usual – much like the tobacco industry use to do, with uncanny similarity.

Education and careers that touch livestock production have been built and hard won. Research at universities, the investment they need and the jobs dependant on it result in science more fit for industry than the health and well-being of the nation. There is no long-term profit and job security in bad news, and what bad news there is, is buried.

The overwhelming science on healthy eating is contrary to industry-supported nutritional science. Blame for the West's booming heart disease, diabetes and obesity epidemic is steered towards salt, sugar, partially hydrogenated margarines and even processed plant-based foods. Little or no notice is taken of real lifestyle studies like *The China Study* and the Blue Zones.

According to the International Agency for Research on Cancer (the cancer agency of the World Health Organisation, or WHO), processed red animal meat is a carcinogen, and in its unprocessed form a probable carcinogen. That's before it is cooked, including white meats, adding mutagenic heterocyclic amines (HCAs) and polycyclic aromatic hydrocarbons (PAHs), also strongly associated in studies with cancer. This doesn't mean everyone who eats processed meat and meat in general will get cancer; it simply increases your risk the more you eat and the less of the good stuff you don't.

From Givens (2018),[8] a study used by the NFU to counter 'myths' about meat, the author states that the 'significant risk of colorectal cancer from consumption of red and particularly processed meat has been reinforced by more recent data' – a fact that does not feature in the NFU literature.

Even if we eat these foods within the UK limit of 70 g per day, it does not stop them having a probable cancer risk, or causing one of the many other diseases associated with eating animal-based foods. 70 g per day is merely a *safety* line in the sand, and a poor one at that.

This is all easily said, but as an alternative can we actually grow enough plant protein and calories to feed ourselves?

This was given some thought back in 1975 by Scottish ecologist Kenneth Mellanby, who published a variety of dietary approaches in his book *Can Britain Feed Itself?* Mellanby took a simplistic approach, and whilst open to enquiry, it was clear the UK has enough land to support a WFPBD. Mellanaby's work has more recently been examined and complemented by Simon Fairlie (2010).[9]

Using a 'Chemical Vegan' scenario, Fairlie showed that Britain's ration could be met using only three million hectares of arable land, freeing up 15.6 million hectares for other uses. Fairlie's recalculation substituted animal protein and fat with pea protein and rapeseed oil. Going 'Organic Vegan' raised the amount of arable land needed to 7.3 million hectares, with 11.2 million spare hectares, due to the extra land needed to rotate the land with nitrogen-fixing crops such as beans, clover or lucerne, given that organic systems do not use synthetic fertilisers.

Fairlie also explored 'Vegan Permaculture' to 'see whether the UK could become more self-reliant, not only in food, but in fibre and fuel'. Combining food, a substantial proportion of our textiles (e.g. by increasing the amount of hemp and flax grown), and the energy for cultivating fields, he calculated a need for 13.4 million hectares, with 8.4 million spare hectares. The difficulty with this model, as he explained, 'is that it results in a lop-sided economy, with almost all the activity concentrated in the arable area; and overall, it appears to provide less employment on the land than the livestock system'.

Taking a 'Net Zero Carbon Britain' approach in mind, with some milk, beef and sheep ('Low Meat and Bioenergy'), Fairlie estimated a need for 7 million hectares of woodland, 10 million hectares of arable, 3 million hectares of pasture, and leaving 2.2 million spare hectares for other opportunities. Such a scenario would need to provide those areas in the UK with a limited arable opportunity a greater share of the animal-food market.

Like Mellandby's approach, the scenarios are simplistic and each would require a transformative approach. They do not include further opportunities either, such as lab-grown meat, urban agriculture and other novel plant-based proteins which would free up even more land.

The savings on land are achieved simply by directly feeding humans, rather than inefficiently putting the food grown on our arable land through animals first. It is not rocket science, though moving away from our existing livestock land/food reliant economy would be complicated. Regardless of government intervention, which would be needed in bucketloads to help us transform, there is an inevitability swelling given the rise and quality of plant-based alternatives, the climate crisis, antibiotic resistance, the aftermath of the Covid-19 pandemic and the need to reverse habitat declines and improve biodiversity on our land and in our seas.

We want rather than need to eat animal-based foods. We will not fall apart at the seams if we stop eating them, provided we eat a properly balanced WFPBD.

Animal-based agriculture is on notice. If we can eat and enjoy foods that do not involve any form of animal unkindness, give us all the calories and nutrition we need, are comparable/cheaper in price, reduce pollution, allow the environment to recover and improve our health, what is not to like and want? The future may not be completely whole-food plant-based, as consumers will still search out less healthy, processed plant-based foods, but nevertheless it is one of hope for us, our children, animals and the Earth.

Citations:

1. Defra (2017), *Agriculture in the United Kingdom.* National Statistics at Gov.uk
2. Ibid.
3. British Summer Fruits (2017), *The Impact of Brexit on the UK Soft Fruit Industry.*
   http://www.britishsummerfruits.co.uk/media/TheAndersonReport.pdf
4. Ibid.
5. British Growers Association.
   http://www.britishgrowers.org/british-growing/
6. Ibid.
7. My Food Data, *Lentils.* https://tools.myfooddata.com/
   nutrition-facts.php?food=172421&serv=wt1).
8. Givens, D. I. (2018), 'Dairy Foods, Red Meat and Processed Meat in the Diet: Implications for Health at Key Life Stages'. *Animal*, 12(8), pp. 1709-1721
9. Simon Fairlie (2010), *Meat: A Benign Extravagance.* Permanent Publications, Bristol, UK

7. The Calorie Count

Another common argument for animal-based foods, and against whole-food plant-based diets, is that plants cannot provide enough calories to feed an increasingly hungry world. Commonly cited are comparisons between calorie-dense meat and less calorie-dense plants.

For example: 'If you stop eating beef, you can't replace a kilogram of it, which has 2,280 cal, with a kilogram of broccoli at 340 cal. You have to replace it with 6.7 kg of broccoli.'[1]

So how does this argument stack up?

A beef cow may use between half to one hectares of land per year, depending on the system and production efficiency.

About 60 per cent of a beef cow's original live weight is not normally eaten. What's left is known as the dressed or carcass weight. The dressed weight excludes internal organs, the head and other body parts, but still includes bones, cartilage and other bits. According to Defra slaughter statistics the average dressed carcass weight for beef cattle is 370 kg (June 2018).[2] A further 30 per cent needs to be removed to reflect the actual retail consumption weight of the meat. This leaves around 260 kg of meat per head of cattle.

Converted to calories, 260 kg of meat eaten multiplied by 2,280 calories provides approximately 0.6 million calories (between 0.6 to 1.2 million calories per hectare, depending on the beef system).

Broccoli grown on an organic farm achieves a yield of around 4.6 tonnes per hectare (4,600 kg). 4,600 kg multiplied by 340 calories provides just under 1.6 million calories per hectare. A conventional broccoli crop has a higher average UK yield of six tonnes per hectare.[3] Multiplied up, providing just over two million calories/hectare.

Case closed for broccoli? Not quite, as there are other factors to take into account.

Still missing from the calculation is the average time it takes to produce a finished beef cow, which ranges from twelve to thirty months depending on the production system. For an average dressed weight of 360 kg the time taken to achieve that weight can be up to twenty-four months. Furthermore, you have to have a calf, which takes up nearly another year (283 days in gestation). Taken together, up to three crops of broccoli can be grown compared to the time it takes to finish a beef cow at twenty-four months. Multiplied by three, that's 4.8 million calories for organic broccoli and six million calories for conventional broccoli, or between four to five times the calories than you could get from beef on the same land in the same time. This wouldn't be quite right as you would include other crops in rotation, and maybe leave a year fallow. However, other crops grown on the same land would provide protein and calories too. You would also need to educate people to eat the extremely healthy and tasty broccoli stalk.

In the UK there is a counterargument that many beef cattle are fed grass produced on pastureland (directly grazed or preserved as silage) that cannot be economically used for other crop types. However, even beef cattle finished on grass can be fed a concentrate ration of up to 0.5 kg per head per day towards the end, produced from a variety of arable crops grown at home and abroad (up to 2.5 kg for a 500 kg steer). Most other cattle are regularly fed other feeds to provide the energy, protein, minerals and vitamins cattle need to meet their target performance (e.g. cereals, soya bean meal/hulls, peas and beans, distillers' grains, rapeseed meal and palm kernel).

My example is ultimately intended as a croppable land-use comparison, given that 55 per cent of croppable land in the UK[4] is used to directly and inefficiently feed livestock. However, some pastureland could be used to grow highly nutritious and calorific foods like nuts and other foods too. I also acknowledge that more parts of a cow can be eaten than the 260 kg of meat per head of cattle, which if included would somewhat close the gap.

Ultimately, the UK livestock sector would have a better argument if they stuck to feeding livestock what nature intended, including herb-rich grassland. However, we desperately need to prioritise rewilding to secure our future and the best place to do that is on the grasslands they graze.

Another advantage for broccoli is that the crop does not burp methane, excrete slurry or consume soya bean meal and palm kernel meal from deforested rainforests.

Citations:

1. Tobias, M. and Morrison, J. (2016), *Anthrozoology: Embracing Co-Existence in the Anthropocene.* Springer, Cham, Switzerland

2. Defra (2018), *United Kingdom Slaughter Statistics.* National Statistics at Gov.uk.
    https://assets.publishing.service.gov.uk/government/uploads/system/uploads/attachment_data/file/734095/slaughter-statsnotice-16aug18.pdf.

3. Richard Soffe (2013). *The Agricultural Notebook: 20th Edition.* Blackwell Science, Oxford, UK

4. Harwatt, H. and Hayek, M. (2019), *Eating Away at Climate Change with Negative Emissions: Repurposing UK agricultural land to meet climate goals.* Harvard Law School. https://animal.law.harvard.edu/wp-content/uploads/Eating-Away-at-Climate-Change-with-Negative-Emissions%E2%80%93%E2%80%93Harwatt-Hayek.pdf

8. Mealtime

With time on my hands, I thought I'd look at the nutritional value of all the food I ate one Saturday. It's a good exercise, whether you eat a WFPBD or a diet with animal-based foods. I've done this several times before since changing my diet, to better understand what I may or may not be getting. Back then, I was glad to learn what I had considered to be a healthy diet was lacking and what was associated with my stroke.

That Saturday I had a couple of Richmond Meat-free sausages and 25 g of a coconut-based vegan cheese, which I bought to do other comparisons (not my normal diet). Saturdays are a run and weight training day for me, but I underate that Saturday by a few hundred calories as I was on a leaning programme.

The food I ate during the day was as follows:

Breakfast: Pea milk, oats, Nestle Shredded Wheat, walnuts, ground flaxseed, blueberries.

Lunch: Home-made wholewheat spelt bread (self-fortified with iron, B12 and iodine), meat-free sausages, vegan cheese, home-made vegan coleslaw.

Post-exercise: Raw hemp protein and banana smoothie.

Evening Meal: Chilli (mung beans, green lentils, kidney beans, haricot beans, tomato, spinach, flaxseed, kale, wild garlic leaves, red pepper, pumpkin seeds, onion, smoked paprika, turmeric, onion powder, chilli powder).

Supper: Mandarin, dark chocolate (two pieces).

Drinks: Herbal teas, espresso coffee, water.

Table 13: Personal nutritional information, Saturday 25 April 2020

Nutritional Information	Personal	PHE England (19-64)[1]	Nutritional Information	Personal	PHE England (19-64)[1]
Energy (kcal)	2,214	2500	Trans Fat (g)	0.014	<2.0
Carbohydrates (g)	334.3	333	Cholesterol	-	
Fibre (g)	69.6	30	Calcium (mg)	907	700
Fat (g)	71.1	<97	Iron (mg)[3]	21.5	8.7
Saturated Fat (g)	13.6	<31	B12 (ug)[3]	1.3	1.5
Omega-3 (g)	7.6		Sodium (g)[4]	1.88	<2.4
Omega-6 (g)	17.3				
Protein (g)[2]	105	55.5			

Table notes:

[1] Public Health England (19–64 years): Government Dietary Recommendations (2016).

[2] Exceeded essential amino acid requirements, with the exception of Lysine (92%).

[3] B12, excludes fortified B12 in home-made bread.

[4] Damn those meat-free processed sausages!

From my personal experience, my daily energy figure is around 2,200 kcal per day. If I eat more than that without exercise to compensate, I would put on weight. However, you really have to be careful with calories as the impact of those calories depends on how they are packaged. If you eat a diet of fast food and high refined sugar, the impact of those calories is totally different to a diet of high fibre and complex carbohydrates. In the latter, fewer calories will be taken up because digestion of fibre uses more energy and some calories remain in the fibre too. How your body digests food, and what your body takes up from it and is able to use for optimum health, also depends on the foods you eat and the gut bacteria those foods influence.

My plant protein intake is less than shown for my non-training days. In any week I run around 20 miles and weight train for 3–3.5 hours, when, on average, I would normally consume around 90 g per day protein. On non-training days I would eat significantly less.

Saturday is always heavier on the protein given it is normally my most active training day. During the week I would eat more whole-food carbohydrates, though on this occasion my carbohydrate intake was nearly bang on the Public Health England (PHE) guideline of 333 g per day.

The only time I would normally have processed sugar is on *pancake* Sunday, a family tradition, when I use two teaspoons. On the odd times I make a cake, using whole grains and tahini for oil, I add less than 50 per cent of the typical amount of sugar. I do not add salt to the food I make either, including bread (I bake bread every other day), and as I cook most of the food from ingredients, my normal salt intake is well below what my results show. Damn that processed meat-free sausage, but still below the UK recommendation of 2.4 g per day. Most of the other sweet treats I have are sweetened with dates or bananas, but I'm partial to the odd chunk of dark chocolate.

That Saturday I comfortably hit my nutritional needs. I certainly was not found wanting in calcium (PHE: 700 mg per day) and iron (PHE: 8.7 mg per day), and I had a high fibre content of nearly 70 g (PHE: at least 30 g per day). My fats were all below the PHE upper limits, with my fats mainly from whole, plant-based foods and not processed oils. My B12 was just below the PHE recommendation but excludes the extra B12 I add to my home-made bread.

My cholesterol intake was zero, which I'm happy to leave it to my liver to make that. Having had a stroke, the last thing I want is an inflammatory diet and the *added* saturated fat/cholesterol from animal-based foods floating around in my blood. I was lucky once and do not intend to chance my luck again!

After doing this Saturday meal snapshot, I also took time to compare cow's milk with the plant-based milks my family and I drink. Not just in terms of nutrition, but also in terms of their environment and other impacts.

Table 14: A comparison of the nutritional, environmental and humanitarian footprints between two plant-based milks and cow's milk.

Measure	Dairy Cow's Milk Verses Plant-based Milks		
	Pea (Sproud)	Oat (Oatly)	Dairy cow
GHG (CO2e/kg)	0.37[1] (Full LCA)	0.31 (Full LCA)	1.60 (Full LCA, incl. c. 25% post-farm-gate emissions)
Energy (kcal/100 ml)	42	46	47/65 (Semi-skimmed/whole)
Protein (g/100 ml)	2.0	1.0	3.5[2]
Total fat (g/100 ml)	3.0	1.5	3.7 (1.8)[2]
Saturated fat (g/100 ml)	0.3	0.2	2.4 (1.1)[2]
Trans fat (g/100 ml)	-	-	0.09 (0.045)[3]
*Sugars (g/100 ml)	1.8	6.7	4.6/4.7
			(Semi-skimmed/whole)
*Free sugars (g/100 ml)	1.8	4.1 (Natural sugars from oats)	-
Lactose (g/100ml)	-	-	4.6/4.7 (Semi-skimmed/whole)
Calcium (mg/100ml)	120	120	120–124
B12 (mcg/100ml)	0.384	0.384	0.4–0.9[4]
Iodine (mcg/100ml)	-	22.5	32
Nitrogen Fertiliser Use	Legume, no nitrogen fertiliser	Low nitrogen use[5]	Medium to high nitrogen use (grass and arable feed crops)[6]

Measure	Dairy Cow's Milk Verses Plant-based Milks		
	Pea (Sproud)	Oat (Oatly)	Dairy cow
Environment (UK)	Low environmental impact, including carbon footprint. Can help improve soil health as part of a rotation.	Low environmental footprint, including carbon footprint. Can help improve soil health as part of a rotation.	High environmental footprint. More reported pollution incidents than any other livestock sector, most involving slurry and silage, with an increasing incident trend (five-year average/ pollution per dairy farm)[7]; up to 95% non-compliance with pollution prevention storage regulations (e.g. SSAFO regulations), and in particular slurry and silage stores[8]; frequent cause of other unreported slurry and diffuse water-related pollution, (e.g. nitrate-N, ammonium-N, phosphorus, soil erosion/ sediment, BOD, microbial pathogens and antibiotics)[9]; 28% of the 87% UK total air ammonia emissions come directly from dairy cows, with a further 20% from beef cattle (dairy beef making up c. 50% of the beef herd)[10]. cows are major contributor to UK agriculture methane and nitrous oxide emissions[11]. Organic manure can be used to help improve soil health and drought resistance, but there is limited soil carbon sequestration potential.

Table 14 cont: A comparison of the nutritional, environmental and humanitarian footprints between two plant-based milks and cow's milk.

Measure	Dairy Cow's Milk Verses Plant-based Milks		
	Pea (Sproud)	Oat (Oatly)	Dairy cow
Environment & humanitarian (offshore)	Low	Low	High. Estimated 150,000 hectares in South America, associated with soya bean meal used for UK dairy cows as a protein feed.[12] Deforestation, with disastrous ecological impacts. Exponential increase in, and uncontrolled use of, herbicides. Chemical water pollution. Agrochemical drift damaging and killing traditional crops. Human health, from agrochemical drift over populated areas, including schools, and contaminated drinking water. Soil degradation and erosion. Enforced displacement/eviction of small farmers, agricultural labourers and indigenous families from their customary land. Violence, including the killing of indigenous activists. Changes to cultural reference. Poverty, and hunger from the transference of traditional and diverse foods to exported animal feeds for use in faraway countries. Enforced migration towards cities and shanty towns, often with little job opportunity and/or transferable skills to find work. Human rights violations.

Measure	Dairy Cow's Milk Verses Plant-based Milks		
	Pea (Sproud)	Oat (Oatly)	Dairy cow
Health	Good. Good source of protein (from processed pea isolate). Useful source of vitamins and minerals. Good alternative for people who are lactose intolerant. Added processed vegetable oil (rapeseed).	Good. 10% oats. Source of fibre, including soluble Beta-glucan fibre, which is associated with reducing cholesterol and providing protection against 'bad' LDL cholesterol damage. Reduced risk of coronary artery disease and colorectal cancer. Can contribute to reduced blood pressure. Contains antioxidants, such as avenanthramides and ferulic acid. Can help lower blood sugar levels. Encourages the growth of good gut bacteria. Fortified with essential nutrients, such	Low. Animal protein and saturated fat have been shown to lower good Short-Chain Fatty Acid (SCFA) producing-bacteria and encourage bad inflammation triggering bacteria. This in turn aids the production of trimethylamine N-oxide (TMAO), associated with increased risk of heart disease, stroke, Alzheimer's disease, type 2 diabetes and other ailments. Can raise 'bad' LDL cholesterol levels. Lactose intolerance in many people, which beyond an individual's level of tolerance can trigger irritable bowel syndrome (IBS) (symptoms include cramping, abdominal pain, bloating and gas, constipation and diarrhoea). Can contribute to dysbiosis, which increases the risk of intestinal permeability ('leaky gut') and the release of bacterial endotoxin into the blood stream. Presence of mammalian cow hormones,

Table 14 cont: A comparison of the nutritional, environmental and humanitarian footprints between two plant-based milks and cow's milk.

Measure	Dairy Cow's Milk Verses Plant-based Milks		
	Pea (Sproud)	Oat (Oatly)	Dairy cow
		as calcium, B12 and iodine, though not in all oat-based milks. Useful source of other vitamins and minerals. Good alternative for people who are lactose intolerant. Added vegetable oil (rapeseed).	such as oestrogens. The contribution of milk, and other dairy products, to these ailments and to others, is influenced by an individual's wider diet, in particular fibre/range of fibre intake, and the health of their gut microbiota. Calcium is an essential nutrient and milk is a convenient source, along with other vitamins and minerals, but it comes with a lot of baggage which can otherwise be avoided. No added processed vegetable oil.
Land Footprint Per Litre (m²/litre)[13]	c. 0.2	c. 0.2	c. 1.0
Water footprint[14]	Unknown	5.3	8 + (1.3-2.5)
Waste footprint (tonnes)[15]	Limited	Limited	330,000

Measure	Dairy Cow's Milk Verses Plant-based Milks		
	Pea (Sprould)	Oat (Oatly)	Dairy cow
Animal Welfare	No issues	No issues	Calf removal, mastitis, lameness, cow infertility, disbudding/dehorning, castration (male calves), acidosis, low social interaction, low daylight/natural light (factory cows), continuous pregnancy, physical exhaustion, transport/slaughter. Subject to physical abuse on some farms.
Packaging	Tetra Pak, recyclable and less plastic than plastic bottles (c. 75% card).	Tetra Pak, recyclable and less plastic than plastic bottles (c. 75% card).	Plastic milk bottles, but 100% plastic and more energy intensive to make than Tetra pack.
Cost (£/litre)	1.80	1.55–1.80	<0.50

Table Notes

1 Climate footprint, full life-cycle assessment (LCA) (CO_2e/kg), information courtesy of Sproud International.

2 Based on Tesco whole milk and semi-skimmed milk. Tesco whole milk has a total CO_2e/pint carbon footprint of c. 900 g (LCA) including, production to farm-gate, post processing, distribution and retail. Post-farm-gate carbon footprint is c. 25% of total. The average farm-gate footprint for milk is around 1,232 CO_2e/litre. https://www.greenerpackage.com/metrics_standards/carbon_footprint_be_displayed_tesco_uk_milk

3 Trans fats 2.5-5% depending on diet and season: https://www.wjert.org/admin/assets/article_issue/11102015/1443683684.pdf / According to ARBRO Analytical Division 100 g of milk contains 0.085 g trans fat: https://testing-lab.com/2017/11/natural-trans-fats-milk/

4 B12 RDA of 2.4 mcg/d.

5 The amount of nitrogen spread to grow oats depends on the soil nitrogen supply (kg N/ha), which will be influenced by factors such as the soil type, previous crop and annual rainfall. Crop requirement may range from 0 kg N/ha to 140 kg N/ha. For example, following a legume around 70 kg N/ha and following a cereal around 110 kg N/ha.

6 The total amount of nitrogen spread in any season depends on several factors, such as expected yield, whether the grass is cut or grazed and the stocking rate. For example, 180 kg N/ha to 340 kg N/ha. Much less nitrogen would be needed where a low grass yield is expected, on swards with an appreciable clover content (e.g. up to 50 kg N/ha), or a bit more or less depending on the soil nitrogen supply (SNS). In a nitrate vulnerable zone the maximum permitted amount of nitrogen from manufactured fertiliser and the amount of available nitrogen available for crop uptake from organic manure) is 300 kg N/ha, with an additional 40 kg N/ha to grass that is cut at least three times a year. Most dairy farms will apply in any calendar year close to and between 170 to 250 kg of total nitrogen per hectare from their livestock manure, of which 40% is readily available for crop uptake.

7 Regulating for people, the environment and growth, 2018 (Environment Agency, October 2019). There are many more slurry-producing dairy farms than slurry-producing pig farms. On beef farms, farmyard manure is still the dominant type of manure produced. By number, and by producing the most amount of slurry, dairy farms are statically more likely to cause a reported pollution incident. Similarly, for silage. Pollution incident statistics do not take account of reduced regulatory visits (from 1% to less than 0.5%) and with a potential reduction in self-reported incidents[8]. Many less Category 3 incidents are also visited/substantiated today than in the past by the Environment Agency due to cuts in funding. Unsubstantiated incidents are not reported in publicly published pollution incident statistics. 5-year rolling averages are used to give a more realistic trend than relying on single years.

8 River Axe Natura 2000 (N2K) Catchment Regulatory Project Report (Environment Agency, November 2019). Nearly 95% of 86 farms visited during the project failed to meet the requirements under the SSAFO Regulations, with the majority found at high risk of causing pollution. 49% of the farms had evidence of a polluting discharge into the River Axe. In another project, North Devon Priority Focus Area, the Environment Agency found that 87 per cent of farms did not comply with environmental regulations and 66 (65 per cent) of the farms were polluting at the time of the farm inspection.

9 The dairy sector has a high potential to release sediment, nitrate, phosphorus and other pollutants to rivers and groundwater. Releases include run–off during rain events from bare/post-harvest stubbles (e.g. after maize), which are often sacrificially used for winter slurry spreading and where the soil is often compacted from harvesting/trafficking the land when wet. Runoff of sediment and slurry may also occur on poorly established grass reseeds and from land poached through stock trampling. Unseen diffuse pollution from nitrate-nitrogen leached from slurry applied post-harvest and into winter, and preferential drainage of pollutants from slurry applied in winter and into spring following significant rainfall. Pollution risk may be increased following intensification, through land-use change (e.g. increased reliance on maize on land not suitable for the crop), increased stocking pressure, increased slurry production and a lack of respective investment in waste management infrastructure.

10 Clean Air Strategy (Defra 2018). Dairy-bred dam and beef sired calves, as well as other male dairy calves, are common in high-input beef systems. Ammonia can travel long distances and when combined with industrial and transport emissions form fine particulate matter ($PM_{2.5}$) that is harmful to human health. Ammonia emissions can cause soil acidification and nitrogen-enrichment, degrading habitats and biodiversity.

11 2016 UK Greenhouse Gas Emissions, Final Figures (Department for Business, Energy and Industrial Strategy, February 2018).

12 From Chapter IV.

13 A LCA undertaken by RISE found that Oatly's oat milk used 79% less land than cow's milk to produce. Pea milk is likely to be similar. Cow's milk footprint is based on dairy's total farmed area of 808,868 ha (England, Defra June Survey of Agriculture), plus the land area used to provide soya bean meal (England, c. 91,500 ha) and then divided by the amount of milk produced per year of c. 9 billion litres (England). The figure does not include other land outside of the dairy farmed area used to produce food for dairy cows, e.g. cereal and oilseed rape meal.

14 Oatly milk is a water use cradle to grave LCA. Down from 6.4 litres to 5.3, through additional efficiency measures. Future target is 3.0 litres. Dairy cows' milk water footprint is up to the farm-gate (c. 8 litres of water /1 litre of milk), with between 1.3-2.5 litres of water used in post-production. There is no available figure for Sproud pea milk.

15 Data from WRAP http://www.wrap.org.uk/food-drink/business-food-waste/dairy. About 90% of milk waste is produced in the home, with around 30,000 tonnes of wastes in the supply chain and 13,000 tonnes during processing. This amounts to 490 million pints of milk as a nation (7% of all milk produced) or 18.5 pints per household. There are no waste figures for plant-based milk, but would be expected to be considerably less than cow's milk.

From my assessment, there are many good reasons to reduce, and ideally stop, drinking cow's milk. Individually they are convincing and together highly persuasive. Other plant-based milks also have a much lower overall impact than cow's milk, though I personally choose plant milks that I enjoy, have the lowest footprint and that are also beneficial as part of a crop rotation. There is a small amount of processed rapeseed oil added to these milks, though this is an extremely small amount of my overall diet and much less of a potential issue than the animal fat present in cow's milk.

I could just as well have included hemp milk, which I use from time to time and also use as a protein supplement to support my training regime.

Hemp is a tremendously versatile crop, with all the top growth having a potential commercial use (food, drink, fabrics, paints, cosmetics, as a building material, biofuel and much more). Hemp requires very little fertiliser and little to no pesticides to grow, enriches soil, reduces soil erosion, needs very little water, effectively sequesters carbon dioxide and is attractive to wildlife. As a food, hemp is packed with fibre, protein, omega-3, 6 and 9 fats, and vitamins and minerals. Nothing to dislike, and one of the best crops used for plant milk, given its even wider environmental credentials.

If it weren't for my preference for oat and pea milk, hemp would feature more regularly within my diet. Nevertheless, I use hemp in other foods such as protein shakes and as an ingredient for savoury and dessert baking.

I do not use almond milk, which has a water footprint the UK dairy industry is deeply concerned about and want you to know about. Most of the world's almonds are grown in California, where unsustainable water use is of a concern, *but* where the amount of water used in the production of Californian cow's milk is an even bigger concern. According to the NFU's Myth 2 in *The facts about British red meat and milk* (2020), a litre of Californian almond milk has a 158-litre 'blue' water footprint – nearly twenty times as much as dairy milk.

The 'myth'-buster as presented gives the impression that all British consumers are drinking Californian almond milk. However, almonds used in plant milks sold in the UK and throughout Europe are more likely to be sourced from Mediterranean countries like Spain, supplied by smaller-scale and traditional farms where most of the production is rain-fed, with some targeted irrigation orchards. The livestock industry complains about being unfairly stigmatised by world-average footprint figures but is more than happy to use them when it suits its interests.

Compare, yes, but compare with what the UK is actually drinking, and include all the other footprints between plant-based milks and cow's milk too, not just water.

If you enjoy almond milk, you can manage your water footprint by avoiding almond milks imported from the US.

Whole almonds are a fantastically nutritious food with many attributable health benefits, including an association with lowering the risk of heart disease and Alzheimer's. So, enjoy them sustainably.

Soya milk is another good option, which compares very well against the environmental footprint of cow's milk.

A 2020 study by Wilkinson and Young[1] published in the *Journal of Applied Animal Nutrition* challenged the notion that soya milk made directly from soya beans is a more efficient process than producing milk from dairy cows:

> *Based on information from the UK Department for Environment, Food and Rural Affairs (Personal communication, 2017), it has been estimated that between 92,000 and 173,000 tonnes of soya bean meal is used in dairy cow diets in the UK (Young, 2017). Total milk production in the UK was 14,713 million litres in 2017/18 (Statista, 2019b), which can be calculated as the yield of dairy cow milk being 85 litres per kg soya bean meal consumed.*

The authors divided 14,713 million litres (c. 1 kg equivalent) by 173,000 to give 85 litres per kg soya bean consumed. If they used the lower value, 92,000, it would calculate at around 163 litres per kg soya bean meal consumed. From this, according to the authors, the less soya bean meal fed to a dairy cow, the more cow's milk you can make, which is nonsensical.

The amount of cow's milk produced from soya bean meal is better represented as a percentage of dry matter consumed. A dairy cow will produce an average of 28 litres per day over a ten-month period, with more or less being produced depending on the lactation stage.[2] For instance, a high-yielding cow may produce as much as 60 litres in a day during peak lactation.[3] It is impossible for a dairy cow to produce 85 litres or more milk from 1 kg soya bean meal, when on average a dairy cow only produces 28 litres of milk a day. Furthermore, a dairy cow does not consume 100 per cent soya bean meal.

In the UK most of the dry matter a dairy cow consumes comes from grazed grass, grass silage, maize silage and cereals. As a percentage of dry matter, the amount of soya bean meal fed may range from between 7 per cent to just over 9 per cent.

Using a UK industry example:[4] assuming a high-yielding dairy cow of around 500 kg in weight produces 35 litres of milk per day, the cow would have a daily feed intake containing around 21 kg of dry matter (DM), of which about 1.5 kg would be from soya bean meal. Each kg of DM would therefore produce 1.67 litres of milk (35 divided by 21). Not quite the 85 litres per kg claimed by the authors.

In contrast, domestically you may get a yield of around 5 litres of soya milk per kg of whole soya beans and commercially squeeze out up to 7.5 litres per kg. The dairy cow example uses soya bean meal, which is around 80 per cent of whole soybeans.

Unfortunately, the authors' errant calculation was picked up and used in the farming press, with *Farmers Weekly* headlining: 'Cow's milk better for the planet than plant-based alternatives.'[5] The *Farmers Weekly* article title is misleading in another way too, effectively lumping all plant-based alternatives into the story when the research only considered soya milk. Ho hum.

The soya bean meal predominantly used in the UK for livestock production comes from GM soya beans grown in South America, mostly Argentina. Most of the soya bean meal is derived from GM crops. A small amount of non-GM soya beans are grown segregated from GM soya beans and sold a premium price (e.g. for direct human consumption). Non-GM soya beans which are not segregated are mixed with GM soya beans. GM soya beans and GM soya bean meal is not directly used in UK food production and instead is indirectly consumed through livestock.

In the UK there is no evidence to suggest GM soya beans or soya bean meal is used in soya milk production (or tofu). Most soya used in UK plant milks originates from Europe, where, as a legume, soya beans add to soil fertility, help reduce the need for manufactured fertiliser and help to rebuild dysfunctional soils. Soya bean milk in the UK is not associated with deforestation, but dairy cow milk is.

Furthermore, according to the Soil Association,[6] the soya bean equivalent used to produce a UK citizen's annual intake of meat and dairy products is about 54 kg per person (1.7 kg for milk directly consumed). This is predominantly sourced from the destruction of rainforests in Latin America. A person regularly eating soya bean-derived products like soya milk and tofu could consume 9–18 kg of soybean equivalent per year that is not sourced from the destruction of rainforests. There are many other healthy plant-based that can be used instead of soya too.

The dairy vs. soya drink calculation used by Wilkinson and Young was challenged by Bailey (2020),[7] which the authors accepted whilst emphasising that most of a UK dairy cow's diet is not associated with soya and deforestation. They reiterated – incorrectly, again – that a top figure of 173,000 tonnes of soya bean meal is fed to UK dairy cows. This appears to be based on a reported soya use in Great Britain animal feed data sets of 1.13 MT (2018–19). Soya use for livestock in the UK is vastly underreported as it only includes soya used by compound feed producers.[8] Add in unreported soya and the amount doubles. On a UK-wide basis the total amount used would be close to 300,000 tonnes according to Fraanje (2020), a figure which my own research agrees with.[9] This amount of soya bean meal requires over 150,000 hectares of deforested land, which however much a dairy cow eats, compared to other feed, is 150,000 hectares too much rainforest.

The authors also now accept that soya drank in the UK is made from soya grown in Europe, freeing UK soya drink from association with deforestation in South America – unlike UK dairy milk. In defence the authors then *imply* that 16 per cent of 53 million acres of species-rich grassland in the Great Plains region of the USA (thus not in the UK) were ploughed between 2009 and 2016 to grow soya beans used in making soya drink. Some of this soya may well be used to make soya drink, but the vast majority is grown to feed livestock. They provided no evidence to back their words. Little thought is given here, too, to the 97 per cent reduction of species-rich meadows which have been destroyed since World War II to largely feed UK livestock, including dairy cows.

The authors conclude that there is more potential to reduce soya bean meal in dairy cow diets without affecting productivity, compared to the use of soya beans in the production of soya drink. In respect of the UK the soya bean meal used to feed livestock is nearly all associated with deforestation, of which about 300,000 tonnes is used to feed dairy cows, according to Fraanje (2020).[10] In the UK, and across Europe, most of the soya beans used to make soya drink are not associated with deforestation, making this a rather head-scratching assertion.

This is no different to saying how if we replaced all UK dairy milk with plant-based milk made from Californian almonds instead of using rain-fed almonds grown in Europe, we would use sixteen times more *blue* water per litre. All this is hypothetical and has no basis in reality, not least as we do not all drink soya or almond milk, but alternatives like oat milk, hemp milk, etc. too. Taking a reversed stance, we could say getting rid of all dairy cows and rewilding the land they occupied would massively increase biodiversity, drastically cut cow-burp methane emissions and ammonia pollution, remove most erosion-ready maize fields and substitute a whole host of other less sustainable, unkind and less healthy food practices associated with drinking cow's milk and eating butter, cheese, etc., if we used soya milk instead.

Dairy cows and other livestock unequivocally contribute to the use and growth of soya plantations in South America and associated deforestation, not people drinking soya milk or eating tofu in the UK and Europe.

Citations:

1. Wilkinson, J.M. and Young, R. H. (2020), 'Strategies to reduce reliance on soya bean meal and palm kernel meal in livestock nutrition.' *Journal of Applied Animal Nutrition* 8(2), pp. 75-82

2. Compassion in World Farming. *Food Business: Standard Intensive Milk Farming.*
https://www.compassioninfoodbusiness.com/awards/good-dairy-award/standard-intenstive-milk-production/

3. Ibid.

4. AHDB Nutrition in dairy cows, *Managing your Feeding.*
https://ahdb.org.uk/knowledge-library/nutrition-in-dairy-cows

5. Clarke, P., 'Cows milk better for the planet than plant-based alternatives', *Farmers Weekly*, 17 August 2020.
https://www.fwi.co.uk/livestock/dairy/cows-milk-better-for-the-planet-than-plant-based-alternatives

6. Soil Association (2010), *Feeding the animals that feed us.*
https://www.soilassociation.org/media/4959/policy_report_2010_feeding_the_animals.pdf

7. Wilkinson, J. M. and Young, R. H. (2020), 'Response to Tim Bailey.' *Journal of Applied Animal Nutrition* 8(3), pp. 103-4

8. Fraanje, W. (2020), *Soy in the UK: what are its uses?* Food Climate Research Network, UK.
https://tabledebates.org/blog/soy-uk-what-are-its-uses

9. Ibid.

10. Ibid.

9. Herbivores or Omnivores

I grew up believing humans have to eat meat; after all, I was told, we had canines.

Many people eat animals. We have since our hunter-gatherer days, and largely from that, and a few adaptions along the way, we are considered omnivores. A few people even believe we are carnivores, either to annoy vegans, because they actually believe a carnivore diet is optimum for health, or a bit of both. The truth is, whether we consider ourselves herbivores or omnivores rests entirely on what particular criteria we want to apply and argue over. Ultimately it is our health that matters the most and given that eating anything other than a small amount of meat, fish, dairy and eggs over time can contribute to a variety of serious illnesses, whilst eating wholesome plant foods do not, is a pretty strong argument for a plant-based diet.

As omnivores, we can live without animal-based foods provided we have enough other food (if not, it's advantageous as an alternative to starvation and death!), but we can't live without eating plants. Furthermore, the food most of us eat today, animal and plant, our lifestyle, circumstances, and the long trouble-free life we want to live is vastly different to hunter-gatherers, and as such it is not helpful to compare things like for like.

With the easy availability of plant foods in modern society, there is no particular advantage to being omnivorous; we are better off being much more herbivorous. Quite simply, the more we displace wholesome plants from our diet in favour of animal-based foods, the more likely we are to need medical interventions and to suffer a shorter life. Not a problem when the prime driver for a species' existence is to reach sexual maturity and reproduce, but it is a problem if we want to enjoy a long and trouble-free life thereafter once the business has been done and we have survived long enough to raise our children to sexual maturity. Hope for a long and trouble-free life is undermined by eating animal-based foods (how much depends on many more factors than eating meat, dairy and eggs alone), probably from the moment we are conceived. We are on the path to a poor health state and lower mortality, hastened by eating the cheap, low-quality animal foods churned out by the billion in factory farms, a scenario a million miles adrift from the wild game our hunter-gather ancestors ate.

The evolution of our brain has allowed us to take advantage of different foods in different environments and do many other things, which has allowed us to spread, reproduce and survive in most places across the world in a way most other animals wouldn't be able to. We have taken advantage of animal-based foods, and if we hadn't, we may not be here now. Even in our world today, eating animal-based foods is still necessary for some people to survive within their environment and circumstances.

When we spread from Africa, we took advantage of eating meat to survive in the places we chose to settle, but the change happened so fast that our bodies have not been able to fully adapt to this new diet and lifestyle. Our ancestors' short lifespan simply did not allow enough time either for the selective shift needed in our biology to enable us to eat animal-based foods without long-term consequences.

Today we now expect to live a long and healthy life and we eat meat that is a poor substitute for what our ancestors ate, and in excess. The animals we eat today no longer eat the foods or live the life best suited for them either, only what is best for profit and is necessary to produce the volume of animal-based foods we are choosing to eat today, which further aids and abets our poor health. In the last few decades alone the quality of the meat most of us eat has drastically diminished, in our drive for ever cheaper food. We have also added other animal-based foods we are not naturally selected to prosper on, and furthermore, unnaturally processed and chemicalised food, making it potentially more harmful to us – meat and even some plant foods. Rather than revert to a diet which is healthier to eat, a predominantly whole-food, plant-based diet (WFPBD), we rely instead on technology and modern medicines to prolong our lives and provide us with a level of comfort in our illnesses.

It's also worth bearing in mind that the ratio of the human population (eight billion) to the number of land animals we slaughter each year for food (78 billion) is massively in excess of Mother Nature's carnivore–herbivore ratio. In nature there are many more herbivores to the animals which eat them, and for good reason: to keep nature in balance. In our hunter-gatherer days, the same would be said of us. Is it any wonder then that defying the balance of nature with such scant regard and arrogance that the *human* predator (who mostly predates on animal reared in factory farms, and to a lesser extent reared in enclosures) is destroying Planet Earth? Worse still when you consider that those of us living in so-called developed countries are eating most of the animals, and that we are NOT carnivores. This alone should tell those people who consider themselves *only* moderate meat, dairy, fish and egg eaters that something about the way they are eating is drastically wrong for our planet, their health and for the well-being of the animals. The only consistent thing we share with carnivores and *non-human* omnivores is that we largely predate on the most vulnerable, unfortunate and sick animals (including farmed fish).

Our brain intelligence, and the flexibility of the human vessel, has served us very well in the past and has made us the dominant species we are today. However, what served us well in our ancestry is our

Achilles heel today and is contributing towards our extinction now. We eat meat, dairy, fish and eggs well beyond our body's adapted tolerances, and the rest of Planet Earth suffers as a consequence.

For most of us on Planet Earth, eating meat, fish, dairy and eggs is no longer a necessity. Our way of eating has become a convenience, a desire, a craving, a question of lust and an extravagance we can no longer afford. The way we have decided to produce and make money from food is a reflection of that and has made us reliant on animals for our food security. We are not stuck with that model and need to change rapidly to, at the very least, a predominant WFPBD system to secure our future.

Regardless of whether we believe we must eat animals and drink their fluids, most of us are beginning to understand that we need to eat a lot less animal foods, eat many more plants, be kinder to our health, our environment, to all animals and other life and regenerate our relationship with nature.

For what it's worth, I consider Homo sapiens to be omnivores and cannot find scientific consensus to think otherwise. We probably would not be here today if we had not, in our past, made animals a part of our diet and developed the means to efficiently catch/farm, prepare and cook them. But what was once a necessity, and remains so for certain people in the world, is no longer an advantage today.

10. We Are Not Carnivores

Humans have small mouths, which are not designed to bite into the hides of the animals our ancestors hunted or the livestock we farm today. Seizing prey requires a wide-opening, stable, clamp-like mouth, with a powerful jaw and muscle configuration. A set of mighty claws is essential too.

Human teeth are compact. We have two very short, blunt, canines, which cannot penetrate an animal's tough hide and bite through into its flesh. To do that requires long, sharp, dagger-like canines. Even if we could, we would want to remove any fur and thoroughly clean the animal of dirt and faecal matter first. A nose peg and a strong stomach to avoid throwing up would also be helpful.

Even if we could bite into an animal hide, we would have to chew the poor creature to death. Carnivores, meanwhile, have triangular, blade-like teeth, designed for a quick kill and to cut into animal flesh and slice off chunks of meat, which they then swallow whole.

When we eat, our lower jaw flexes sideways. As we chew, we use our tongue to help move, turn and blend food with our digestive saliva. Chewing is important to break down the cell walls of food and promote digestion. Perfect for eating plants. Well-chewed food is also easier to swallow, passage and digest in our stomachs. We need tools to hunt animals, tools for slicing and macerating their meat, heat to cook and soften it, heat to kill pathogens and heat to make the meat palatable.

If our natural ability to eat animal foods was as strong as many people want to believe, we would be able to safely drink milk straight from the udder of a cow without pasteurising it and be able to eat any meat without having to clean and/or cook it first. Even the production of raw milk requires some precautions, including disease-free dairy herds, a hygienic milking environment, sterile equipment and regular bacterial testing.

The meat we eat damages our health. It builds up harmful levels of gut bacteria, such as Bacteroides spp., Alistipes spp. and Bilopila spp. Conversely, meat-eating lowers health-promoting bacteria, such as Bifidobacterium spp., lactobacillus spp., Roseburia spp., Eubacterium spp. and Ruminococcus ssp.[1]

We have not evolved the ability to eat animal-based foods as we do throughout our lives without it causing us harm. If we had, we wouldn't suffer intolerances to foods like milk, and eating meat, fish, eggs and dairy would not harm our health in the ways that I have described already.

For the first 18 million years of our evolution, we mainly ate plants. We have only been hunter-gatherers for around two million years. Research suggests that the animal protein that hunter-gatherers ate from this time helped them to reach sexual maturity quickly. With a very short lifespan, this would have been beneficial for survival and the many health-harmful effects of eating meat would not have had time to fully manifest. With such a short lifespan, around twenty-five years, there would have been little opportunity for evolutionary change in response to eating meat.

This begins to unpick why the prolonged consumption of meat and other animal-based foods can cause so many health problems today, and why eating a WFPBD, or at least a predominant one, can enable us to live longer and healthier lives.

Eating meat would have helped us reproduce quickly and have more children. Eating plants would have slowed our development, and although it probably would have helped us live longer, we would have had more opportunity to die for other reasons before reaching sexual maturity. If only we could test such a hypothesis.

In the immortal words of Dr. Michael Greger, M.D., Raubenheimer and Simpson (2020)[2]: 'put it to the test!'

In an experiment the researchers noted a protein and carbohydrate competition within the diet of locusts, with protein winning out. They noted that:

Locusts had a target mix of protein and carbs for best growth and survival [(target diet), but] when their diet didn't allow them to reach the target, protein was prioritised over carbs but at a cost to growth and survival.

In the experiment each locust was fed a single food. In the first phase they received a high-protein and low-carbohydrate diet and in the next a high-carbohydrate and a low-protein diet.

To get the right proportion of protein they needed while being fed a low-protein and high-carbohydrate diet, the locusts instinctively over-fed on the carbohydrates to meet their protein requirement. The diet also delayed the locusts reaching winged adulthood. This would be a problem in a natural situation, as any delay to adulthood would increase the risk of being eaten before achieving reproduction. The diet also caused them to put on fat.

When fed the reverse diet, the locusts became overly lean, short of energy and less likely to reach adulthood compared those fed the target diet.

The locusts were then offered two food choices containing a different balance of protein and carbohydrate, and instinctively managed to eat the two in the right proportion to hit their *target diet*. They naturally achieved the best balance to support their survival, ignoring one nutrient over the other to achieve their optimum diet.

The researchers evidenced that protein and carbohydrates were the main nutrients driving a locust's choice of nutrients.

A similar study was then performed with cockroaches, a species known to scavenge a wide range of foods. Given a choice, cockroaches were also found to intuitively balance their protein and carbohydrate intake. The same was found when they studied three predator species: ground beetles, wolf spiders and web-building spiders.

Experiments were then conducted with cats and dogs. The researchers found that working cats, used as a pest control, selected a diet with 52 per cent of energy coming from protein; dogs selected a diet of only 25–35 per cent protein, well below that of wolves.

It appears that during the process of domestication from wolves, dogs switched their diet to fit their shared environment with humans. However, cats that were kept and left to feed on the pests they caught maintained a target diet similar to that of their ancestors.

In a natural system the target diet of a species would be expected to change across its life cycle for a variety of reasons. For instance, a caterpillar turning into a butterfly and a baby weaning from its mother's milk to the food it eats as an adult.

Throughout a year, nature does not always provide the same foods and in the same abundance too. As the authors put it, to maintain a nutritionally balanced diet, 'animals would have to have a plan B'.

Next up for the authors were humans.

For two days participants were allowed to eat what and as much as they wanted from a wide selection of animal and plant-based foods. The individuals' food intake and amount of protein, carbohydrate and fat they consumed was carefully measured. On the next two days they were split into two groups. One group was provided with a high-protein buffet of mainly animal-based foods; the other a low-protein but high-carbohydrate and high-fat buffet, without meat, fish and eggs. Again, they ate as much as they wanted. On days five and six they were returned to the initial buffet.

In the first phase the participants ate an anticipated number of calories, at a ratio of around 18 per cent protein. In phase two, both groups consumed a similar amount of protein, with the low-protein but high-carbohydrate and high-fat group eating 35 per cent more total calories to achieve the same protein intake. In contrast, the high-protein group ate 38 per cent fewer calories than before. A simple, but limited, experiment showing a similar outcome to locusts and other species. From this the authors hypothesised:

In a protein-poor but energy-rich food environment, humans will overeat carbs and fats to try to reach their protein target. However, when the only available diet is high in protein, humans will under consume carbs and fats.

Like the locusts, humans ate more calories to get the same amount of protein, which if left unchecked would result in weight gain. The 'leverage [to] everything else we eat' was protein. An interesting observation in relation to the human obesity epidemic.

The authors later conducted a blind trial along similar lines, with three diets containing 10, 15 and 20 per cent protein. The participants were kept and fed in a controlled environment. The same outcome was shown, with 12 per cent more calories eaten than on the week spent on the diet containing the lowest protein, with most of the extra calories consumed through snacking.

The authors concluded:

1. *Humans, like locusts, prioritise eating a target amount of protein.*
2. *In a protein-poor but energy-rich world, we overeat carbs and fats to try to reach our protein target, risking obesity.*
3. *When the diet is high in protein, we under consume carbs and fats to avoid overeating protein.*
4. *This is why high-protein diets can help you lose weight. But why would we risk eating too few calories to avoid eating too much protein?*

   It seems, then, as though our appetites are telling us we're better off eating too few calories, at the risk of running out of energy than we would be ingesting too much protein. This alone suggests that there is something seriously undesirable about excess protein intake.

Other research has shown that overeating protein (not all protein, but animal-based protein) is not good for us in other respects than obesity.

For reasons I describe elsewhere, humans are likely to develop a variety of health issues, and with them a lower quality of life in older age, and a lower average lifespan, from overeating animal-based proteins. The same associations have not been found with plant-based foods, which instead have been found to be protective.

The authors turned their attention next to whether there is a trade-off between lifespan and offspring, with previous studies suggesting living longer is associated with having fewer offspring, where energy and resources are spent 'on living longer or on having and raising babies'.

In experiments on the fruit fly *Drosophila* they found that 'lifespan had virtually nothing to do with total calories consumed and everything to do with the ratio of protein to carbohydrate [eaten].' Flies fed a higher amount of protein produced more eggs, whilst a diet richer in carbohydrate led to a longer lifespan.

When protein was further increased, egg production fell, showing that 'there was such a thing as too much protein'.

They surmised:

The message for protein was clear – eat a small amount and you will live a long time but not produce many offspring; eat a bit more and you will leave lots of offspring but not live as long; eat even more and you will neither live long nor produce many offspring. At least if you happen to be a fruit fly.

In a five-year experiment on mice, they found the same relationship, with mice kept on a low-protein, high-carbohydrate diet living longer. The authors noted that it was a combination of low protein and high carbohydrate that promoted the longest lifespan, not protein alone. In human terms:

This would be the equivalent for us eating less meat, fish, eggs but more healthy carbs, like low-calorie vegetables, fruits, beans, and whole grains.

A low-protein, high-fat diet didn't give as long a lifespan as the low-protein, high-carbohydrate diet either.

Where the high-protein diet had the advantage was in reproduction. The extra protein produced a more *endowed* male mouse and females with a bigger uterus to carry a larger litter.

The authors termed the two dietary approaches the 'longevity pathway' and the 'growth and reproduction pathway', with the later meaning 'make hay while the sun shines and to hell with the consequences'.

Making hay comes with a cost, which in us, in our desperate attempt to 'have our cake and eat it', often results in years of health struggles in later life, a lower lifespan, and the associated, but avoidable, consequences on our health services.

The same positive outcomes of a low-protein and high-carbohydrate diet in humans are seen in Blue Zones and in *The China Study*.[3] The authors specifically cite the Okinawan diet, with the zone's number of centenarians five times that of the rest of the developed world. A diet containing on average 'nine per cent protein, 85 per cent carbs and six per cent fat ... precisely the same ratio that [the] mice ate to support maximal lifespan.'

However, unlike the mice, who became obese, the Okinawans were lean, thanks to the high level of fibre they ate.

Consistently, the evidence leads us towards a WFPBD.

Our hunter-gatherer ancestors chose to eat meat and it was probably important in maximising reproduction and survival. We may not be here if they hadn't. They had a very short lifespan. A longevity diet would have been less important and may have decreased reproduction and survival.

Hunter-gatherers would have eaten a diet of whole, fibre-rich foods from the vegetables, tubers and fruits available to them, supplemented by wild meats, so dissimilar to the plant-based and animal-based foods we eat today. Even a WFPBD is thought only likely to match the lower end of the fibre they ate.

Since our hunter-gatherer days, human anatomy has not substantially changed. Change has mostly been forged through culture, since we learned to control fire and cook.

Cooking is likely to have been essential. Our mouths and teeth are not designed to eat the meat we caught; tools and cooking were necessary to make it more available, smaller, safer and softer to eat, and more digestible. Similarly, cooking would have expanded the range of plant-based foods available to our ancestors; including plant foods, which before may have been less palatable and/or made us sick. In both cases cooking allowed us to maximise the nutrition from a much wider range the foods. As suggested by Richard Wrangham of Harvard University:[4]

The extra energy gave the first cooks biological advantages. They survived and reproduced better than before. Their genes spread. Their bodies responded by biologically adapting to cooked food, shaped by natural selection to take maximum advantage of the new diet.

The fire we cooked with would also have been important in the way we lived our lives, allowing us to live on the ground by providing protection, warmth and comfort.

For 18 million years we were herbivores, perhaps except for the odd insect or grub. For us and the health of our planet, we need to return to how we started.

Citations:

1. Lawrence, D. et al. (2014), 'Diet rapidly and reproducibly alters the human gut microbiome.' *Nature* 505, pp. 559–563.
2. Raubenheimer, D. and Simpson, S. J. (2020), *Eat Like the Animals: What Nature Teaches Us About Healthy Eating*. William Collins, London, UK
3. Campbell, T. Colin and Campbell, Thomas M. (2016), *The China Study: Revised and Expanded Edition: The Most Comprehensive Study of Nutrition Ever Conducted and the Startling Implications for Diet, Weight Loss, and Long-Term Health*. BenBella Books, Dallas, TX
4. Wrangham, Richard (2010), *Catching Fire: How Cooking Made Us Human*. Profile Books, London, UK

11. Tricks of the Trade

1. Njike, V. et al. (2010)[1]

It is generally recommended we reduce our consumption of eggs, which are high in cholesterol, to reduce cardiovascular disease. This study looked to test this against recent industry evidence suggesting that dietary cholesterol has a limited influence on cholesterol or cardiac risk.

The study found that 'single dose egg consumption had no effect on endothelial function compared to [sausages/cheese]; and the daily consumption of egg substitute for six weeks significantly improved endothelial function compared to eating egg and lowered serum total cholesterol.'

The study concluded 'Egg consumption [as compared to sausages/cheese] was found to be non-detrimental to endothelial function and serum lipids in hyperlipisemic adults, whilst egg substitute consumption was beneficial.'

The study was 'positively' published, neglecting to mention that egg consumption was compared to eating sausages and cheese!

2. Thorning, T. K. et al. (2015)[2]

This study compared the following three diets:

1) a high-cheese (96–120 g) intervention (i.e. intervention containing cheese [CHEESE]),

2) a macronutrient-matched non-dairy, high meat control (i.e. non-dairy control with a high content of high-fat processed and unprocessed meat in amounts matching the saturated fat content from cheese in the intervention containing cheese [MEAT]), and

3) a non-dairy, low-fat, high-carbohydrate control (i.e. non-dairy, low-fat control in which the energy from cheese fat and protein was isocalorically replaced by carbohydrates and lean meat [CARB]).

The study, supported by the dairy industry, found that total cholesterol, LDL cholesterol, ApoB and triacylglycerol were similar between the three diets. How did they do it?

As explained by Michael Greger M.D. on NutritionFacts.org,[3] this was done by making sure each of the three diets contained the same amount of saturated fat. For instance, in the low-fat diet (number 3 – CARB) they added a large amount of chocolate, coconut milk, coconut fat and sweetened biscuits!

Table 15: Contributions of different foods in percentage of total content of saturated fatty acids (SFAs)

| | SFAs | | |
	CHEESE diet	MEAT diet	CARB diet
Cheese	53	-	-
Chocolate	14	17	24
Meat	11	48	28
Coconut milk	7	11	20
Coconut fat	-	11	6
Sweetened biscuits	6	6	10
Almonds, peanuts	2	-	0
Canola oil	1	-	1
Sunflower oil	-	1	-
Bread, pasta, rice	4	5	7
Other[1]	2	1	4
Sum	100	100	100

Table notes:

[1] Fresh fruit, dried fruit, vegetables, eggs, cornflour and saccharose.

Table 16: Nutrient composition

	CHEESE diet	MEAT diet	CARB diet
SFAs, % of fat	49.9 (52.1)	49 (50.8)	49.8 (51.6)

3. Demmer, E. et al. (2016)[4]

This study was designed to show that cow cheese caused less inflammation than a vegan alternative. The vegan cheese had 2 g of saturated fat compared to 6 g in the cow cheese. So how was this done?

The researchers added lots of palm oil to the vegan cheese diet, which has been shown in other research to cause more inflammation than milk fat. With the addition of palm oil, the vegan cheese meal contained 17.2 per cent saturated fat. The cheese meal came out slightly lower at 16.8 per cent saturated fat.

4. Ashton, E. and Ball, M. (2000)[5]

In this study the beef industry compared meat with tofu, adding butter and lard to the tofu diet to match the fat intake with the meat diet! Unsurprisingly, they found a similar LDL cholesterol and HDL cholesterol ratio between the two diets.

5. Roussell, M. A. et al. (2102)[6]

The objective of this study, supported by the American Beef Checkoff program, was to see beef's effect on LDL cholesterol by using cholesterol-lowering diets with varying amounts of lean beef.

In the study beef was added to the subject's diet and the percentage of total and LDL cholesterol went down. From this, they concluded that low saturated fat, heart-healthy dietary patterns containing beef produce favourable effects on cardiovascular disease risk.

This miracle was achieved by cutting out so much high-fat dairy, poultry, pork, fish and eggs in the beef diet that they cut the amount of saturated fat in the diet by 50 per cent, from 12 g to 6 g![7]

6. Fernandez, M. L. (2012)[8]

In 2012 Professor Mari L. Fernandez, a proponent of the egg industry, reviewed contemporary evidence which challenged the dietary restrictions of that time regarding cholesterol in the US (300 mg per day), whilst presenting some beneficial effects of eggs (an icon for dietary cholesterol) in healthy individuals. The review received funding from the American Egg Board, but despite this the professor declared no conflict of interest.

Professor Fernandez accepted that eggs raised LDL 'bad' cholesterol, but argued that HDL 'good' cholesterol also rose, resulting in the maintenance of the LDL–HDL ratio, a key marker in cardiovascular heart disease (CVD). Part of her evidence relates to an American Egg Board and Egg Nutrition Center study, which she was also a part of.[9]

The study concluded:

> *[...] postmenopausal woman and men >60 years old with healthy lipoprotein profiles may consume eggs as part of their regular diet. For those subjects who experienced an increase in LDL cholesterol in response to the dietary cholesterol challenge, this was countered by an increase in HDL cholesterol and an increase in the size of the LDL particle (antiatherogenic). In addition, the susceptibility of LDL oxidation is not enhanced by egg intake.*

However, evidence in a 2001 meta-analysis of seventeen trials actually showed that the rise in LDL was much greater than the rise in HDL after egg consumption.[10] The 2001 meta-analysis concluded:

> *[...]the favourable rise in HDL cholesterol with increased cholesterol intake fails to compensate for the adverse rise in total and LDL cholesterol concentrations and, therefore, that increased intake of dietary cholesterol may raise the risk of coronary heart disease.*

Digging deeper into Professor Fernandez's review on egg intake, regarding the formation of less atherogenic lipoproteins in large fluffy LDL being better than small dense LDL, this is true.

Large LDL was found only to raise CVD risk by 44 per cent compared to 63 per cent for small LDL! However, both have a considerable impact.

It has long been established, over many decades of extensive research, that both large and small LDL cholesterol are causal agents in atherosclerosis.[11] However, Professor Fernandez's review chose to relay the message in a misleading way:

The formation of large LDL is considered atheroprotective [in relation to cardiovascular disease] *relative to small LDL.*

As Michael Greger succinctly put it, it's like saying 'getting stabbed by a knife is protective relative to getting shot.'

Eggs do contain some beneficial components; for instance they are a source of lutein and zeaxanthin, two antioxidants which have been shown to protect against macular degeneration and cataract formation and also in the reduction of LDL oxidation.

However, the levels of these phytonutrients are so small it would require the consumption of many eggs to see any benefit.

In respect of eggs reducing the susceptibility of oxidised LDL, the exact opposite has also been found (Yishai Levy et al., 1996):[12]

[...] egg consumption, in addition to its hyper-cholesterol effect, increases plasma and LDL oxidizability, a phenomenon which was shown to enhance the progression of atherosclerosis.

The same research concluded that eggs, as with other research, raise LDL 'bad' cholesterol levels.[13]

Citations:

1. Njike, V. et al. (2010). 'Daily egg consumption in hyperlipidemic adults: Effects on endothelial function and cardiovascular risk.' *Nutr J.* 2010 9(28)
2. Thorning, T. et al. (2015), 'Diets with high-fat cheese, high-fat meat, or carbohydrate on cardiovascular risk markers in overweight postmenopausal woman: a randomized crossover trial.' *Am J Clin Nutr.,* 102(3), pp. 573-81
3. Dr. Michael Greger M. D., NutritionFacts.org
4. Demmer, E. et al. (2016), 'Consumption of a high-fat meal containing cheese compared with a vegan alternative lowers postprandial C-reactive protein in overweight and obese individuals with metabolic abnormalities: a randomised controlled cross-over study.' *J Nutr Sci.* 5:e9, pp. 1-12
5. Ashton, E. and Ball, M. (2000), 'Effects of soy as tofu vs meat on lipoprotein concentrations.' *European Journal of Clinical Nutrition* 54, pp. 14-19
6. Roussell, M. et al. (2012), 'Beef in an Optimal Lean Diet study: effects on lipids, lipoproteins, and apolipoproteins.' *Am J Clin Nutr.* 95(1), pp. 9-16
7. Michael Greger M. D. (2018), NutritionFacts.org. *How the Dairy Industry Designs Misleading Studies.* https://nutritionfacts.org/video/how-the-dairy-industry-designs-misleading-studies/

8. Michael Greger M. D. (2014), NutritionFacts.org. *Does Cholesterol Size Matter?* https://nutritionfacts.org/video/does-cholesterol-size-matter/

9. Greene, C. M. et al. (2005), 'Maintenance of the LDL Cholesterol: HDL Cholesterol Ratio in an Elderly Population Given a Dietary Cholesterol Challenge.' *Journal of Nutrition* 135(12), pp. 2793-98

10. Weggemans, R. M. et al. (2001), 'Dietary cholesterol from eggs increases the ratio of total cholesterol to high-density lipoprotein cholesterol in humans: a meta-analysis.' *Am J Clin Nutr.* 73(5), pp. 885-91

11. Robinson, J. G. (2012), 'What is the Role of Advanced Lipoprotein Analysis in Practice?' *Journal of the American College of Cardiology*, 60(25), pp. 2607-15

12. Levy, Y. et al. (1996), 'Consumption of Eggs with meals Increases the Susceptibility of Human Plasma and Low-Density Lipoprotein to Lipid Peroxidation.' *Ann Nutr Metab.* 40(5), pp. 243-51

13. Ibid.

CHAPTER III

Livestock Farming's Longer Shadow

I think we've become very disconnected from the natural world. Many of us are guilty of an egocentric world view, and we believe that we're the centre of the universe. We go into the natural world and we plunder it for its resources. We feel entitled to artificially inseminate a cow and steal her baby, even though her cries of anguish are unmistakable. Then we take her milk that's intended for her calf and we put it in our coffee and our cereal.

We fear the idea of personal change, because we think we need to sacrifice something; to give something up. But human beings at our best are so creative and inventive, and we can create, develop and implement systems of change that are beneficial to all sentient beings and the environment.

Joaquin Phoenix, speech at the 2020 Oscars

12. Introduction

In this chapter I deal with livestock's longer shadow in varying degrees of detail, and in particular the history of slurry pollution. History is particularly important as it provides a better understanding and appreciation of the legacies of the past which manifest in the present. My intended readership is also wide and diverse, so I've aimed to satisfy most of what I'm commonly asked and feel necessary to record, and which may otherwise be lost to time. If parts of what follow are too exhaustive, please skip, and if not, it is there for you to indulge in, and I hope you do.

When it comes to livestock and pollution, don't just take my opinion. According to Emma Howard Boyd, chair of the Environment Agency, 'pollution problems are rife in the dairy and intensive beef farming sectors'.[1] To that I would add pigs and poultry too.

Livestock farming once coexisted in relative harmony with nature. Today that relationship has been broken and overwhelmed with livestock, to lavish our plates and taste buds with a surfeit of meat, eggs, dairy and farmed-fish foods we do not need to eat.

The lion's share of the UK's productive agricultural land is used to keep and nourish these animals, with what's left to nature subservient to livestock farming.

According to a study by Henri de Ruiter et al. (2011):[2]

Only 15% the UK's land footprint [total land footprint associated with UK food supply, excluding cropland and grassland 'embodied' in exports] in 2010 was associated with crops that are directly grown for human food, while 22% of the total land footprint was used for growing feed crops, with the remaining 63% due to grazing area.

At the same time, this 85% of the land footprint used to produce animal products only contributed about 32% to total calorie supply and 48% of total protein supply. This illustrates the relative inefficiency of producing livestock products.

The authors point out that often grasslands are not suitable for crop production, and so they do not always compete with food for human consumption and especially milk production. Whilst this is somewhat true, grasslands used for livestock have greatly harmed ecosystems (and with them ecosystem services), have reduced soil health and the soil carbon stock, and support livestock which are often fed supplementary feed (sourced from home and overseas) which competes with available cropland for direct human consumption. The extensive monocultured grasslands which carpet the UK are biodiversity deadlands, and woe betide any plant or animal that competes, or otherwise inhibits, livestock farming – for which the livestock industry has an 'unfortunate' solution.

Specific to the UK territory, a study carried out by Harwatt and Hayek (2019)[3] showed that animal agriculture occupied 115,900 km² out of the UK's total land area of 241,930 km² (48 per cent of the total). The 115,900 km² consists of 84,000 km² of permanent pasture and around 31,900 km² (3.19 million hectares) of cropland used for animal feed (55 per cent of the current cropland area). When permanent pasture is combined with cropland used for animal feed, the livestock share of the UK's agricultural land is 81.5 per cent. The 55 per cent of UK cropland used to inefficiently feed livestock, rather than to directly feed people, is staggering, hard to digest and a huge waste of food.

According to WRAP, around two million tonnes of food surplus and waste occurs in primary production on farms,[4] and the Food and Agriculture Organization of the United Nations suggests that 8.4 million people in the UK struggle to afford a meal. 3.19 million hectares of cropland can grow a staggering amount of food, well in excess of 10 million tonnes, which is currently wasted by being fed to livestock and otherwise inefficiently converted into meat, fish, dairy and eggs. A good part of the total food surplus and waste (around 3.6 million tonnes[5]) is also fed to livestock and anaerobic digesters, which is perfectly good to feed to people and should be prioritised. Fantastic organisations like FareShare (www.fareshare.org.uk), and others/volunteers across the UK, are leading the way to fight hunger and tackle this food waste, but we must also do this by stopping wasting crops by feeding them to livestock. Grow healthy/wholesome food on our croppable land for people, not livestock. How much more land does the livestock industry want to occupy? They already have over 80 per cent. However, even 3.19 million UK croppable hectares is not enough for the livestock industry, as they use well over one million hectares more in the rainforests of South America too, primarily used to feed chickens and pigs, and to a lesser extent cattle and sheep, soya bean meal (see chapter IV). There is a further danger that a greater amount of croppable land will be consumed by livestock if the UK becomes more self-sufficient in livestock protein crops (e.g. lupins and forage beans); and you can be sure that if croppable land is built on (e.g. roads, housing and industrial premises), it won't be livestock feed crops that give way. Factory farms who can pay the most for food grown on a piece of land find the means to hoover it up.

Harwatt and Hayek (2019) also looked at scenarios for repurposing some of the animal feed cropland for human-edible fruit and vegetable production, 90 per cent of which by value is currently imported. They created a list of seventeen fruits and vegetables currently grown as field crops in the UK and assigned them 1/100th each extra of the cropland currently used by animals. Grossing the numbers to repurpose a third of this land, the authors found it was sufficient to provide 62 million adults with the UK's Eatwell dietary guide recommendation of five servings of fruits or vegetables a day. They identified another list of fruits and vegetables also currently grown in the UK but not yet in any significant quantity, like lentil, sunflower, sesame and sweet potato, to repurpose more land and add dietary variety. This still excluded crops like quinoa and hemp, which are also grown in parts of the UK, and other opportunities to become more self-sufficient in salad crops grown in greenhouses and through planting pastureland with fruit and nut trees. Ultimately, what is possible to grow commercially in any specific area or region of the UK would be variable, but it would be a much better use of land to improve the UK's health, environment and food self-sufficiency with plant-based foods.

The transition from livestock farming within nature to one of nature eking out an existence within livestock farming has gone almost unnoticed. Much of this change happened after World War II, in a march towards cheap food, led by corporate agricultural interests and its increasingly urbanised supporters. Farms began to become bigger and less diverse, halving in number by the late 1970s.[6] Hay meadows fell to the plough, hedgerows were dug up, peatlands were drained and biodiversity slayed. Crop rotations became less varied, livestock numbers grew and the animals were increasingly placed in intensive industrialised units, and continue to be today.

Livestock stock density doubled between 1930 and 1990, from 0.83 head per hectare to 1.77 head per hectare respectively.[7] By the end of the 1970s, courtesy of increased fertiliser and pesticide use and technological advancements, a dairy cow could be sustained on 0.6 hectares of land compared to one hectare just prior to World War II.[8]

Today, at any one time, the UK is home to: [9]

- 9.9 million cows and calves;
- 5 million pigs;
- 34 million sheep and lambs;
- 188 million poultry.

This is over three and a half times the UK human population.

Keeping and feeding this livestock requires millions of hectares of land in the UK, and over million hectares of deforested land in South America and Asia. The UK's millions of cows and sheep produce huge amounts of greenhouse gases and, along with associated deforestation, furthers our planet's climate emergency. Livestock also contributes to soil damage, erosion and loss, and produces huge amounts of excrement, all of which needs to be carefully managed to avoid causing harm such as ammonia, nitrate-nitrogen, ammonium-nitrogen, phosphorus and microbial pathogen pollution.

When it comes to nutrient pollution, and in particular nitrogen and phosphorus, a good-news story about the reduced quantity of fertilisers used by farmers on crops, since the early 90s, sometimes surfaces. However, this trend relates to the overall use of manufactured fertiliser, which has levelled out since 2008.[10] In this book I concentrate on the pollution from nutrients arising from the animals farmed (handled or directly deposited on land), not manufactured fertiliser, which is managed/controlled in an entirely different way.

For instance, handled organic manures all too often:

- are spread/disposed of at the wrong time of the year, when there is no soil and crop need and the risk of causing pollution through leaching, preferential drainage and run-off is significant;
- are spread less efficiently than manufactured fertilisers, still all too frequently using inefficient and less accurate technology (e.g. vacuum tankers and splash plate); and
- contain a mix of nutrients that can give rise to an exceedance of soil and crop need for a particular nutrient (e.g. phosphate), unlike manufactured fertilisers which can be well-targeted/fine-tuned using straights or a ratio of nutrients and with a much better use efficiency/consistency than organic manures.

Furthermore:

- Livestock farms, and in particular pig, dairy and chicken, are continually concentrating and intensifying, along with their manures, in different parts of the country; for instance: pig farms to the drier east; dairy farms to the wetter west; and chickens within the counties of Herefordshire, Shropshire and Powys.[11]

Added to this are the organic manures/wastes from the food producers/plants which follow them (e.g. slaughterhouses, anaerobic digesters, renderers, dairy processors, etc).

This situation creates sprawling manure hotspots within the UK, concentrating spreading/disposal close to the places of production, beyond soil and crop need and at a higher risk of causing water pollution and advancing ammonia emissions.

Livestock may also be kept out on land at high-risk times of the year and generate pollution directly from faeces, urine and through trampling wet soils.

Whilst it is only right to show reductions in the use of manufactured fertiliser, it cannot be used as evidence of a downward trend of pollution or good agricultural practice in relation to organic manures – far from it, as you will go on to see.

Citations:

1. Emma Howard Boyd (2021), *The Green Industrial Revolution needs Green Industrial Regulation* [Speech].
   https://www.gov.uk/government/speeches/the-green-industrial-revolution-needs-green-industrial-regulation
2. de Ruiter, H. et al. (2017), 'Total global agricultural land footprint associated with UK food supply 1986-2011.' *Global Environmental Change* 43, pp. 72-81
3. Harwatt, H. and Hayek, M. (2019), *Eating Away at Climate Change with Negative Emissions: Repurposing UK agricultural land to meet climate goals.* Harvard Law School.
   https://animal.law.harvard.edu/wp-content/uploads/Eating-Away-at-Climate-Change-with-Negative-Emissions%E2%80%93%E2%80%93Harwatt-Hayek.pdf
4. WRAP (2019), *Food surplus & waste in Primary production costs UK more than £1 billion.*
   https://www.wrap.org.uk/content/food-surplus-waste-primary-production-costs-uk-more-%C2%A31-billion
5. Ibid.
6. Royal Commission on Environmental Pollution (1979), *Agriculture and Pollution, Seventh Report.* HMSO, London, UK
7. Edwards, A. and Withers, P. (1998), 'Soil phosphorus management and water quality: A UK perspective.' *Soil Use and Management* 14, pp. 124-130
8. Royal Commission on Environmental Pollution (1979), *Agriculture and Pollution, Seventh Report.* HMSO, London, UK
9. Defra (2017), *Agriculture in the United Kingdom.* National Statistics at Gov.uk
10. Ibid.
11. Defra, *Maps of livestock populations in 2000 and 2010 across England.*
    https://assets.publishing.service.gov.uk/government/uploads/system/uploads/attachment_data/file/183109/defra-stats-foodfarm-landuselivestock-june-detailedresults-livestockmaps111125.pdf

13. Poo Sticks

Livestock manure has been used for millennia to maintain and improve soil fertility, typically as:

- a solid mixture of excreta, urine and a bedding material, derived from livestock housed in buildings or kept on yards.
- a semi-liquid or liquid slurry voided by livestock, kept on, or mixed with, a limited amount of bedding material.
- excreta and urine directly voided by grazing animals.

In its solid well-rotted form, it is easy to handle, treat, store, transport and spread on land reasonably safely when needed. In a liquid form, it is much more problematic and expensive to manage, transport and return safely to land without causing pollution.

When livestock manure is allowed to escape into the environment, as is too often the case, whether in water, soil or air, it has the potential to cause serious harm or contribute to harm. Livestock manure managed inappropriately may be harmful to human health and the quality of aquatic ecosystems, or terrestrial ecosystems directly dependent on aquatic systems.

Gases which escape directly from livestock, and from their manure, contribute to:

- the gathering climate emergency, for instance, methane and nitrous oxide.
- ammonia gas volatilising from manures, which when mixed with nitric acid from industrial and vehicle nitrogen emissions is linked to respiratory and cardiovascular diseases, cognitive decline, low birth rates and increased mortality.
- ammonia emissions, which drift and deposit on sensitive plant communities in keystone ecosystems, causing unnatural soil enrichment and acidification which in turn damages the natural flora and fauna dependent on it.

Given the harm livestock emissions can cause, you would be forgiven for thinking the UK has robust legislation and regulatory presence to protect us and the natural environment from them. Sadly, that is not the case, and what legislation has been provided bears the scars of politicians, corporate interests and a woeful lack of resource to advise and regulate the livestock industry effectively.

Our legislative history is surprisingly short and only goes back in any significance to the Control of Pollution Act 1974. Even then, it wasn't until after the Water Act 1989 came into force that much was done about livestock pollution.

The urgency to legislate livestock pollution was primarily an outfall of post-World-War-II intensification. Before then, in the intervening war years, concern was not with on-farm pollution but with pollution from factories processing sugar beet and dairy products.[1]

This was not too surprising. Back in 1930 there were many more farms, and each had a much lower stocking density compared with today. For example, 0.83 head per hectare in 1930 compared to 1.77 head per hectare in 1990.[2]

In terms of farm numbers, in 1955 there were 141,399 holdings with dairy cows (2,475,610 cows) in England and Wales. By 1986 the number of dairy farms had reduced drastically to 41,055 (2,573,222 cows).[3] Around 85 per cent of the 141,399 kept fewer than 29 dairy cows, while by 1986 around 75 per cent of the 41,055 had 30 dairy cows or more. The milk yield in 1955 was also low at 3,144 litres per cow per year, compared to 4,935 litres by 1986. This is important, as the more milk a cow produces, the more slurry it produces, and the bigger the pollution risk.[4] Interestingly, between these years, the number of dairy cows remained similar, at around 2.5 million. In England it is about half that number today, but the average milk yield is now a heady 8,000 litres per cow per year, which more than makes up for it. Beef, pig and poultry production was equally less intensive and more spread out across the UK before World War II.

More farms in the past also kept a mixture of livestock, whilst today many farms specialise in one or two types, with most livestock being kept on strawed yards and within straw-bedded sheds, producing farmyard manure (FYM). In 1953, around 35 million tonnes of FYM a year was produced on English and Welsh farms.[5] The amount of slurry was not recorded, or at least I can find no record, but it would have been a smaller percentage in comparison. What slurry farmers did produce was either collected in a tank or allowed to drain *somewhere*. The quantity of slurry lost would have been widely spread out compared with today, and the consequences on the environment would have been less concentrated, obvious and noxious.

Back then, usually only the most concentrated, nutrient-rich slurry was stored. Rainwater and wash water contaminated with slurry, from open yards and the washing down of livestock buildings, was usually allowed to run onto open land, drain directly/indirectly into surface waters or disposed of in a soakaway.

A government advisory leaflet from 1960 touches on the handling of wash water:

If much of the water used for washing down the cowstalls is allowed to run into the storage tank, the liquid manure will be made weaker. Arranging a simple flap valve so that the water can be diverted into the drains is worthwhile. After washing down is finished, the valve can be turned so that the urine from the standing flows into the tank from which it can be pumped into a liquid manure distributor or irrigation system.

In open yards it is not always possible to provide enough litter to absorb the urine and the rain, and consequently the losses of nitrogen, phosphorus and potash are generally large. Sometimes, however, the dark liquid which drains away from the heap [farmyard manure] can be collected in the tanks into which the liquid manure from the cow sheds is channelled.

Source: Ministry for Agriculture, Food and Farming, Advisory Leaflet 435. Making the Most of Farmyard Manure, 1960.

Storage capacity, though, was minimal, with slurry regularly spread to land throughout the high-risk autumn and winter months.

As livestock farms have consolidated and grown in size, farmers have increasingly turned towards slurry-based systems. By 1976 only around 45 per cent of dairy farms were keeping their cows on straw bedding, with around 32 per cent adopting a slurry-based system and a further 20 per cent a part slurry-based system. The remaining 3 per cent overwintered their cows outside.

More recently, a 2008 study found that the amount of dairy and beef solid manure and slurry spread to land in England and Wales was estimated at 25.6 and 34.2 million tonnes respectively.[6] The lion's share was dairy, with solid manure and slurry spread amounting to 9.6 and 25.5 million tonnes respectively.[7] Beef was estimated at 16 and 8.7 million tonnes, indicating a much greater reliance on straw-bedded systems.[8]

Inclusive of all livestock, the total amount of solid manure spread was 33.8 million tonnes and slurry 37.7 million tonnes, with a further 73 million tonnes of excreta deposited during grazing, of which 16.4 million tonnes is dairy, 25.8 million tonnes beef and 30 million tonnes sheep, goats, deer and horses combined.[9] For the UK as a whole, the total amount spread is of the order of 90 million tonnes.

Post-war intensification was focussed on growing and getting more out of crops and livestock to provide food security, and not on the swell of agricultural pollution which grew from it.

This was aided and abetted by the NFU, and to a lesser extent the Country Landowners Association (CLA). The NFU and CLA enjoyed a cosy relationship with the Ministry for Agriculture, Food and Farming (MAFF), who had responsibility for agricultural pollution. Together, they fought fiercely to protect farmers from having any pollution prevention policy and regulation foisted on them, whenever it reared its 'ugly' head.[10] However, a tide of pollution was rising, and their ability and success in trying to contain it as a non-issue began to wane.[11] But, whenever their success did dwindle, together they still managed to ensure whatever policy and regulation transpired was diluted and dispersed.

MAFF was created in 1889 and was originally known as the Board of Agriculture. The ministry enjoyed a long and productive existence until it was eventually dissolved in 2002, in the wake of a tarnished reputation over its handling of the 2001 foot-and-mouth disease outbreak and the earlier bovine spongiform encephalopathy ('mad cow disease') outbreak. MAFF responsibilities were reborn into a new government department, as part of a merger with the Department of the Environment, Transport and the Regions in 2000. The new department was named the Department for Environment, Food and Rural Affairs (Defra).

After World War II, as farmers increased their dairy and pig herds and increasingly housed them on slurry-based systems, pollution incidents began to escalate. Throughout the 1960s slurry pollution became increasingly commonplace. In acknowledgement of the problem, and despite ministry and industry corporate resistance, existing pollution prevention law was strengthened by the introduction of the Control of Pollution Act 1974 (Part II).

The 1974 act re-enacted and expanded upon earlier land, water, noise and air pollution legislation, which the Rt Hon Margaret Thatcher (her prime ministerial role yet to come) hailed as 'the most comprehensive attempt for many years to bring pollution under control.'[12] However, despite such optimism, many provisions of the act were slow to come into force.

In February 1984 the *Tenth report of the Royal Commission on Environmental Pollution: tackling pollution – experience and prospects* took stock of progress made over the previous ten years, and commented: 'so far as the majority of the Act's provisions relating to water pollution are concerned, that claim [by Rt Hon Margaret Thatcher] remains to be tested'.[13]

In fact, by 1984 major elements of the act had still not been laid – for example, Part II, Section 31, concerning the prevention of pollution of water. MAFF set out the current state of affairs: 'The main provisions of Part II of the Control of Pollution Act, which is concerned with pollution of water, are being introduced in stages from July 1983 to July 1987.'[14]

The act's delay was largely caused by the economic cost to industry. It was also hindered by financial pressures on the regional water authorities, who had been made responsible for regulating the water pollution requirements of the act.[15]

Water authorities were a division of ten publicly owned regional water authorities (RWAs). RWAs were privatised in 1989 and responsibility for pollution and fishery regulation was transferred to the National Rivers Authority (NRA) under the Water Act 1989. The NRA was granted additional pollution offence powers under Sections 85 and 86 of the Water Resources Act 1991.

The Environment Agency. The NRA was merged on 1 April 1996 (under the provisions of the Environment Act 1995) with Her Majesty's Inspectorate of Pollution (NMIP) and county council waste regulation authorities to form the present-day Environment Agency (EA).

The EA is a non-governmental department sponsored by Defra, covering the majority of England. Each of the devolved nations has its equivalent body and powers.

Under the Environment Act 1995, the EA has a duty to exercise its pollution control powers for the purpose of preventing or minimising, or remedying or mitigating the effects of, pollution of the environment and to such extent as it is considered desirable, generally to promote the conservation and enhancement of the natural beauty and amenity of inland and coastal water and of the land associated with such waters and the conservation of flora and fauna which are dependent on an aquatic environment.

Today, in respect of livestock pollution, the EA's role includes its water pollution control functions under the Water Resources Act 1991, the Water Resources (Control of Pollution) (Silage, Slurry and Agricultural Fuel Oil) (England) Regulations 2010, the Nitrate Pollution Prevention Regulations 2015, the Environmental Permitting (England and Wales) Regulations 2016 and the Reduction and Prevention of Agricultural Diffuse Pollution (England) Regulations 2018. The Environmental Permitting (England and Wales) Regulations 2016 retain similar water pollution offences to those previously used in Section 85 of the Water Resources Act 1991.

Ultimately, the EA is responsible for securing compliance with the Water Environment (Water Framework Directive) (England and Wales) Regulations 2017, or WFD for short. Compliance with WFD includes preventing deterioration in status or the potential of a water body from discharges and aiming to achieve good status or potential in water bodies.

Despite remaining committed, the water authorities urged ministers to continue to defer full implementation on grounds of cost. However, ministers reasserted the commitment:

Ministers, however, reaffirmed a commitment to the full implementation of COPA II, not least because it would ease the necessary implementation in the United Kingdom of various European Community directives on environmental protection and public health, but they decided that decisions on ways of phasing in the outstanding provisions would depend on the availability of resources in both the public and private sectors.[16]

The offence provisions of Part II of the act were eventually enacted in 1984, with certain transitional measures applying to existing uncontrolled discharges to water.

Part II of the act was important to agricultural pollution management, with Section 31 making it an offence for a person to cause or knowingly permit:

(a) any poisonous, noxious or polluting matter to enter any stream or controlled waters or any specified underground water;

(b) any matter to enter a stream so as to tend (either directly or in combination with other matter which he or another person causes or permits to enter the stream) to impede the proper flow of the water of the stream in a manner leading or likely to lead to a substantial aggravation of pollution due to other causes or of the consequences of such pollution;

(c) any solid waste matter to enter a stream or restricted waters.

However, Section 31(2)(c) of the act also provided a defence that a person would not be guilty of the offence of causing water pollution if the entry in question was due to an act or omission which was in accordance with good agricultural practice. What was meant by good agricultural practice was described in Section 31(9) of the act:

> *[...] any practice recommended in a Code approved for this purpose by the Ministry for Agriculture, Food and Farming shall be deemed to be good agricultural practice (but without prejudice to evidence that any further practice is good agricultural practice).*

This made it extremely difficult to prove an offence and is an example of how political and industry corporate interests were able to influence and water down the potential impact of the act on the farming community.

What constituted good practice was ill-defined, and therefore by extension bad practice was also hard to define. To aid the defence of farmers MAFF produced the 'Farm Waste Management' series of booklets, including Booklet 2200, *Advice on pollution and other slurry wastes,* which farmers could rely on as evidence against a pollution offence (i.e. by following that advice). There was little urgency for anything more comprehensive, given the offence provision of the act did not come into force until 1984. However, enactment of Part II of the act in 1984, and the risk of a case being brought against the UK by the European Court for that inaction, forced the issue.[17]

MAFF's first Code of Good Agricultural Practice, arising out of the Control of Pollution Act 1974, after eleven years of intransigence, was finally published in January 1985.

The legal defence against causing pollution remained in place until the enactment of the Water Act 1989, four years later. Even after its removal, a defence of compliance with the Code of Good Agricultural Practice 1985 (and later revisions) remained a mitigating factor against a water pollution charge. Unless the offence was very obvious, a lot of technical skill would be needed to prove a case and enable it to proceed to court. Mitigation can also significantly affect the outcome and fine received on conviction. Fines were very low at the time, making prosecution an unattractive option for all but the most serious pollution incidents.

When it became known that the NRA was to inherit water pollution enforcement powers from the regional water authorities, MAFF took exception, fearing it could threaten its own existence. Of particular concern was the threat to freedoms ('exceptionalisms') from regulation they, backed by the NFU and CLA, had secured for the farming community since World War II. They attempted to seize control of, and marginalise, the NRA's powers, but ultimately failed.[18]

Despite that failure, MAFF remained a powerful actor in agriculture: as a statutory consultee, administrating research and development and through the provision of guidance and advice to farmers on pollution management.

For example, through their advice programme, led by their advisory unit ADAS, MAFF continued to hold great sway on whether an offence of pollution by a farmer would actually occur under Section 31 of the 1974 act.

The intransigence of MAFF, supported by the likes of the NFU and CLA, unfortunately led to a deluge of farm pollution incidents, as shown in Chart 1.

Chart 1: Farm Pollution Incidents (England and Wales)[19]

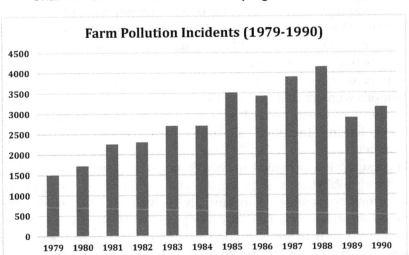

The majority of these incidents were related to: slurry, typically slurry stores (e.g. 881 in 1985 and 632 in 1990); yard/parlour washings, commonly known as *dirty water* slurry (e.g. 695 in 1985 and 763 in 1990); land 'slurry' run-off (e.g. 237 in 1985 and 395 in 1990) and treatment system failure (e.g. 123 in 1985 and 130 in 1990). Solid manure store (i.e. farmyard manure) incidents amounted to 185 in 1985 and 118 in 1990. Another large contributor, and specifically in dairy farming, was silage liquor (e.g. 1,006 in 1985 and 470 in 1990). Reported incidents in Scotland rose from 252 in 1981 to 572 in 1988.[20] Across Great Britain the increase in reported incidents was considered in part due to the public's greater awareness of pollution and willingness to report incidents.

It is also worth noting that these were *reported* incidents, and many more would have gone unreported, and in particular those not involving a fish kill. Dairy farms registered by far the majority of slurry-related incidents.

In 1990, 72 per cent of the 3,147 reported pollution incidents were related to cattle farming (dairy and beef).

The slurry emergency was to be a major test for the newly appointed NRA, and the weakened resistance to change from MAFF, NFU and CLA.

Much-needed change came about in the form of the Control of Pollution (Silage, Slurry and Agriculture Fuel Oil) Regulations 1991 (SSAFO), which had its origins in the Water Act 1989 (sections 110 and 185). These regulations required farmers with custody or control of a crop being made into silage, livestock slurry or certain fuel oil to carry out works and take precautions for preventing pollution of waters.

For the first time, installations of storage for slurry, silage and agricultural fuel oil were to be constructed to the relevant British Standards, have a twenty-year design life (generally with maintenance) and remain effective whenever the installation was used. The regulations were framed in a way that would allow the NRA to deal with the majority of situations with minimal resource implications. However, the regulations included an unfortunate exemption from the requirements.

The exemption provided in SSAFO, negotiated between MAFF and agriculture lobbyists, was to ensure the regulations only applied to installations constructed, substantially reconstructed or substantially enlarged *after* 1 March 1991 (or 1 September 1991 where a contract was entered into or construction commenced before 1 March and where work would be completed before 1 September 1991).

Rather than an exemption, it would have been much better to have included a transitional arrangement to bring all stores up to the required standard. It was perhaps *reasonably* envisaged that the pre-1991 exemption for existing stores would be revisited in subsequent years. However, despite a review of SSAFO prior to the 1997 SSAFO amendment regulations (1 April 1997) no changes were made to SSAFO, nor in two subsequent revisions in 2010 and 2013.

Removal of the SSAFO 1991 exemption has been a crusade for myself and others for many years. Representations have been made, including from environmental non-governmental organisations (ENGOs), for its removal. Each time it has been thwarted by the Knights Templar of the agricultural industry, who fiercely protect the exemption like a Holy Grail – and, by extension, the pollution that goes with it – to protect industry profits and viability. If you support the SSAFO exemption, you support poor agricultural practice and pollution. It is as simple as that.

The SSAFO regulations do contain a notice provision, which can be served on exempt stores where, in the opinion of the Environment Agency, works, precautions or other steps are appropriate to the requirements of the regulations for reducing to a minimum any significant risk of polluting waters. Failure to comply with a served notice would remove the exemption, the reason for this requirement being to ensure pre-1991 polluting stores were not allowed to continue unchecked. However, it can be immensely difficult to prove a structure is unsound, unless obvious.

To help make the new regulations palatable to industry, and to encourage farmers to comply with SSAFO, a capital grant scheme was made available in Great Britain.[21] Under the scheme, which became operational on 20 February 1989,[22] a 50 per cent grant was made available to eligible farmers for the provision, replacement or improvement of facilities for the handling, storage and treatment of agricultural effluents and waste (including safety fencing) and fixed disposal for such effluents and wastes. The scheme provided for other kinds of works too, including poultry manure stores, hedges and shelter belts (including trees for shading livestock and associated protective fencing).

When the scheme was introduced, the then Minister of Agriculture, the Rt Hon Earl Howe, announced a provision of up to £50 million over three years in recognition of the need to tackle the problem of farm waste pollution.[23] The deadline for this part of the scheme was 30 November 1994.

MAFF, aided by the industry soldiers, also benefited farmers with a concessionary opportunity to build less than a 'maximum quantity of slurry likely to be produced in any continuous four-month period' set out in SSAFO, provided it could be demonstrated to the NRA that a lesser storage period is required.

Where a farmer considered that their arrangements for storing slurry required less than four months' storage capacity, the proposal was only to be accepted by the NRA on the production of an evidenced-based 'professionally' prepared farm waste management plan which could demonstrate a farm had safe winter-spreading land. In reality, there is no safe time to spread slurry in the winter without causing or contributing to pollution, which would have been known by MAFF and the NRA at the time.

The NRA produced guidance on what they required from farmers.[24] In order to accept such a proposal, the following factors needed to be taken into account:

- Effluent volumes, including information on its nature and the production period of the slurry.
- Land available for spreading slurry, including its area, soil type, topography, proximity to watercourses/aquifers, field drainage arrangements, cropping patterns, use by livestock and field capacity period.
- Growing season/cropping patterns.
- Spreading period(s).
- Methods of spreading.
- Contingencies (e.g. for adverse weather, etc).

This requirement was to be made clear to any farmer considering such a proposal. An explanation on dealing with such proposals was to be provided in a field guide known as NRA Policy Implementation Guide Note No. 5, Farm Waste Regulation, 91/18.

To ensure NRA field staff adopted a common approach, an internal training guide was produced by the South West region in 1992.[25]

This provided farmers with an opportunity to produce separate plans for slurry types with different characteristics, i.e. slurry and 'dirty water' slurry.

Dirty water is a dilute slurry, arising from lower polluting water (e.g. milking parlours), rainwater falling on soiled yards, liquid removed from slurry stores (e.g. weeping wall structures) and drainage from farmyard manure stores.[26] The nature of 'dirty water' slurry has altered over time, but at this time was described as generally having less than 3 per cent dry matter.[27]

Requirements for farmers wishing to install a settlement tank and distribution system to handle 'dirty water' slurry:[28]

1. Details of stocking and volumes of waste entering the system.
2. Calculations showing:
   a. Yard areas and other wastes draining to [Low-Rate Irrigators].
   b. Infiltration of clean water to the system.
3. Calculations showing capacities of the settlement and buffer storage area.
4. Pump output, rates of application and type of irrigator, i.e. mobile or static.
5. Farm maps for the area to be irrigated, showing risk aspects (preferably in colour).
6. Diagrams of buildings showing clean and foul drainage routes and exits, and to individualise contaminated and clean areas.
7. Assessment of land showing soil type and slope.
8. Nitrogen loadings.
9. Contingencies.

For farmers wishing to install a slurry storage system of less than 4 months' capacity, a plan showing:

1. Details of stocking and wastes entering the system.
   a. Including run-off from contaminated yards.
   b, Rainfall on the store.
   c. Provision to deal with liquid drainage.
2. Disposal methods for waste.
3. Calculations showing nitrogen loadings and application rates.
4. Farm map showing area to be used for disposal including risk and no-go areas.
   a. Drained land.
   b. Slope.
   c. Watercourses.
   d. Trafficability and Porosity.
5. Contingencies.

Should a farm's available land not be sufficient to take both categories of waste independently, then a full farm waste management plan was required, showing a combination of both.

This picks up on the waste disposal nature of FWMPs, which the NRA elaborated on in 1992:

> *The disposal of a whole range of agricultural waste and by-products to land is generally advocated as the soundest option. This is based on the principle that the soil is a suitable medium for retaining, breaking down or releasing substances in a controlled manner within a capacity of the environment [...] The process is currently dependant on voluntary compliance with a series of detailed codes and guidelines, the results of which are still far from satisfactory to farmers and the NRA alike.*[29]

The NRA's 1992 publication appended an example of a colour-coded FWMP 'traffic light' risk assessment prepared by MAFF/ADAS to support a new, and more substantial, 1991 Code of Good Agricultural Practice for the Protection of Water (CoGAP 1991). CoGAP 1991 introduced a five-stage planning process for FWMPs.

Stage 1: Areas where waste should not be spread at any time

For example, within at least 10 metres of all watercourse; and to reduce the risk of polluting groundwater within at least 50 metres of a spring, well or borehole that supplies water for human consumption, or was to be used in farm dairies. Recognition was given for a need for a bigger distance in higher-risk situations. These buffer areas were to be indicated in red. A slope greater than 12 degrees may also be considered too risky to spread on without causing run-off.

Stage 2: Matching land area to nitrogen in waste

A calculation of how much land was needed to limit the amount of *total nitrogen* in livestock waste, and other wastes, that may be applied to less than 250 kg per hectare per year (excluding any dung or urine deposited while livestock is grazing). However, a lower amount may be appropriate in sensitive catchments such as nitrate vulnerable zones.

Stage 3: Estimating the risk of pollution from spreading

A field-by-field assessment, taking into account field capacity; proximity to all watercourses, springs, well or boreholes; field slope; soil permeability, including any compaction that would restrict infiltration;

soil depth over fissured rock; the presence of a land drainage system, and subsidiary drainage such as mole drains and subsoiling in the last twelve months.

The assessment broke the risk down into three parts:

Very high risk (orange areas)

1. Fields likely to flood in the months after waste is applied.
2. Fields that are frozen.
3. Fields next to a watercourse, spring or borehole,
   a. where the surface is severely compacted,
   b. that are waterlogged,
   c. that have a steep slope [8–12 degrees] and the soil is at field capacity,
   d. that have a moderate slope [4–7 degrees], a slowly permeable soil [within 40 cm] and the soil is at field capacity.
4. Fields that are pipe- or mole-drained and the soil is cracked down to the drains or backfill.
5. Fields that have been pipe- or mole-drained in the last twelve months.
6. Fields that have been subsoiled over a pipe or mole drainage system in the last twelve months.
7. Fields where the soil depth over fissured rock is less than 30 cm and the soil is cracked.

Very high-risk status is generally only applied during the winter and field capacity period, or other periods when the soil is deeply cracked (e.g. summer) and/or disturbed within the previous twelve months so as to create preferential drainage conditions (e.g. by mole ploughing and subsoiling). In other periods when the risk of causing pollution was less, the risk would be lowered to a 'high-risk' area status.

Frozen was taken to mean soil frozen for more than twelve hours. Days when the soil was frozen overnight but thawed out during the day were not included.

Field capacity is described as the period when the soil is fully wetted and more rain would cause water loss by drainage. The typical field capacity period falls between October and March and when potential transpiration and/or evaporation is relatively small. Year-to-year variation is usually small, though in wet seasons the period can start earlier and end later.

High risk areas (yellow areas)

These are areas not otherwise ruled out, where waste could be spread 'safely' most of the year (but not really) at a reduced rate (i.e. at a maximum application rate of 50 m^3 per hectare, leaving at least three weeks between each application to stop the surface sealing and to let the soil recover).

1. Fields next to a watercourse, spring or borehole.
   a. that have a moderate slope and the soil is at field capacity.
   b. with a slowly permeable soil and the soil is at field capacity.
2. Fields where the soil depth over fissured rock is less than 30cm.

Like very high-risk areas, the risk status would generally be lowered outside of winter and field capacity conditions.

All other areas (green areas)

These are lower risk areas of causing point-source pollution, where higher amounts of waste can potentially be applied but without exceeding 250 kg per hectare of total nitrogen from organic manures. Care was still required not to apply waste at a rate that would most likely result in drainage or run-off beyond crop use.

An unfortunate consequence is that the traffic light system implied that 'green areas' meant that that slurry could be spread safely all year round and particularly on grassland. CoGAP 1991, and versions since, were mainly directed at the avoidance of direct pollution. Whilst the guidance has *somewhat* contributed to lowering diffuse pollution, in respect of such pollution it was more about controlling releases within the capacity of the soil and water to dilute nutrients like nitrogen and phosphorus to an apparently safe level. This is a concept that was never realistic at the time, based on the available science, and which has led to thirty years of diffuse pollution since, which could otherwise have been avoided. It's a legacy that still impacts water quality today. You cannot spread slurry, and many other manures, safely all year round, including on grassland, without causing a significant risk of agricultural diffuse pollution. Even in parts of the UK, like Cornwall, where grass has a longer growing season, you will cause nitrate-nitrogen, ammonium-nitrogen and phosphate pollution through leaching, drainage and run-off; as I go on to explain later.

Stage 4: Choosing the size of storage facility

The size of store required is based on the amount of waste produced during winter months minus the amount that can be *safely* spread over the same period, the biggest surplus of slurry that could not be spread giving the size of the store needed. Due to separate handling on most farms, the calculation for slurry and 'dirty water' slurry would normally be estimated separately.

If the type of store needed could not be emptied in stages, then the farmer also needed to take account of the animal housing period, with the amount depending on:

1. Number and type of livestock.
2. Volume of dirty water and rainwater going into the store.
3. Amount of bedding used.

Stage 5: Choosing a storage system

The system chosen depended on the type of livestock, how they were to be housed, the amount and type of bedding material and whether the waste was diluted with 'dirty water' slurry. The system chosen would also depend on winter rainfall – for instance, the surface area of different stores can have a significant bearing on the amount of rainfall entering a store, thereby increasing the capacity requirement and amount of slurry needed to be disposed of.

To further help farmers, MAFF/WOAD ADAS produced a separate *Farm Waste Management Plan, Step by Step guide for farmers* in 1994, republished in 1998 and 2003. The last iteration is found within *Defra Guidance on complying with the rules for Nitrate Vulnerable Zones in England for 2013–2016.*

Ask a farmer how much of the above they have in place today and the answer would be very little.

Nitrate vulnerable zones are rules (Nitrates Action Programme measures) originally brought in under the European Community Nitrate Directive.

The Nitrates Directive requires member states to designate as vulnerable zones areas of land that drain into waters affected by pollution and waters which could be affected by pollution if actions were not taken. Polluted waters are defined as:[30]

- surface freshwaters and ground waters with nitrate concentrations of greater than 50 mg/l, and

- natural freshwater lakes, other freshwater bodies, estuaries, coastal waters and marine waters that are eutrophic.

The directive requires member states to establish voluntary codes of good agricultural practice throughout their territory and to set up Action Programmes in the designated nitrate vulnerable zones (NVZs). In the UK, different Action Programmes operate in England, Northern Ireland, Scotland and Wales.

Nitrates Action Programmes are required to take account of available science and relevant technical data, mainly in respect of nitrogen contributions originating from agriculture and other sources.

The directive states that Action Programmes must include the measures in the code of good agricultural practice and the regulatory rules relating to:

- periods during which the application of certain types of fertilisers is prohibited.
- the capacity of storage vessels for livestock manure, which 'must exceed that required for storage throughout the longest period during which land application in the vulnerable zone is prohibited', unless it can be demonstrated that the excess will be disposed of in a manner that is not environmentally harmful.
- limitations on the application of fertilisers, consistent with good agricultural practice and taking into account: soil conditions, soil type and slope; climatic conditions, rainfall and irrigation; and land use and agricultural practices. There must be a balance between the nitrogen requirement of the crops and the nitrogen supply from the soil and from fertilisation.
- the amount of livestock manure applied to the land each year, which must not exceed 210 kg nitrogen per hectare during the first four-year Action Programme and 170 kg nitrogen per hectare thereafter.

The first twenty-two compulsory NVZs were introduced in July 1994, without compensatory payments to farmers.[31] Today, 55 per cent of England falls within an NVZ.

The original concept, or 'hope', for FWMPs was that farmers knew their land well and with some technical guidance could develop simple, practical waste management plans which could ensure sound 'disposal' regimes for all wastes and by-products to land, or elsewhere, and hence prevent pollution.[32]

In November 1994, when the 1989 Farm and Conservation Grant Scheme 1989 came to an end, it was deemed a success. In 1994, the minister stated in Parliament that the number of serious incidents had reduced markedly, from 940 incidents in 1988 to 63 in 1993.

The minister concluded that the time was right for farmers to fully accept responsibility for the costs associated with waste-handling facilities. In total the Scheme paid out £150 million in grants to farmers in the UK, representing a total investment of about £300 million including the farmers' contribution. In all, the scheme benefited 11,500 farmers in England and Wales.

Unfortunately, the information fed to the minister on the number of serious pollution incidents was not correct.

Shortly after the grant was launched, the terminology and criteria used to record incidents had changed. Previously, those incidents having the greatest impact on aquatic ecosystems were lumped together into one category, 'Serious' incidents. After 1990, 'Serious' incidents were subdivided into two categories – 'Category 1' and 'Category 2' incidents.

In 1988, the 940 'Serious' incidents related to slurry were more or less equivalent to a combination of 'Category 1' and 'Category 2' incidents. The sixty-three incidents reported were simply just 'Category 1' incidents. Including 'Category 2' incidents the fall would have been much less, and probably around 350 (my estimate). Nevertheless, the overall number of incidents did fall, with *Category 1* and *Category 2* incidents averaging out at around 350 until 1996 (again my estimate).

Although the grant was withdrawn, a commitment to provide free pollution advice and help in the preparation of FWMPs continued to be available. Thanks to these plans, many farmers didn't have to go to the full expense of building slurry stores with four months' storage capacity. While the NRA, and after 1996 the Environment Agency, increasingly toughened their stance in this respect, some stores were still built with less than four months' storage capacity up until around 2010.

To accommodate NVZ legislation, and to avoid the stigma associated with the term 'waste', FWMPs were later renamed manure management plans (MMPs). This move was driven by the farming community.

The adoption of a FWMP/MMP was essential in raising awareness and has helped lower the number of *direct* slurry run-off-related pollution incidents. What they didn't do was adequately address diffuse pollution. FWMP/MMPs allowed farmers to carry on spreading their slurry to land in the autumn and winter months, aided and abetted by a lack of slurry storage when there was little or no soil and crop need for nutrients, soils were wet, it rained a lot and the risk of diffuse pollution was high. For example:

- When slurry is spread in the autumn, the nitrogen is quickly converted into nitrate-nitrogen, a form of nitrogen which if not used by a crop is easily washed out of the soil during the rainy winter months. This process is known as leaching. Slurry spread in the first half of the winter is also at risk of leaching.
- When slurry is spread on a wet soil during the winter months, and significant rainfall follows, nutrients in slurry, like ammonium-nitrogen and phosphorus, and microbiological pathogens, can be washed downwards and into any underlying field drains or groundwater. This process is known as preferential drainage, and it continues into the spring until the soil dries sufficiently. If the application rate is too high, some slurry can drain directly down and out of a soil.
- When slurry is spread on a wet soil, provided the application rate is low, the land's slope is slight and the soil is not compacted, it should not easily run into a watercourse. However, if a significant rainfall event follows, any slurry present at the surface, along with any loose nutrient-enriched soil particles, remain at risk of being washed away to water.

To ensure slurry is used in a manner that doesn't exceed the needs of soils and crops or give rise to a significant risk of agricultural diffuse pollution, as a minimum you need around six months' slurry storage capacity. Sometimes more, but not less than six months. In my opinion, nine months' storage on many farms would not be gold-plating. This is a volume calculation with the emphasis on reducing the volume needing to be stored to a minimum, through limiting the amount of rainfall, washings or other liquids that may otherwise enter the store by being waterwise and following best farming practice. Extended grazing can reduce the amount of slurry which needs to be stored too, but can give rise to soil damage and water pollution in other ways.

Unfortunately, even for farms in an NVZ, who are required by law to have five months slurry storage for cattle slurry, and six months for pig and poultry, many farms do not even have this capacity – thirteen years after the 2008 regulations came into force. The five months for cattle was a concession made to industry, based on cattle slurry having less readily available nitrogen than pig slurry and more potential opportunity to spread 'lightly' (e.g. <20 m³/hectare) up until the end of

October onto grassland (medium/heavy soil), with a lower risk of nitrogen loss to the environment in some years. However, there are many more dairy farms, and associated slurry, than pig farms, raising the overall risk. Autumn can also be awash with rainfall, filling slurry stores before winter and returning the soil to field capacity earlier in some years, increasing the risk of pollution. The Environment Agency suggests at least 50 per cent of farms have inadequate slurry storage arrangements.[33]

An NFU survey completed by over 150 dairy farmers also highlighted the lack of slurry storage capacity.[34]

Of the 150 questioned:

- 45 per cent of farmers said they did not have enough storage to comply with the five-month NVZ storage requirement (9 per cent didn't say). 46 per cent said they had sufficient or more capacity.
- A fifth of farmers said they were not planning to invest in measures to comply with the regulations.

This was a very honest appraisal of the situation, which has been confirmed by other studies I refer to later.

Farmers, and others, often get their slurry storage capacity calculation wrong for a variety of reasons. One notable error is an historic failure to take into account wetter-than-average rainfall years, which is calculated as the wettest year in the last five years. This results in a lot of rain falling on slurry-contaminated yards, which then drains to a slurry store, and rain falling on the storage itself not being accounted for in any *reasonable* year. The more cattle, the more yard area, the bigger a slurry store's surface area, the bigger a store's potential storage capacity discrepancy.

On dairy farms, as milk yield has increased through intensification, so has the amount of slurry excreted, which has also put more pressure on existing slurry storage arrangements. Given most dairy farms are found in the wetter parts of England, which are better suited to growing grass, this can add considerably to slurry storage capacity and pollution pressures. Many farmers are oblivious to how much capacity they actually have.

In 2010, in England alone, the amount of slurry produced was 29 million tonnes a year. The actual amount needing to be stored, for instance produced during winter housing, is much less than this, as livestock like cattle voids a large amount of slurry whilst grazing. This figure includes dung, urine and wash water. Many pigs, though, are kept indoors on slurry-based systems all year, and produce a particularly ammonium-nitrogen-rich, and therefore potentially more polluting, manure if used when there is no soil and crop need, or in high-risk spreading situations (i.e. if applied to winter cereal land prior to sowing in the autumn).

Over a six-month full occupancy (1 October to 1 April) collection period, I estimate farms in England have a slurry storage capacity shortfall of around 7.25 million tonnes, excluding rainfall. The average cost of a cubic metre of slurry storage is £40–50. Using £45 as an average, the cost of investment is around £325 million. Further capacity is required for rain. Given most slurry-producing livestock farms are found in the wetter parts of England, and most stores are uncovered, this can be a considerable extra expense for those farms. Dairy cows today have a very high milk yield of 8,000 litres per cow per year, and more milk means more slurry, so that has to be added in too. On top of this, some slurry stores would need to be replaced entirely.

Overall, I would suggest a total investment of over £500 million would be needed to bring most farms in England up to at least six months' (excluding complete rebuilds), uncovered slurry storage capacity. In many situations it is better to have at least nine months' storage capacity and sometimes even twelve. Add in amortised costs and we are getting close to £750 million. Add in slurry covers, to reduce ammonia pollution emissions, and the cost would rise to more than £1 billion. Nine months' slurry storage capacity would bring the cost to well over £1.5 billion. Similarly, anaerobic digestion (AD) facilities may need nine months' or more digestate storage to maximise nutrient efficiency and avoid pollution (autumn through to early spring).

7.25 million tonnes, plus up to two million more to cater for rainfall, is a lot of slurry which is not being used as well as it could and contributing to pollution, most of which is found in the wettest parts of the country, to the west and south west. For the whole of the UK, also assuming a 50 per cent capacity shortfall (three months), and excluding rainfall, the slurry storage deficit could be in excess of 11 million tonnes.

So how have farmers found themselves in such a predicament? The blame does not rest entirely on farmers' shoulders. The farm-gate price livestock farmers receive does not cover the full cost of production needed to comply with environmental law. Instead, farmers externalise their pollution costs, and we benefit by buying meat, eggs and dairy products in supermarkets at a price well below their true environmental cost. Very little public pressure exists to hold meat, egg and dairy corporations, and retailers who sell their products, accountable for the slurry pollution footprint which they are largely responsible for.

So, what are these polluting emissions like today, and how do they translate into the number of pollution incidents we now see?

Unfortunately, the true number of pollution incidents and diffuse pollution is unknown. In terms of direct pollution there are not as many reported incidents as in the 1980s, but pollution, dairy pollution in particular, like the sea level, is on the rise.

Using England as an example, based on a rolling five-year average, Category 1-2 (most serious) dairy pollution incidents between 2012 (2008-2012) and 2017 (2013-2017) rose from around 30 to 40.[35] There was a dip in 2018 data, which the Environment Agency attributed to the dry weather in that year and a substantial drop in the number of registered dairy farms (fewer farms means fewer farms that can pollute).[36]

This rise has not been reflected in industry publications. In fact, in Dairy UK's report *The Dairy Roadmap 2018 – Showcasing 10 Years Environmental Commitment*, the industry actually celebrated a 23 per cent *drop* in pollution incidences in the dairy sector between 2008 and 2014.

Here are the actual figures I obtained for England from the Environment Agency (2008 to 2020), with incidents rising to a five-year high in 2020, made up of four Category 1, 20 Category 2 and 138 Category 3 slurry-related incidents.[37]

Table 17: Substantiated pollution incidents (2008 to 2019)

Year	Substantiated Pollution incidents Category 1-3 on dairy farms	Number of slurry incidents on dairy farms
2008	246	152
2009	231	128
2010	204	121
2011	191	109
2012	261	181
2013	314	205
2014	321	184
2015	267	137
2016	245	153
2017	230	145
2018	235	142
2019	221	150
2020	245	162

In 2008, the total Category 1–3 pollution incidents for dairy was 246, dropping to 191 in 2011 (based on a calendar year). Incidents began to rise again then, jumping by 70 in 2012 to 261 and rising still further to 321 in 2014. Out of all the substantiated dairy pollution incidents reported (2008 to 2019), slurry-related incidents represented 61 per cent of the total.

From further data, the most serious incidents, Category 1–2, totalled 33 in 2008, 44 in 2013, 36 in 2014, 47 in 2015, 33 in 2016 and 42 in 2017. Isolating slurry only, there were twenty-two Category 1–2 incidents in 2008 and 26 in 2014 (32 in 2012, 37 in 2013 and 34 in 2017).

It is also odd the report should choose to use 2014 in a 2018 publication when 2017 data was available. Given most UK dairy farms are in England, there would need to have been a massive drop in incidents in the rest of the UK to give the drop of 23 per cent they reported. The devolved administrations are not without their own dairy pollution issues.

An important omission in incident data, historically published by the Environment Agency, is that in 2008 there were 10,077 registered dairy farms but only 8,820 in 2014 (7,074 in 2018). This means, fewer farms causing even more pollution on a per dairy farm basis. When pollution incidents are reported it is important to ensure the data reflects changes within the industry, like the number of dairy farms. The Environment Agency have, to their credit, addressed this somewhat in their 2019 pollution report by using five-year averages to smooth out weather variations between years which impact on pollution numbers across dry to wet years.[38]

For comparison, the average number of reported pollution incidents in England for pig farms over the last decade is about one per week. About a third of those are directly related to slurry. This is less than the totals for dairy, but there are about a third less pig farms in England (c. 2,200). Many more pigs are housed outdoors or in strawed barns too, compared to dairy, and therefore do not produce slurry, or only produce very little.

There will be other omissions in the current number of incidents the Environment Agency substantiates each year, due to cuts in funding. These include environment officers attending fewer reported pollution incidents per year (you can't substantiate what you do not see) and less self-reporting during routine farm visits (on average, environment officers self-report around 11 per cent of all farm pollution incidents).[39]

In 2019–20 environment officers only visited 308 out of around

106,000 farms (a once-every-344-years inspection rate). This is around 600 fewer farms visited than in 2014–15 and, based on an 11 per cent shortfall in self-reported incidents, could have resulted in sixty-six fewer incidents being recorded.[40]

As a consequence of austerity, including a sharp reduction in grant-in-aid (GIA) funding, the Environment Agency, like other government-funded bodies, does not have the funds and the inspectors available as it once did. Cuts are nothing new and the Environment Agency, like other regulators and advisers, have to continually adapt to these pressures and find new ways of working to maintain a credible presence to deter poor farming practices and ultimately to maintain public confidence as an effective regulator. The farming community cannot complain about being over inspected by the likes of the Environment Agency, and the result appears to show up in non-compliance and pollution data.

In the 1990s most reported farm-related incidents would have been followed up, unlike today where attendance is now prioritised based on the potential severity and impact of a reported incident. Reported incidents still followed up, according to the Environment Agency, include all likely Category 1 and 2 incidents (subject to health and safety concerns and Covid-19 restrictions), but not all suspected Category 3 incidents.[41] In my experience, a significant number of reported incidents judged to be Category 3 at the desk but not visited would be upgraded to be Category 2, and in a few cases to Category 1. This would mean the incident numbers reported annually cannot be considered a true representation of what pollution is happening on our livestock farms today compared to the past.

Suspected Category 4 incidents, the lowest incident category, would not normally be attended either. If they were, it would also be reasonable to suggest some would be upgraded to at least a Category 3. Furthermore, many incidents in all categories will inevitably go unreported, being out-of-sight-and-mind, but which have the potential to cause chronic diffuse pollution. In recent times, physical evidence from catchment-based studies which back this opinion is coming to light, which I return to later in 'Slurry Fears', the next section of this chapter.

In all, information obtained from the Environment Agency shows that between 2015 and 2020 only 37.95 per cent of total incidents reported to be emanating from agricultural premises were actually attended.[42]

When it comes to farm pollution it is not just England of course; the devolved administrations have their own unfortunate share of livestock pollution. In Wales farm pollution got so bad, three incidents a week, that the Welsh government, after giving Welsh farmers four years to clean up their act, felt they had no choice but to place the whole of Wales, in 2021, under the nitrate vulnerable zone regulatory regime. The decision was labelled as 'draconian and unimaginable' by NFU Cymru, who had been advocating for a voluntary approach to improve water quality.[43] Welsh farmers have little room to move, given their buyers do not price pollution prevention into what they are prepared to pay.

Welsh rural secretary Lesley Griffith, in response to the union's criticism and after a lengthy period of engagement with the industry, said: 'I have given the industry every opportunity to demonstrate a change in behaviours through voluntary action.' Furthermore, he added: 'Some progress has been made over the past four years but not enough to demonstrate the scale, rate and commitment to change needed.'[44]

Many pollution incidents reported in Wales, as in England, are dairy related, with most farms likely to be members of Red Tractor Assurance. In the next section ('Slurry Fears'), I further discuss Red Tractor Assurance, a British farm standard scheme which includes environmental protection as one of its four key principles.

The *Dairy Roadmap* also states that pollution incidents are strongly influenced by weather conditions, as poor weather puts pressure on slurry store capacity and limits spreading opportunities. This is true, but poor weather conditions in a normal, non-exceptional year would be much less of an issue if dairy farms at least had the legal minimal slurry storage capacity, better still six months or more storage.

For many years after the mid-1980s farmers have been advised that they can operate with less than four months' slurry storage capacity on the production of a professionally prepared FWMP/MMP. A change of direction came with the introduction of the Nitrate Pollution Prevention Regulations 2008, which required farmers to have five months' slurry storage capacity for cattle and six months for pig and poultry. However, the regulations only applied to parts of the country which fell within a designated NVZ. The regulations also included slurry spreading closed periods, ranging from 1 August to the end of the following January; a whole farm *livestock* manure total nitrogen limit (170 kg per hectare); a per hectare maximum *organic* manure total nitrogen limit (250 kg); and nitrogen planning and spreading restrictions. I have emphasised *livestock* and *organic*, as in the first instance farmers can add a further 80 kg per hectare of total nitrogen to their whole farm limit from other *organic* manures and thereby defeat this NVZ measure.

Farms outside NVZ carried on as before, but after this time most farmers building new stores were required to have a minimum of four months' slurry storage capacity. Those who had less than four months and/or who had a store older than 1991 have steadfastly refused to substantially enlarge or substantially reconstruct their stores for fear of losing their SSAFO exemption. What slurry storage they may have had before, and even after 1991, has been further reduced by years of intensification, with their livestock producing more slurry than ever before.

Ask a farmer if they know about the SSAFO regulations and many will say no, but question their slurry storage capacity and most have at least heard about the SSAFO exemption. Of even greater concern is the (still quite a few) farms that do not have any slurry storage, and who verdantly still claim they have 'safe' all-year-round spreading land. They do not; they are causing pollution. In these situations, slurry may be allowed to:

- run off the farmyard and onto the adjacent land;
- drain to a soakaway;
- drain to a sump, which may be irrigated to land or be allowed to drain onwards to a surface water;
- drain to a low point in a farmyard, where it may be trapped by a structure, then regularly sucked up by a slurry vacuum tanker and frequently spread to land;
- drain into a series of settlement ditches, before discharging to water;
- drain to a sewer;
- drain somehow and somewhere else.

Another point the *Dairy Roadmap* makes is that a prolonged period of low milk price in recent years has restricted farmers' ability to invest in necessary slurry store improvements and spreading equipment that helps to reduce pollution incidents. Again, this is plausible, but dairy farms have still managed to increase the average herd size from 119 in 2008 to 151 in 2018. Money spent is invested in new or extended dairy parlours, buildings and cattle collecting yards (the bits that provide a financial return and keep the banks happy). In my experience, only a fraction, the odd farm excluded, is spent on waste management to compensate for this growth. The rise in dairy pollution incidents in England is a testament to that.

The Dairy Roadmap further promotes Defra statistics from 2015, that:

- 90 per cent of dairy farmers have a manure management plan (MMP); and
- 73 per cent of dairy farms have a nutrient management plan (NMP).

Both are all well and good statistics, so long as you have at least six months' slurry storage capacity, and can spread slurry without exceeding soil and crop need, or without causing a significant risk of diffuse pollution. Until then they are unfortunately meaningless statistics. The statistics are also hollow, unless you know how good or bad the MMPs and NMPs farmers have are. In the livestock sector I've seen many poor plans and very few I'd call comprehensive.

The slurry emergency extends beyond dairy farming. Dairy gets highlighted, because there are many more dairy farms than slurry-producing pig farms. Beef cows are still much more likely to be kept on straw and the average stock number is much smaller, at twenty-eight cows. Slurry-producing pig farms are just as likely to cause a pollution incident as a dairy farm; and slurry on beef farms often drains somewhere other than where it should.

Whilst most beef herds are small, a few have grown to seismic proportions. The growth of these farms was highlighted in research undertaken by the Bureau of Investigative Journalism, which was reported in the *Guardian* newspaper.[45] The report highlighted the rise in the number of industrial-scale, unpermitted fattening units of up to 3,000 cattle held on barren feedlot pads, the largest fattening up to 6,000 cattle a year. These units, and their smaller cousins, are commonly known in the UK as woodchip or straw-bed corrals, or out/overwintering pads (OWPs). In America they are known as concentrated animal feeding operations (CAFOs).

Corrals or OWPs can be a pollution hazard. When they first arrived on our shores the potential impact to water quality was passed over, with the exception of Ireland, who from 2007 provided good pollution prevention advice.[46] Good practice got lost somewhere in translation when they came across the Irish Sea and landed in Scotland, from where they steadily descended into England and Wales.

The reason for investing in OWPs is that they are much cheaper than installing a roofed livestock building. Out-wintering pads can also provide animal health and performance benefits over indoor accommodation.[47]

Many of the initial OWPs installed were not lined, with enthusiasts believing that a deep bed, of up to fist-sized pieces of chipped wood, was sufficient to retain and treat slurry and protect groundwater. More unsettling was the removal of any protective topsoil, which was used to help form a bank around the pads. Needless to say, such enthusiasm was not met with reality. The effluent draining through the OWPs remained highly polluting, despite the woodchips retaining some of the nutrients. A few other large unlined pads were built and promoted in England, using straw rather than woodchips.

In recognition of pollution concerns, the Agriculture Development and Horticultural Board (AHDB) took the issue by the horns when it was brought to their attention. AHDB Beef & Lamb invested money in research on woodchip and straw-bed OWPs and published best-practice guidance in 2011[48] and 2016 respectively.[49]

Unfortunately, several early birds escaped the net. The oddest design I have come across is a saucer-shaped OWP, reminiscent of some mound systems used in American feedlots. It was circular in shape, with excavated soil from the outer half of the OWP piled up in the centre, creating a moat-and-mound structure. The moat was filled, and the mound covered, with copious amounts of straw. Extra straw was added daily.

The idea was that the beef cattle would spend most of their time loafing and feeding on the dry, cleaner mound, only venturing down to the lower moat area for water. In the moat, the farmer hoped the straw would soak up all the dung, urine and rainfall. It didn't work, and despite the farmer's best efforts, the moat became a quaking bog of super-saturated straw and slurry, with a few deep pools of slurry around the drinking troughs. Cattle could be seen standing up to their bellies in slurry and avoiding other deeper areas, perhaps fearing drowning.

Three other examples I have come across, all woodchip pads, were lined but still externalised their pollution costs.

One used a pad as an extension to a barn. The cattle preferred the outdoor pad area to the barn, in all but the worse weather. Effluent was collected through a drainage system resting on a lined, impermeable base, which drained to a sump. The sump was connected to a low-rate irrigation system, which on reaching a certain level automatically switched on and discharged the effluent onto a sacrificial piece of sloping land in an adjacent field. Essentially whenever it rained, it resulted in slurry running down the slope.

Another simply drained to a hole in the ground, which never overflowed! The farmer of the other OWP collected all the drainage but then disposed of it to groundwater through a herringbone soakaway system, essentially defeating the objective of lining the pad.

Slurry pollution occurs on most farms and compliance with environmental law and codes of good agricultural practice is generally poor. I provide more information on this in the next section. Slurry pollution is growing, with reported pollution statistics no longer representative on their own of the actual pollution happening. It is not true in my experience that the majority of livestock farms are high environmental performers, as the industry wants to, and does, promote. They are not; the majority cause and/or contribute to pollution.

Until farms put in the right infrastructure to prevent all forms of pollution, slurry pollution will continue to grow in proportion to livestock intensification. Until then, we are well and truly in a slurry emergency.

The time is ripe to challenge the livestock farming industry's Knights Templar and put an end to the SSAFO exemption and slurry pollution they know goes with it.

Citations:

1. Michael Winter (1996), *Rural Politics: Policies for Agriculture, Forestry and the Environment*. Routledge, London, UK
2. Edwards, A. and Withers, P. (1998), 'Soil phosphorus management and water quality: A UK perspective.' *Soil Use and Management* 14, pp. 124-130
3. Marks, H. F. (1989), *A Hundred Years of British Food & Farming: A Statistical Survey*, edited by D. K. Britton. Taylor & Francis, London, UK
4. Ibid.
5. Section of Comparative Medicine (1953). 'Discussion on Organic Manures and Fertilizers and the Production and Composition of Food for Man and Animals.' *Proc R Soc Med*. 46(9), 791-8
   https://journals.sagepub.com/doi/pdf/10.1177/003591575304600912
6. Defra, Project WQ0103 (ADAS 2008). *The National Inventory and Map of Livestock Manure Loadings to Agricultural Land*. MANURES-GIS.
7. Ibid.
8. Ibid.
9. Ibid.
10. Michael Winter (1996), *Rural Politics: Policies for Agriculture, Forestry and the Environment*. Routledge, London, UK
11. Ibid.
12. Royal Commission on Environmental Pollution (1984), *Agriculture and Pollution, Tenth Report*. HMSO, London, UK
13. Ibid.

14. Ministry of Agricultural Fisheries and Food (1983), *Advice on avoiding pollution from manures and other slurry wastes,* Farm Waste Management Booklet 220, p. 2

15. Royal Commission on Environmental Pollution (1984), *Agriculture and Pollution, Tenth Report.* HMSO, London, UK

16. Ibid.

17. Michael Winter (1996), *Rural Politics: Policies for Agriculture, Forestry and the Environment.* Routledge, London, UK

18. Ibid.

19. NRA (1992), *The Influence of Agriculture on the Quality of Natural Waters in England and Wales.* Water Quality Series, No. 6

20. National Audit Office (1989), *Grants to Aid the Structure of Agriculture in Great Britain*

21. MAFF (1989), *Farm and Conservation Grant Scheme*

22. MAFF ADAS (1989), *Water Pollution from Farm Waste: England and Wales*

23. The Parliamentary Secretary, Ministry of Agriculture, Fisheries and Food (Rt Hon Earl Howe, 1992-1995). House of Lords debate 19 January 1995, Hansard vol. 560, col. 809

24. NRA (1991), *Control of Pollution (Silage, Slurry and Agricultural Fuel Oil) Regulations 1991,* Policy Implementation Guidance Note No. 5.

25. Newport, S. B. (1992), *Agriculture Training Control of Pollution Regulations Guidance Notes.* NRA South West Region, PC/W/I/92/002

26. MAFF (1985), *Dirty water disposal on the farm: Farm waste management.* Booklet 2390

27. MAFF (1991), *Code of Good Agricultural Protection for the Protection of Water.*

28. Newport, S. B. (1992), *Agriculture Training Control of Pollution Regulations Guidance Notes.* NRA South West Region, PC/W/I/92/002

29. NRA (1992), *The Influence of Agriculture on the Quality of Natural Waters in England and Wales.* Water Quality Series, No. 6

30. Council of the European Communities (1991), *Council Directive of 12 December 1991 concerning the protection of waters against pollution caused by nitrates from agricultural sources.* 91/676/EEC, Annex I

31. National Audit Office (1995), *River Pollution from Farms in England*

32. NRA (1992), *The Influence of Agriculture on the Quality of Natural Waters in England and Wales.* Water Quality Series, No. 6

33. Stacey, G. (2018). *Farm Inspection and Regulation Review: Summary and Recommendations.* Independent Report for Defra

34. NFU (2011), *Dairy NVZ Survey.* Report by the Dairy, Membership and Economics Team

35. Environment Agency (2019), *Regulating for people, the environment and growth, 2018.*
https://www.gov.uk/government/publications/regulating-for-people-the-en vironment-and-growth

36. Ibid.
37. Environment Agency (3 December 2020), via Freedom of Information Request NR192233
38. Environment Agency (2019), *Regulating for people, the environment and growth, 2018.*
     https://www.gov.uk/government/publications/regulating-for-people-the-environment-and-growth
39. National Audit Office (1995), *River Pollution from Farms in England*
40. Via Environment Agency Freedom of Information Request, NR192233 (3 December 2020)
41. Ibid.
42. Ibid.
43. Debbie James, 'Farm Pollution breaches trigger all-Wales NVZ decision, *Farmers Weekly*, 27 January 2021.
     https://www.fwi.co.uk/business/compliance/farm-pollution-breaches-trigger-all-wales-nvz-decision.
44. Ibid.
45. Wasley, A. and Kroeker, H. (2018), 'Revealed: industrial-scale beef farming comes to the UK.' *The Guardian*, 29 May 2018.
     https://www.theguardian.com/environment/2018/may/29/revealed-industrial-scale-beef-farming-comes-to-the-uk.
46. Department of Agriculture and Food and the Marine (2007) *Minimum Specifications for Out Wintering Pads*. S132.
     https://www.agriculture.gov.ie/
47. AHDB (2011), *BRP+ Improved design and management of woodchip pads for sustainable out-wintering of livestock.*
48. Ibid.
49. AHDB (2016), *BRP+ Guidelines for managing outdoor straw pads for beef cattle.*

14. Slurry Fears

According to a 2013 Environment Agency study,[1] around 45 per cent of slurry stores in use today predate 1991, which calls into question their reliability over 30 years on. Age, though, is not always an indicator of a store's integrity, as some were built like air-raid bunkers and some have been well maintained. However, the age of a slurry store is usually indicative of a lack of storage, and many stores are found in a poor state of repair. Given the fact that some of these stores would have been built well before 1991 and the intensification of the livestock industry which has happened since their construction, many are likely to be woefully short on the capacity needed to ensure slurry is only spread when there is a soil and crop need, or without causing pollution.

Common structural problems associated with slurry stores include:

- Concrete stores: deterioration/defects of concrete/sealants and consequent leakage, in particular around joints (bases and walls); surface pitting and internal honeycombing; corrosion of steel reinforcement (e.g. exposure to acidic liquids and gases); tree damage (mainly older stores); impact damage.
- Masonry (concrete block/brick): insufficient load reinforcement; corrosion of steelwork; cracks and deterioration of mortar between blocks and consequent leakage; tree damage; impact damage.
- Above-ground cylindrical tanks: deterioration/defects of protective enamelling/epoxy coatings and sealants, in particular around the edges of steel sheets and bolts; insufficient reinforcement to support slurry and wind loads; formation of pinholes and associated slurry spurting; impact damage.
- Earth banked: ground stability; insufficient permeability; incorrect wet slope angle for geology/soil (internal embankment); erosion/failure of internal and external embankments; animal burrowing and tree/shrub root damage; and deterioration/defects in plastic liners used.

As part of the 2013 study 250 catchment sensitive farming (CSF) infrastructure audit reports were reviewed, most of which were on dairy farms.[2] Launched in 2005, CSF is a partnership between Defra, the Environment Agency and Natural England. It works with farmers and others to enhance water and air quality in priority areas. Through CSF, farmers have access to free training, advice and support for grant applications.

Eighty-four of the farms audited under CSF were in a designated nitrate vulnerable zone (NVZ), but despite most of them being in an NVZ since 2009 only 32 per cent had the required slurry storage capacity (i.e. five months for cattle and six months for pigs and poultry), with a further 6 per cent unclassified. A further 149 farms were audited outside of NVZ, of which 69 per cent were assessed as not having four months' slurry storage. While four months is the legal minimum, some of these farms would have benefited from the SSAFO pre-1991 exemption.

Between 1 January 2012 and the beginning of 2013, a record was also kept of 218 livestock farm visits conducted by the Environment

Agency.[3] In NVZs, 37 per cent were confirmed as volume compliant; 52 per cent were confirmed as non-volume compliant; and for the remaining 11 per cent, storage was either uncertain or there was no data. Outside of NVZs, 51 per cent of slurry stores constructed before or after 1991 had less than four months' storage (including exempt structures); 33 per cent were confirmed as volume compliant; and of the remaining 15 per cent, storage was uncertain. In all cases, visiting environment officers judged 36 per cent of the slurry stores to be causing environmental harm.

The 2013 report concluded that as CSF visits were not random, and a few EA visits would have been in response to a pollution incident, there may have been a degree of reporting bias. Notwithstanding that, the level of compliance is similar to my experience, which stretches back thirty years. It is also similar to an NFU survey I mentioned in the previous section, which found 45 per cent of farms asked did not have enough storage capacity.

In recent times potential reporting bias has been 'put to the test' by the Environment Agency (England).

Between 2016 and 2019 the Environment Agency carried out a campaign on the River Axe catchment, Devon, following a judicial review instigated by the World Wildlife Fund and the Angling Trust.[4]

The River Axe catchment covers an area of 308 km² across Devon, Somerset and Dorset. The lower reaches are designated as a Site of Special Scientific Interest (SSSI) and Special Area of Conservation (SAC). 237 km² drain to the SAC, which is in unfavourable and declining status owing to nutrient enrichment and sediment pollution. The majority of the pollution and run-off issues are associated with the catchment's intensive dairy farms, of which there are around 125.

Pollution prevention campaigns in the catchment are nothing new. An extensive five-year campaign in the River Axe catchment began in 1991, with seventy-four farms visited in the first year. Pollution was identified, remediation measures requested and follow-up visits undertaken to assess their effectiveness.[5] Several other advice-led initiatives have been carried out since 2004, providing farmers with free advice and guidance on diffuse pollution, including funded meetings, workshops, demonstration events, one-to-one guidance and grant aid to deal with diffuse pollution.

All these were light-touch, advisory campaigns, which is the agricultural industry's preference in regulatory-led initiatives. However, despite all these interventions, poor regulatory compliance and pollution in the catchment was suspected.

The Environment Agency's 2016–2019 campaign set out to see just how bad things were, and to use their regulatory powers to put things right.[6]

By the end of 2019:

- 49 per cent of eighty-six farms visited had evidence of a polluting discharge (two Category 2 and 40 Category 3 pollution incidents).
- 95 per cent of farms failed to meet requirements under SSAFO regulations, with the majority at high risk of causing pollution.
- The presence of sewage fungus – evidence of chronic, persistent, pollution – was found in many tributaries to the River Axe.
- 40 per cent of the farms visited had less than four months' slurry storage, with some benefiting from the SSAFO pre-1991 exemption.

According to the report's findings, 'most farmers were aware of the requirement for four months' slurry storage but often admitted to taking a business risk by not investing in infrastructure because there was little regulatory presence of the Environment Agency in the catchment and the lack of payback'. Promotion of forage maize cropping by agronomists and nutritionists, to support increased milk yield, was also cited as leading to pollution.

The Axe is typical of many dairy-dominated catchments. Increases in herd size have led to higher stocking densities (about 3.6 dairy cows per hectare), higher milk yields (more milk more slurry), and a larger forage maize crop area to feed them (more maize equals more run-off pollution).

Intensification requires considerable investment in new dairy parlours, housing, collecting/loafing yards and other infrastructure. Unfortunately, investment in waste management – to cater for the increased volume of excreta, parlour washings, silage effluent and run-off from contaminated yards – often falls by the wayside. The difference in investment is considerable, with most of the added pollution cost externalised.

Compliance with environment law and pollution prevention makes very little money and is not adequately provided for in cost-of-production models used by milk buyers. As a result, the UK is littered with creaking, failing and inadequate slurry pollution prevention infrastructure. Most spreading is also carried out using outdated splash plate technology, increasing the risk of run-off and ammonia emissions.

Intensification of livestock herds can:

- escalate soil compaction, through livestock treading of wet soils, increasing the amount of contaminated run-off entering watercourses.
- increase riverbank erosion and the amount of excreta and urine voided into the water, where dairy cows and other cattle have access to unfenced rivers.
- put further pressure on slurry storage capacity.
- result in more slurry being applied to land in exceedance of soil and crop need, giving rise to an increased risk of diffuse pollution through leaching.
- result in more slurry being applied to land in unfavourable soil and weather conditions, increasing the risk of run-off and pollution through preferential drainage.
- lead to the planting of more maize forage, which, being a late harvested crop, often results in soil compaction and soil erosion (slurry is often spread onto the maize stubble too, to ease pressure on slurry storage, exacerbating the risk of pollution).
- increased ammonia emissions, which can harm human health and vulnerable habitats.

The Axe catchment is prone to all these risks, but despite this, it isn't in an NVZ, and so farm slurry storage capacity would have been judged against the minimum SSAFO four-month requirement.

Another concerning example is a corresponding 2016–2019 study in North Devon (North Devon Priority Focus Area), where during the associated winter periods the Environment Agency carried out 101 farm audits.[7]

The allied Taw–Torridge Estuary has protected status for Shellfish Waters, is a bass nursery, migratory water for salmonids, eels, elvers and shad, a SSSI, the focus point of the Taw Estuary NVZ, is in an Area of Outstanding Natural Beauty and a UNESCO (United Nations Educational, Scientific and Cultural Organisation) Biosphere Reserve. The estuary is failing water quality targets for nitrogen.[8]

According to the Environment Agency, agricultural practices in the Focus Area have contributed to elevated nitrogen due to intensification of dairy farming and associated maize-growing for fodder. The soils in the catchment are vulnerable to compaction and erosion and a significant proportion of land is unsuited to growing maize or winter manure spreading, but these activities are nevertheless widespread.[9]

In all, the Environment Agency found that 87 per cent of farms did not comply with environmental regulations and 66 (65 per cent!) of the farms were polluting at the time of the farm inspection.

The majority of farms visited were Red Tractor assured, showing, according to the Environment Agency, that 'Red Tractor is not effective at assuring farms are meeting environmental regulations.'[10] The same conclusion was made by the regulator in the Axe report.

Red Tractor Assurance works with around 46,000 British farmers to champion care and attention for the British countryside and animals.[11] Assured Food Standards (AFS), the not-for-profit company behind Red Tractor, is owned and created by the British farming and food industry. AFS has six guarantors: NFU, Dairy UK, Ulster Farmers' Union, AHDB, British Retail Consortium (BRC) and NFU Scotland. Whilst AFS own Red Tractor, it is operated independently. AFS claims: 'Red Tractor is the only scheme that integrates a range of issues, food safety, animal health and welfare and controls on environmental pollution into a single set of standards and a single inspection.'[12]

According to Red Tractor, if 'you want food and drink that's responsibly sourced, just look for [their] logo',[13] with environmental protection being one of the four key principles of the Red Tractor Standards. But alas, not really for the environment at this time, according to the findings of the Axe and North Devon studies, pollution incidents, other factors and my own experience. Pollution has become an 'inevitable' part of livestock farming, and window-dressing it with a standards logo is not changing that. Change requires the injection of conspicuous advice and regulatory presence, from well-trained, capable and willing people. Farmers also need to have the means to make the right investments, using the best available advice and techniques. But even that wouldn't be enough without a drastic cut to the amount of livestock farmed and animal products eaten. Unfortunately, the goodwill to do the right things, which exists in good measure, is met with silence.

As will become apparent later, what basic commercial interests AFS has for the environment, when it comes to livestock farming, stop at the British border and do not extend into the rainforests of South America and Asia. Red Tractor's principal standard on animal health and welfare is also a misnomer, as it is based around (my emphasis) livestock farming not causing *unnecessary suffering and injury* to the animals as embedded in UK animal welfare law. Red Tractor does have some standards above UK regulatory requirements, but it is nevertheless a basic animal welfare assurance scheme, and one that the retailers will wilfully hide behind and use to gaslight their customers that they care for animals, and those same customers are only too willing to be *reassured*, given the inconvenient truths of the *necessary* suffering and painful injuries served up on their dinner plates (for some people, three times a day). It is important not to leave out other schemes, such as RSPCA Assured, which is claimed to be an ethical food label dedicated to animal welfare. This is another indefensible misnomer, which although intended to raise the animal welfare bar significantly in contrast to UK animal welfare law and Red Tractor assured, to further improve the lives of farm animals, still turns a blind eye to animal cruelty in relation to the animals we want to eat and in turn itself promotes unethical practices and acts of speciesism. If it is the ultimate mission of the Royal Society for the Prevention of Cruelty to Animals to end the cruelty that animals suffer, then it would be a reasonable supposition for them to promote and fully partake in a whole-food plant-based diet, along with all the rainforest, ocean and climate-change ecowarriors.

When it comes to slurry storage, not all rainfall falling on yards and wash water from dairy parlours (e.g. dirty water or lightly fouled water), which is contaminated with livestock excreta, finds itself included in a farm's slurry storage capacity calculation. Rainfall during winter is expected to increase, climate-change predictions tell us, wash water would also be expected to increase as livestock farms further intensify and expand. In law, both in and out of NVZs, there is no exemption to store less slurry than the statutory minimums, regardless of the slurry's nature. If it's slurry, it's slurry.

Dirty water may be seen by some as low-risk, but on dairy farms it can amount to 37 litres per cow per day, and the volume per dairy farm has increased markedly as herd sizes have grown. Dirty water is often regularly applied throughout winter to a small amount of already wet land in close to proximity to the farmstead, where some inevitably finds its way into inland freshwaters (thus delivering nitrate-nitrogen, ammonium-nitrogen, phosphorus, microbial pathogens and other contaminants). I have walked many a saturated field where dirty water has and is being applied, with varying degrees of slope and poaching (e.g. run-off risk) and land drainage (e.g. preferential drainage). This contributes to diffuse water pollution, which can grow with each added cow.

Farmers and their supporters also often point to the SAFFO four-month calculation as the minimum required by law, so it must be okay, but SSAFO is thirty years out of date. In 2018 new regulations, The Reduction and Prevention of Agricultural Diffuse Pollution (England) Regulations 2018, were laid to fulfil obligations on diffuse pollution under the Water Framework Directive. The Diffuse Pollution Regulations apply across England and require each application of organic manure (and manufactured fertiliser) to be planned so that it does not:

- exceed the needs of the soil and crop on the land, or
- give rise to a significant risk of agricultural diffuse pollution.

Here, 'significant' means the significance of risk and *not* the significance of pollution. Account must be taken of the weather conditions and forecasts at the time of application. This is particularly important in relation to run-off and preferential drainage. In my opinion, this cannot be achieved with four months' slurry storage capacity. Six months is the bare minimum needed UK-wide, and more in some places depending on soil type, slope and rainfall, and in particular where risk factors are stacked.

Pig farmers reliant on arable land to spread slurry in the autumn may need as much as nine months' storage, or even more, given the majority of arable crops grown in the UK do not have a soil and crop nitrogen need until late winter/early spring. Furthermore, pig slurry has a very high ammonium-nitrogen content (70 per cent), which is readily converted into nitrate-nitrogen, a highly leachable form of nitrogen, which when not used by a crop in the autumn can give rise to a significant risk of diffuse pollution. For example, on a clay loam soil a typical application of pig slurry (50 m³ per hectare) can result in a loss of around 40 kg of nitrate-nitrogen on each hectare spread, or up to 70 kg on a silty loam soil. The risk is even higher for pig slurry put through an anaerobic digester (AD), which has an ammonium-nitrogen content of 80 per cent and is more akin to a liquid manufactured fertiliser.

The regulations represent parts of the Code of Good Agricultural Practice for Water, which extend back to 1985. They were brought in due to a countrywide failure of voluntary compliance. For example, in respect of nitrogen, dating back 35 years:[14]

Nitrogenous fertilisers should only be applied at times when the crops can utilise the nitrogen. In autumn and winter applications should be avoided except when there is a specific crop requirement [...]. Seed bed applications of nitrogen for winter cereals are therefore rarely necessary[...]

Diffuse pollution in these regulations, also commonly known now as Farming Rules for Water (FRFW) regulations, means the transport of agricultural pollutants into inland freshwaters or coastal waters, or into a spring, well or boreholes, where:

- the transportation occurs by means of soil erosion or leaching; and
- the agricultural pollutants may be harmful to human health or the quality of aquatic ecosystems or terrestrial ecosystems directly depending on aquatic ecosystems.

Diffuse pollution is generally associated with an accumulation of minor discharges of polluting materials from disparate sources, with each individual discharge contributing to, rather than causing, obvious pollution alone. However, some distinct sources that are persistent and chronic in nature can also result in pollution of particular concern.
Signs of diffuse pollution are:

- diminished water life.
- declining fish numbers.
- high or rising nitrate, phosphorus and suspended solids levels.
- a build-up of sediment and the presence of sewage fungus or other unwanted growth.

To comply, farmers must take into account any factor which would give rise to a significant risk of agricultural pollution, and then take whatever reasonable precautions are necessary to prevent pollution. For instance, this could include thoroughly composting materials like farmyard manure before spreading in the autumn.

Unfortunately, unlike inland freshwaters, the regulations do not protect all groundwater, just groundwater which transports pollutants into the waters I describe above. At the very least, although untested, this could include an area around features like springs, wells and boreholes with a fifty-day travel time from any point below the water table to the source; or as a minimum a 50-metre radius from the source.

A fifty-day travel time is classified in groundwater protection terms as a Zone I Inner Source Protection Zone, modelled on the basis of microbial 'die-off time' in the subsurface. The orientation, shape and size of the zones are determined by the hydrological characteristics of the strata and the direction of groundwater flow. The zone is not usually defined where the aquifer is confined beneath substantial and continuous covering strata of very low permeability. There are other zones, such as a Zone II Outer Source Protection Zone, defined by a

400-day travel time, and a Zone III, which covers the complete catchment area of a groundwater source.

The Environment Agency models zones around public water supplies and supplies used in commercial food and drink production, amongst some others (e.g. including mineral and bottled water). Domestic supplies are not modelled, and the minimum 50-metre radius from the source is used as a reasonable precaution. However, this may be less than a 50-day travel time, and relying on a minimum 50-metre distance around springs, wells and boreholes may not be sufficient to comply with the diffuse pollution regulations in all cases. This is complicated for farmers but is important in protecting water quality and resources. Something being complicated is not an excuse for business-as-usual activities which cause pollution.

The likelihood of causing pollution would also depend on the soil's leaching potential, determined on the basis of the physical properties of the soil, which affect the downward passage of water, and the ability of the soil to attenuate three types of diffuse pollutant:

- pollutants which under certain circumstances can be retained in the soil;
- pollutants which can readily pass through the soil layer (e.g. nitrate-nitrogen);
- liquids from slurries and manures, etc. (e.g. through preferential drainage).

The highest-risk soils are known as soils of high leaching potential (H). These soils have little ability to attenuate diffuse source pollutants, and as such non-adsorbed pollutants and liquid discharges percolate rapidly through them. There are three categories:

H1: Soils which readily transmit liquid discharges because they are either shallow or susceptible to rapid bypass flow directly to rock, gravel or groundwater.

H2: Deep, permeable, coarse-textured soils which readily transmit a wide range of pollutants because of their rapid drainage and low attenuation potential.

H3: Coarse-textured or moderately shallow soils which readily transmit non-adsorbed pollutants and liquid discharges, but which have some ability to attenuate adsorbed pollutants because of their large organic matter or clay content.

The depth of the unsaturated zone, the part of an aquifer which lies above the water table, and its pollutant attenuating properties all also play a part in determining the vulnerability of groundwater to pollution from surface activities.

Effects of diffuse pollution are often long-term and widespread, and the place(s) of origin is/are less easily identified. Because of this, diffuse pollution and the source(s) is/are less likely to be reported and go unnoticed, in particular nitrate-nitrogen pollution.

When ammonium-nitrogen (ammonium-N) is applied to soil, it is readily adsorbed onto the surface of clay minerals and soil organic matter. However, it is not fixed in that form and in certain conditions ammonium-N is oxidised to nitrate-nitrogen (nitrate-N) in a two-phase process by two groups of bacteria: firstly, Nitrosomonas bacteria, which oxidises ammonium-N to nitrogen dioxide, and then by Nitrobacter, which oxidises nitrogen dioxide to nitrate-N.

When a crop is growing, nitrate-N is taken up quickly, reducing its susceptibility to leaching. When there is no need for nitrogen, or crop growth slows or stops, unused nitrate-N is washed out of the soil beyond any future crop use by winter rainwater. Highly permeable sandy and shallow soils are particularly susceptible to nitrate-N leaching.

How much nitrate-N is lost by leaching is governed by the soil type, crop, amount of readily available nitrogen (RAN) applied, the method/accuracy of application, the time of application (as winter progresses, the soil chills and nitrification reduces) and how much rain falls between the time of the application and the 31 March (c. the end of field drainage). More nitrate-N is leached in higher than lower rainfall areas, though that does not necessarily mean a pollution risk is more or less. For example, there is less dilution in drier parts of the country.

Nitrogen can be immobilised in the presence of a rich carbon source, which is low in nitrogen but has a high C/N ratio (e.g. 80:1). The carbon-rich residues are decomposed by soil microorganisms, which absorb ammonium-N and turn it into microbial biomass. When the microbial biomass dies, it is open to decomposition and ends up as humus and mineral nitrogen. Humus builds up extremely slowly.

Even organic manures with a lower readily available nitrogen content can pose a significant risk of pollution. To prevent nitrate-N leaching, it's important not to dispose of nutrients at a time of the year when they are not needed, and instead find alternative solutions to make the best use of them. Unfortunately, creating liquids and disposing of liquids to land when there isn't a soil and crop need is a cheap way to externalise costs and the problem. Solutions require change and cost money.

Charts 2 to 5 below, give an indication of how much nitrate-N and other sources of nitrogen can be lost over a calendar year. These have been modelled using MANNER-NPK software, using an annual rainfall of 955 mm, an application rate of 50 m³ per hectare (spread using a band spreader) and a typical dairy cattle slurry (6 per cent dry matter, 2.6 kg per m³ total nitrogen and 1.2 kg per m³ readily available nitrogen). In all, a total of 130 kg per hectare of total nitrogen is applied.

Chart 2: Grass / Sandy Loam over Sandy Loam

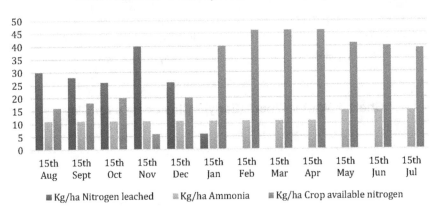

Chart 3: Grass / Clay Loam over Clay

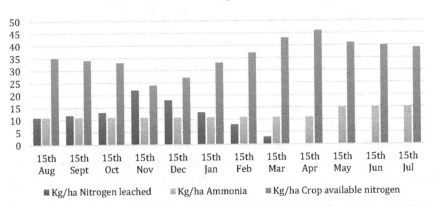

Chart 4: Winter Wheat (Early) / Sandy Loam over Sandy Loam

15th Aug	15th Sept	15th Oct	15th Nov	15th Dec	15th Jan	15th Feb	15th Mar	15th Apr	15th May	15th Jun	15th Jul

■ Kg/ha Nitrogen leached   ■ Kg/ha Ammonia   ■ Kg/ha Crop available nitrogen

Chart 5: Winter Wheat (Early) / Clay Loam over Clay

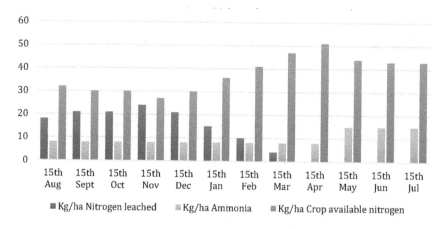

15th Aug	15th Sept	15th Oct	15th Nov	15th Dec	15th Jan	15th Feb	15th Mar	15th Apr	15th May	15th Jun	15th Jul

■ Kg/ha Nitrogen leached   ■ Kg/ha Ammonia   ■ Kg/ha Crop available nitrogen

The results shown are not cumulative; they are based on what is lost after you apply 50 m³ per hectare of cattle slurry on any 15th of a month, ranging from zero to 45 kg per hectare in these examples.

As shown in Charts 2 and 4, nitrate-N loss is highest when slurry is applied to sandy, free-draining soils between the 1st of September and the end of December.

Charts 3 and 5 show significant nitrate-N loss from autumn and winter applications, in particular for winter wheat given the nitrate-N produced is not used by the crop at that time of year. Unused nitrogen in a sufficiently warm and wet winter soil can also lead to significant nitrous oxide emissions, a very potent greenhouse gas.

The amount of nitrate-N lost on grassland is less, despite nitrogen use not normally being productive after September, as grass can use up to 20 kg per hectare of nitrogen from organic manures before the end of October.

Spreading slurry at the right time is constrained in several ways, for example, by insufficient slurry storage and relying on antiquated, heavy broadcast spreading equipment (tankers with splash plates). With enough storage, large enough fields and the availability of precision spreading equipment, in particular trailing hose, it is possible to spread more broadly to growing crops in the spring, such as winter wheat, whilst limiting pollution swapping (e.g. phosphorus pollution) and soil damage. Precision spreading can also be used before a spring-planted crop, and can also allow a winter-cover crop to be grown beforehand to improve soil health, nutrient retention and carbon sequestration.

Significant risk does not end with nitrate-N leaching; it also concerns run-off and preferential drainage. Preferential drainage occurs through old root channels, cracks, wormholes etc. Preferential drainage can occur on both dry and wet soils. On dry soils, risk occurs on cracked soils above land drains or cracked shallow soils over fissured rock. On wet soils (at or close to field capacity), over winter and into spring (November until April), the risk of preferential drainage is high.

The risk of compaction, and associated run-off, e.g. from trafficking slurry-spreading equipment on wet soils, including grassland, can also be high. The risk of loss includes ammonium-N, phosphorus and microbiological pathogens.

Ammonium-N loss is an addition to nitrate-N loss through leaching. Wet soils have limited capillary action, with soil water mainly held by surface tension on the walls of cracks and holes. Ammonium-N in contact with the surface of clay minerals and organic matter is absorbed, but some is not, and if it rains sufficiently within ten to twenty days after application some will be washed downwards and beyond crop use. The risk of loss is greater on drained soils, which encourage drainage.

The high-risk period for spreading liquid organic manures, including lower-RAN manures, starts from 1 August and ends in April (leaching and preferential drainage combined).

The core risk period for both is November to February, inclusive, as shown in Table 18.[15]

Table 18: Diffuse water pollution and soil compaction risks following slurry applications to medium/heavy soils

*Low Risk. **Medium Risk. ***High Risk

Application timing	Nitrate-N cereals/grass & OSR	Ammonium-N	Phosphorus	Microbial pathogens	Soil compaction
Autumn(Aug–Oct)	***/**	*	*	*	*
Winter(Nov–Jan)	**	***	***	***	***
Spring(Feb–Apr)	*	**	**	**	**
Summer(May–Jul)	*	*	*	*	*

Source: Williams, J. *et al. The Contribution of Slurry Management Practices to Diffuse Pollution from Agriculture.* ADAS. Defra Project WT1508, Work Package 1.

Table 19: Diffuse water pollution and soil compaction risks following slurry applications to free-draining soils

Application timing	Nitrate-N cereals/grass & OSR	Ammonium-N	Phosphorus	Microbial pathogens	Soil compaction
Autumn(Aug–Oct)	***/**	*	*	*	*
Winter(Nov–Jan)	**	**	**	*	**
Spring(Feb–Apr)	*	*	*	*	*
Summer(May–Jul)	*	*	*	*	*

Source: Williams, J. *et al. The Contribution of Slurry Management Practices to Diffuse Pollution from Agriculture.* ADAS. Defra Project WT1508, Work Package 1.

Whilst the Agriculture Diffuse Pollution Regulations do not ban the application of organic manures to land, they do, as mentioned earlier, require applications to be planned so as to not exceed the needs of the soil and crop on the land, nor give rise to a significant risk of diffuse pollution.

Table 18 shows that changing applications from the autumn to spring on medium/heavy soils reduces the risk of nitrate-N loss from high to low, but can increase the risk of phosphorus pollution from low to medium. This is a concern which needs to be mitigated. The type of pollution livestock farming causes is not a choice:

- Spread/dispose of slurry in the autumn and cause pollution; or
- Spread/dispose of slurry when fields are wet in the winter and spring and cause pollution.

This kind of pollution conundrum would be greatly eased by the public ditching meat, dairy and eggs and eating plants instead. In the interim, the livestock industry, and others that depend on agricultural land to dispose of their wastes, cannot rely on pushing pollution pebbles around the seasons. Solutions need to be created, working together, and undoubtedly will involve heavy expense. Business-as-usual and pollution-as-usual is not an option under the regulations.

For too long diffuse pollution has been ignored. It may be less obviously seen or appreciated, but it can be/become persistent/chronic in nature, as observed in the River Axe and North Devon catchments. For farmers that have followed the Code of Good Agricultural Practice for Water and have the right storage capacity requirement and spreading equipment to comply, compliance with the diffuse pollution regulations should be less of a problem.

Farmers generally know their land well and know what is necessary to avoid direct pollution. Diffuse pollution, though, is much less well understood and/or appreciated by farmers. Existing manure management plans do not adequately address the risk of diffuse pollution and, like nutrient management plans, are less effective when a farm does not have enough storage and the right spreading equipment. All too often, a lack of storage capacity and/or inappropriate spreading equipment is driving poor practice and the pollution that goes with it.

Farmers in any doubt can check their slurry storage capacity requirements using the AHDB slurry wizard,[16] whilst also taking into account the common calculation mistakes I list below. The wizard benefits from the inclusion of an inbuilt Environment Agency formula to calculate wetter-than-average years (one in five years), and a calculation which takes account of wet slopes on earth-banked lagoons.

Common calculation mistakes result from a farmer's failure to:

- use M5 (five-year wetter-than-average) rainfall, required for SSAFO and NVZ compliance.
- take account of slurry that can reasonably be expected to remain in the store prior to the storage capacity requirement calculation period.
- take account of freeboards, 750 mm for earth-banked lagoons and

300 mm for other stores (earth bank lagoon freeboard must be maintained at all times).

- accurately calculate the yard area draining to the slurry store, including increased yard area due to livestock expansion (e.g. collecting/loafing yards).
- take account of rain falling on the slurry store during the storage capacity period.
- take account of end/side wet slopes on earth-banked lagoons (the smaller the store the greater the disparity).
- include lightly fouled water slurry (LFW) within the farm's total slurry storage capacity requirement calculation. Whilst it is often assumed LFW can have less volume storage than prescribed in SSAFO and NVZ, it is slurry, so is not actually exempt from the calculation. However, it has been allowed to have less capacity in the past.

  Note: Lightly fouled water means a dilute form of slurry produced from any water collected from yards and buildings used temporarily by livestock and where, as far as reasonably practicable, the yard or building is scraped or brushed down immediately after use to minimise contamination with livestock excreta.[17]
- take account of bedding (e.g. straw, sawdust, sand, etc), which may end up, and also build up, in the slurry store.
- take account of rainwater, draining from silage clamps, which enters the slurry store.
- keep clean water out of the store (e.g. from broken guttering, 'clean' yards, etc).
- update the slurry storage capacity requirement and as a result continue to use the old, outdated calculation.

Another issue is that the current calculation used to determine the amount of slurry against a dairy cow milk yield (6,000–9,000 litres per annum) is outdated. A value of 53 litres of slurry per day is used, yet a dairy cow achieving today's average milk yield of 8,000 litres will more likely produce over 57 litres of slurry per day. The closer to 9,000 litres, the greater the disparity – and naturally this only increases for milk yields over 9,000 litres. Therefore, whilst a farm may have the correct calculated storage capacity, if the milk yield achieved is greater than 7,500 litres per year, the slurry store may have insufficient storage in practice.

In some instances, there is a failure to collect and properly manage contaminated wash water and yard effluent, which instead drains *somewhere* (e.g. to a sump or soakaway, barrier ditch, farmland abutting yards, to ground through permeable 'rubble' yard areas or defects in concrete yards and to roads/road drains below sloping yards). In some cases, sumps without a soakaway are drained to surface water.

Whilst some dairy farms have yard effluent that drains 'somewhere', it is perhaps more common practice on beef units with no or inadequate waste management infrastructure. This includes both conventional and organic farms. Just because a farm is organic, it doesn't make it less polluting.

Some farmers claim that they do not spread during the legal minimum storage capacity periods and use that as evidence of compliant slurry storage capacity. But, as alluded to above, this does not mean a farmer is compliant and/or is not causing pollution, unless *all* the slurry generated ends up in the slurry store and is accounted for. Sadly, there are too many farms that do not collect and store all their slurry, allowing a proportion to drain to that 'somewhere else' place and end up directly/indirectly in watercourses or groundwater.

The cost of slurry storage can be substantial in relation to profit margins, which farmers can't easily pass on to the consumer. Cost-of-production models used to determine the price paid for meat and milk at the farm-gate need to factor in the cost of compliance with at least the regulatory minima and pollution control. If not, then farmers will continue to sell more than milk. Processors, supermarkets and others buy livestock pollution and regulatory non-compliance on a daily basis, and then they sell it on to us.

In England around 10 billion litres of milk is produced each year (15 billion for the whole UK). A price increase of a single penny per litre of milk would yield the princely sum of £100 million. Three pence a litre over ten years would be enough to sort out most of the slurry emergency – a very low price to pay to minimise pollution and for a clearer conscience, unless you are supermarket.

The story does not stop directly with slurry. ADs may use slurry as a feedstock, food waste, field crops (e.g. maize and grass) or combinations of these and other feedstocks (e.g. blood and rendering waste). What goes into AD comes back out the other end and is usually spread back to farmland. It doesn't go away.

Effluent from AD is known as digestate, and the process of AD creates a digestate with a much higher ammonium-N content than pig and cattle slurry. This makes it a good fertiliser but a *very high-risk organic manure* if applied to land when there isn't a soil and crop need, or when it would give rise to diffuse pollution (e.g. increased risk of nitrate-N leaching). AD plants notoriously do not have six months' or more storage capacity, including those associated with slurry-producing farms – similarly, animal waste rendering plants, whose own effluent is commonly applied to farmland. In fact, these wastes and others do not have a legal minimum storage capacity requirement, with AD digestate storage only being steered by the nitrate vulnerable zone (covering 55 per cent of the UK) closed spreading period rule and whether the effluent contains any livestock manure (i.e. for digestate that has to be spread within an NVZ). AD is a very welcome addition to climate-change mitigation, but it must not be used as yet another excuse for polluting water.

Citations:

1. CIRIA Farm Waste Storage (ADAS 2013), *Potential Pollution Risks Evidence Programme, Slurry Storage Situation Report*. Project CON197 (Prop 2886), via Freedom of Information Request. Environment Agency, NR192233 (3 December 2020).
2. Ibid.
3. Ibid.
4. Cossens, J. (2019), *River Axe N2K Catchment Regulatory Project Report*, Freedom of Information Request. Environment Agency, NR192233 (3 December 2020)
5. National Audit Office (1995), *River Pollution from Farms in England*
6. Cossens, J. (2019), *River Axe N2K Catchment Regulatory Project Report*, Freedom of Information Request to the Environment Agency, NR192233 (3 December 2020)
7. Environment Agency, *North Devon Priority Focus Area Case Study Summary Messages*, via Freedom of Information Request to the Environment Agency, NR192233 (3 December 2020)
8. Ibid.
9. Ibid.
10. Ibid.
11. Red Tractor Certified Standards. Accessed at https://redtractor.org.uk/
12. Red Tractor Certified Standards. Accessed at https://assurance.redtractor.org.uk/who-we-are/company-guarantors.
13. Red Tractor Certified Standards. Accessed at https://redtractor.org.uk/
14. MAFF ADAS (1985), *Code of Good Agricultural Practice*
15. ADAS, Williams, J. et al. (2013), *The Contribution of Slurry Management Practices to Diffuse Pollution from Agriculture*. Defra Project WT1508, Work Package 1

16. AHDB, *Slurry Wizard*.
   https://ahdb.org.uk/knowledge-library/slurry-wizard
17. CIRIA C759F (2018), *Livestock manure and silage infrastructure for agriculture*. Part 1 Selection Guide

15. Silage Tears

Silage is commonly grown as a winter feedstock for dairy and beef cows. It can be defined as any crop harvested while green for fodder and kept succulent by partial fermentation, essentially pickling, in a silo under anaerobic conditions. Most silage is made from grass, rapidly wilted to preserve nutrients.

Silage is commonly made in a silo/silage clamp. To create anaerobic conditions, air is removed by compressing the crop in layers with farm machinery. The clamp is then sheeted to prevent rainwater ingress and to keep it airtight. Once air-free, the acidity increases quickly, preventing any remaining respiration, which would otherwise spoil it. The process can take several weeks and in the initial phase creates large amounts of highly polluting and corrosive effluent, which must be safely contained.

Silage can also be compressed into bales that are then wrapped and sealed into impermeable membranes, or enclosed in impermeable bags. Farmers may also store silage in field heaps, typically done if they run out of clamp capacity. Straw bales may be used to form a rudimentary structure, with a layer of straw underneath advised to help soak up any silage effluent.

Maize is the second most commonly ensiled fodder crop, with the advantage of producing much less potentially polluting effluent than grass.

The making and storage of silage in the UK is also regulated under The Water Resources (Control of Pollution) (Silage, Slurry and Agricultural Fuel Oil) Regulations 2010 (SSAFO). England and the devolved administrations have their own versions but these are essentially the same.

SSAFO contains an exemption, as explained previously. This applies to silos in use before 1 March 1991 too. Many silos still in use today benefit from this long-outdated exemption, ferociously defended by the agricultural lobbyists on behalf of their members. If a pre-1991 silo is substantially enlarged or substantially reconstructed then it must be constructed in accordance with SSAFO, to avoid pollution.

New silos are expensive and it's of little wonder that the exemption is so hotly defended. Preventing pollution can cost a lot of money. As of 2021 the exemption was thirty years old, with many silos now well beyond their design life. Farmers may patch them up, but there is only so long before they degrade and fail. Visit farms with silage stores and the greater majority will either be in a poor state or not built to SSAFO requirements. Furthermore, many silos built since 1991 do not fully conform to SSAFO and/or have not been adequately maintained. My personal experience is that at least 60 per cent of silage stores are unlikely to be built in accordance with SSAFO.

Some common problems are cracked and/or eroded floors (silage effluent is highly corrosive), and worn or damaged floor joints, reducing permeability; wall tilting (concrete, masonry and timber sleeper walls) through overloading and/or under-design; cracks, deformation or other stress-related damage to wall panels; and effluent leakage from the edge of the silo floor slab and exiting the clamp, including undersized perimeter drains or simply blocked or broken.[1] Signs of leaking include scorched/dead vegetation, soft ground and the presence of nitrogen-loving nettles and docks.

Silos are designed to drain into a below-ground effluent tank, subject to a minimum capacity to ensure all potential effluent is safely collected. Some tanks may have an overflow to a slurry store. The effluent can be spread *safely* to a crop, usually grass, but needs to be diluted with water and account taken of the nutrients applied. Poorly drained clamps or blocked drainage channels at the front of the clamp can lead to uncontrolled run-off.

SSAFO applies to any ensiled crop, and not just when used for fodder. Clamps used to store maize, grass or other green crops used for AD feedstock must also conform to SSAFO.

Old or poorly designed clamps are commonplace on livestock farms. Silage effluent is much more toxic to aquatic life than slurry and is a common cause of water pollution incidents.

AD plants themselves, however beneficial towards climate change mitigation, can lead to an increased risk of nitrate-nitrogen and ammonia pollution. Some of the bigger AD plants can produce up to 50,000 m³ of digestate effluent a year, or 10–12 times the amount of slurry produced on an average dairy farm. This is a huge concentration of effluent for single locations, which in turn is much stronger/potentially polluting than untreated cattle slurry. A serious amount of crop land is needed to safely and agronomically service this amount of effluent, which under English diffuse pollution legislation can only be realistically applied to crops for six months of the year. With many AD plants also looking towards maize as a feed crop, this increases the risk of associated soil erosion and water pollution too.

Unfortunately, many AD plants outside of NVZs lack adequate storage capacity, resulting in digest being spread when there isn't a soil and crop need and when it will give rise to significant diffuse pollution. Like slurry, the problem is not with AD per se but with inadequate infrastructure and a sufficient/secure land bank. AD plants are built to provide an income and providing all the necessary post treatment and storage to avoid pollution eats into profit.

AD digestate applied directly to the surface of land results in high ammonia emissions, which can be vastly reduced by injecting the digestate into the soil. However, if done post-harvest the practice greatly increases the risk of over-winter nitrate-nitrogen leaching. Injecting into a wet, clay-rich soil also causes the digestate to concentrate in and around the injection channel, which greatly increases the risk of preferential drainage.

Where silage is made from whole cereal crops, urea may also be added to prevent spoilage. This in turn gives rise to ammonia emissions, to further extend livestock's longer shadow.

Citations:

[1.] CIRIA C759F (2018), *Livestock manure and silage storage infrastructure for agriculture*. Part 2, Design and construction.

16. Ammonia Pollution

Livestock manure produces ammonia pollution, and an awful lot of it. 87 per cent of the UK's ammonia emissions come from agriculture, with 48 per cent from cattle. Stepping a short distance outside the UK, 99.2 per cent of all Irish ammonia emissions come from agricultural activities, with 89.4 per cent from manures.[1]

Ammonia emissions from agriculture cause half of all dangerous $PM_{2.5}$ particulate pollution (fine inhalable particles, 2.5 micrometres and smaller) within the UK. It is lost from slurry, farmyard manure and other nitrogen-containing organic matter used in agriculture; and from manufactured fertilisers, in particular urea-based fertilisers.

Ammonia reacts with nitric acid from industrial and vehicle nitrogen oxide emissions to form $PM_{2.5}$, which in turn is linked with respiratory and cardiovascular diseases, cognitive decline, low birth rates and increased mortality. It is estimated that particulate matter emissions as a whole result in 29,000 early deaths every year in England.[2]

Ammonia emissions are also toxic to keystone ecosystems, as ammonia emissions unnaturally enrich and acidify soil and water. For example, as soil fertility increases, mosses, lichens, liverworts and other plant species, many rare, begin to die off, and with them their associated *bedrock* wildlife communities. Nutrient-loving plants, like fast growing grass, nettles and brambles begin to move in, eventually crowding out what may remain. Furthermore, ammonia pollution indirectly contributes to nitrous oxide, a potent greenhouse gas.

Whilst livestock factory farms are ammonia emission hotspots, smaller livestock farms – and there are many thousands more of them – collectively far outweigh them. Given that ammonia emissions to date are only tackled on permitted intensive pig and poultry units, it is a clear weakness in the fight against the harm caused by ammonia emissions from livestock farms.

However, things are about to improve somewhat. Government intransigence has been replaced with a commitment to permit intensive beef and dairy farms. Pig and poultry farms above a certain threshold already have to be permitted, though there remain many other farms below the threshold pumping out ammonia emissions.

The English plan, to fulfil legal international commitments on air quality, is to reduce ammonia emissions from 2005 levels by 8 per cent, and then by 16 per cent by 2030. The emission factors used to calculate UK farming's requirements for the pig sector are contested by the likes of the Agriculture and Horticulture Development Board,[3] but regardless, the percentage reduction falls way short of what is really necessary to properly protect terrestrial and aquatic ecosystems and human health. What 'permitting' will look like and how effective it will ultimately be we will have to wait and see.

So how do we reduce or stop ammonia emissions to protect us and our environment? The easiest and simplest way is to dramatically reduce or stop farming livestock, with the added benefit of solving all the other countless issues associated with the sector at the same time. The list is very long.

I wish it were all that simple. Whilst we wait for the rest of the world to catch up with a whole-food plant-based diet (WFPBD), it is crucial work is done on reformulating our livestock's diet, keeping livestock housing squeaky clean, stopping livestock urine mixing with dung (because they react to form ammonia), minimising the surface area of livestock floors (without further compromising animal welfare) and keeping manures dry.

Another option is to cover slurry stores. Assuming an impermeable cover is used, it can cut ammonia emissions during storage by as much as 80 per cent. To avoid wasting all that benefit, slurry when spread needs to be placed directly on or into the ground, as opposed to broadcasting it. Solid manures need to be incorporated into the soil within at least twelve hours. Not only is this a good idea, but reducing nitrogen loss leaves more available in the soil for the next crop, thereby reducing the need for manufactured fertilisers and contributing towards solving the climate emergency. Lowering the acidity of slurry, e.g. with nitric or sulphuric acid, can also reduce ammonia loss.

All this can come at an incredible cost. I estimate covering all slurry stores alone could cost at least £300 million, excluding the extra £700 million I estimate is needed to provide six months' slurry storage.

Citations:

1. Buckley, C. et al. (2020), *An Analysis of the Cost of the Abatement of Ammonia Emissions in Irish Agriculture to 2030.* Teagasc, Carlow, RoI
2. Defra (2018), *Code of Good Agricultural Practice for Reducing Ammonia Emissions*
3. ENDS Report, 'UK ammonia targets based on 'outdated' emission factors.' https://www.endsreport.com/article/1683921/uk-ammonia-targets-based-o utdated-emissions-factors

17. Forage Maize and Pollution

Forage maize, grown and ensiled as a livestock feed, has proven a revolutionary crop. In dairy herds it has helped increase forage intake, milk yield and milk protein content and ultimately profit.[1] Unfortunately, this has come at the cost of pollution in many instances.

Maize in the UK is typically harvested from mid-September and well into October, sometimes later, depending on the variety, location and weather conditions. By late autumn, when soil has begun to wet, the heavy machinery used to harvest maize can compact/rut and damage the soil's structure, greatly reducing its porosity. This heightens the potential water erosion risk, as it lowers the infiltration rate and increases the risk of run-off. Nutrient-enriched soil particles on the land's surface may be washed away, and in larger amounts if rills and/or gullies develop, into a watercourse.

Soil entering a watercourse can have a profoundly negative impact on water quality and cause considerable harm to aquatic life. In addition to pollutants such as phosphorus and nitrogen, a coating of sediment forming on the bottom of gravel-bedded rivers renders them unsuitable for fish spawning, e.g. salmon and trout. Sediment too, if allowed to build up, can contribute to flooding.

Run-off can also occur in the early days after planting, typically April to May. Before a protective crop canopy forms, topsoil rich in organic matter, along with fertiliser and pesticides residues from early treatments, can also be washed from the land when it rains sufficiently.

Maize stubbles may also be used as sacrificial land to dispose of slurry in winter, when there is no soil and crop need, or there is a risk of causing diffuse pollution. Crop stubble is an ideal place to relieve the strain on a farm's diminishing slurry storage capacity, before winter fully sets in. Heavy machinery used to spread slurry and other livestock manures can exacerbate soil rutting and compaction. A mixture of soil and slurry can be particularly harmful if it flows into a watercourse.

Soil erosion, including loss from maize fields, has been of concern since the early 1970s.[2] Erosion caused by rainfall can happen in several ways. The impact of rain dislodges soil particles, which are then washed away in a thin film of water. If the run-off shear stress is sufficiently powerful, it can scour the land surface, creating a network of numerous narrow, shallow channels called rills. Where rills form, a considerable amount of topsoil can be lost. In severe cases deep, ditch-like gullies, may be cut into the land. Light, free-draining soils are particularly prone to the formation of rills, whilst clays soils are more associated with soil wash. Common to all soil erosion events is a bare, or otherwise exposed, soil and associated poor soil health.

In 1973 the UK maize crop was around 8,000 hectares and by 2019 around 228,000 hectares.[3] A remarkable transformation, with more being added every year, often in places unsuited to the crop from an environmental perspective.

There are options to curb soil erosion, including under sowing maize after early establishment with Italian rye-grass, or even a legume, between the rows. Oats can also be planted for quick establishment after harvest. A cover crop may also be planted, provided it can be established in time after harvest. Another option is to cultivate the soil sufficiently after harvesting maize to break up any compaction and allow rainwater to soak into the soil. However, leaving land bare in whatever way damages soil biology/health.

There are some UK locations where the soil type means that growing maize causes run-off in most years, and in these locations maize should not be grown. In some situations, an early-harvest maize variety can be grown, which allows a following crop to be sown soon after to lower the risk of winter run-off, but with extremes in weather these days this can't always be guaranteed in all years.

The best option is not to grow it where the pollution risk can't be effectively managed, or better still, until we stop farming cows, to feed them grass from uncompacted land.

Citations:

1. Potash Development Association (2018), *17. Forage Maize – Fertiliser requirements*.
   https://www.pda.org.uk/pda_leaflets/17-forage-maize-fertiliser-requirements/
2. Boardman, J. (2013), 'Soil Erosion in Britain: Updating the Record.' *Agriculture* 3(3), pp. 418-442
3. Defra (2019), *Agriculture in the United Kingdom*. National Statistics at Gov.uk

18. Erosion-ready Livestock

Livestock is not always viewed as the largest source of soil erosion – that award tends to go to winter cereal fields – but contribute they do. A large percentage of winter cereal crops are used to feed livestock, and an expanding hectarage of erosion-ready forage maize fields (8,000 hectares in 1973 and 228,000 hectares in 2019, with the majority used to feed livestock and about a quarter used in AD plants), a crop increasingly used to boost the carbohydrate content of animal feed to help meet industrial-scale production demands. Pastures too, permanent and short-term grass (ley grass, normally less than five years old), contribute to erosion.

Despite winter cereals being a large source, the actual rate of soil erosion per hectare is typically much less than for other crops fed to livestock. For instance, the mean rate of erosion (medians in brackets), based on the National Soil Monitoring Scheme (1982–1986), for winter cereals is 1.85 m^3 per hectare (0.68 m^3 per hectare). For maize it's 4.46

m^3 per hectare (1.00 m^3 per hectare) and ley grass 4.09 m^3 per hectare (1.14 m^3 per hectare). Kale and other fodder crops have a mean erosion rate of 2.10 m^3 per hectare (1.41 m^3 per hectare). There is a marked variation between means and medians as erosion is highly dependent, in particular on the soil's nature (including its health) and condition, ground cover, slope, rainfall and a farm's management practice.

Soil erosion from forage maize production is only one way livestock farming contributes to soil loss and associated pollution. Over-grazing land and soil poaching caused by livestock treading can cause soil structural deterioration, which can result in a significant loss of soil too – especially during wet weather or on poorly drained land. Erosion is not limited to lowland farming and is prevalent on the fragile, sheep-grazed soils of low-fertility upland hill pastures and peat moors too.

Vulnerability can increase with soil wetness, which makes soil less stable and more pliable. Take a dry loam soil and squeeze it between your hands and it will stay in its original state. Do the same with a wet loam and it will compress and hold its new shape, more so with a heavier clay soil. What you are doing is squeezing the air out of the soil, reducing the pore spaces between the soil particles and ultimately decreasing its volume. When it dries, what was once a friable and air-filled soil that breaks apart easily, and takes up water readily, becomes hard, oxygen-limited and resistant to wetting.

Animal treading works wet soil in the same way, creating an uneven, heavily potted land surface. In addition, the action of treading both buries and uproots vegetation, and the resultant damage exposes the soil to unwanted weed species more adapted to the soil condition.

Soil poaching makes the land less productive and reduces a soil's infiltration capacity when it rains, particularly on sloping soils, which in turn increases the rate of surface run-off. Dislodged soil particles are readily carried away with it, along with any nutrients and other contaminants it may contain. The surface of dry soils can become brittle too and turned to dust by treading, therefore encouraging loss by wind erosion.

Compaction ultimately:

- restricts plant root oxygen supply.
- impedes root growth.
- reduces soil warming.
- limits soil nutrient diffusion, causing nutrient deficiency.
- reduces worm populations.
- causes a decline in soil microbial activity.

- increases denitrification, which contributes to climate change.
- lowers a soil's ability to sequester carbon.
- decreases land's resilience to the effects of climate change (e.g. stressing and stunting crop growth in dry years).

The more land is exposed to poaching, the less productive it becomes and the more susceptible it becomes to further degradation. Livestock poaching literally squeezes the life out of a soil.

A worrying dairy trend in terms of soil damage, run-off and associated pollution is the adoption of New Zealand style over-winter grazing in the UK. In this method, cows are grazed outdoors into the winter despite weather, site and soil conditions to make use of the prolonged grass-growing season in the warmer parts of the UK. The system helps lower the amount of slurry storage needed, but there are consequences. Like spreading slurry, the excreta cattle void on the land at this time of year is subject to nitrate-N leaching and ammonium-N and phosphorus loss through preferential drainage. Runoff from poached, churned, muddy grazing paddocks carries manure, pathogens and nutrient-enriched soil into watercourses too. What was presented as a good idea by New Zealand dairy farmers now contributes to their own dairy cow pollution crisis. In winter, and in particular where most dairy farms in the UK are located, the soil is simply too wet for all-year-round grazing. These systems cut costs, but at the expense of the environment and damaged reputations.

Avoidance is the most desirable solution to soil damage and erosion, the most obvious solution being, particularly in the uplands, to only keep enough livestock to maintain a more biodiverse mix of naturally regenerating habitat. Ideally this would be part of a rewilding initiative, with hill farmers paid to perform as conservation managers rather than as livestock farmers per se. At the moment we squeeze conservation into livestock farming when it should, at the very least, be other way around.

The UK, with a total land area of c. 245,000 km^2, is a complex mix of geology and soils, a landscape shaped by glaciation and fluvial processes, with variable climatic conditions across locations and seasons. This and other influences have a huge bearing on what we think might be right or wrong for the land. Not to worry; we can largely let nature make that choice and do most of the work for us if only allowed. However, considerable investment to support change, and the livelihoods of farmers and rural communities, is needed to transition towards a better place for us, livestock and wildlife.

We are already getting an ecological reawakening from rewilding initiatives, such as the one at Knepp Estate, Sussex. Once an intensive dairy farm, set in challenging geography, the Knepp Estate, in little over fifteen years, has made a remarkable transformation by working with, rather than against, nature. The dairy cows are gone, replaced with a mix of free-roaming Old English longhorns, Exmoor ponies, Tamworth pigs, fallow deer, red deer and roe deer. This mix of herbivores and omnivores is more akin to those that once roamed the UK, each having a specific function in creating and maintaining an already, and increasingly, diverse ecosystem.

In her account Isabella Tree sums up the wildlife successes on the Knepp Estate:

- *2009. Five-year monitoring revealing breeding skylarks, woodlarks, jack snipe, ravens, redwings, fieldfares and lesser redpolls; thirteen out of the UK's seventeen bat species, and sixty invertebrate species of conservation importance including the rare purple emperor butterfly.*
- *2012. Thirty-four nightingale territories.*
- *2013. Presence of thirteen birds on the International Union for Conservation of Nature (IUCN) Red List and nineteen on the Amber List; and several extremely rare butterflies and plants.*
- *2014. Eleven male turtle doves recorded; first sightings of short-eared and long-eared owls.*
- *2015. UK's largest breeding population of purple emperor butterflies.*
- *2016. Sixty-two species of bee and thirty species of wasp, including seven bee and four wasp species of national conservation importance.*

The good news goes on, with many more animal and plant species not listed here and regularly being added to the list.

Before this transformation the Knepp Estate was like many dairy farms: lifeless green fields, fragmented with limited biodiversity and hostile to the land's full wildlife potential. At Knepp, the livestock they farm is now at least a functioning part of the land's ecology.

Before being rewilded, Knepp's original farm was not bereft of soil-related issues. The farm's soil was an easily compacted/poached heavy clay, which amongst other consequences of intensification, such as the overuse of chemical fertilisers and pesticides, would have contributed to water pollution. Today, the land's natural regeneration has greatly improved the quality of water that drains from the estate.

Knepp may seem an unrealistic model, but only if we choose to continue our unsustainable ways and diet.

Whether for wildlife, pollution reduction, dietary health, a better life for livestock or as a necessary response to the climate emergency, lifestyle change is needed. We waste vast amounts of our agricultural land's resource to feed livestock, when we can grow more calories on it to feed us directly on a mix of foods that are nutritionally superior to meat, fish, eggs and dairy and much better for our health. We would not need to farm as much land for our food, making it feasible to return a few million hectares back to nature.

19. Tackling Pollution

Having read much of what I have written so far, and given what is to follow, you would be forgiven for thinking, 'What are government agencies like the Environment Agency and Natural England, and their equivalents in the devolved administrations, doing about it all?' The truth is, an awful lot, which is only limited by the resources made available to regulate and advise farmers to prevent pollution. These government organisations and others work tirelessly and have achieved much to be proud of towards protecting and enhancing our shared environment, making it better for people and wildlife despite the limited resources given to them. With resource constraints and arms (and legs) often tied behind their backs, government agencies constantly have to innovate to do more with less, including working with many equally conscientious external people and bodies, but there are inevitable limitations. A lot of good work invariably goes unnoticed.

It is also important to highlight that the success of a regulator is not *solely* determined by the number of prosecutions taken, but by the outcomes delivered for people and the environment. Agencies may report high levels of non-compliance, particularly after new regulations are introduced, many of which can be dealt with most effectively using other enforcement mechanisms, reserving prosecution for the most serious offences and/or against those who refuse to comply with the law. Delivering visible outcomes can also take time, as it can involve considerable investment in infrastructure and difficulties in getting planning permission. Often interim betterment measures are necessary, until a farmer can come into full compliance.

Sometimes it is expressed that taking prosecutions is a failure of the regulator – if you like, a failure of regulatory advice to achieve compliance. Speaking for myself, I have investigated and led on many prosecutions during my career, including acting as an expert witness on several cases. If I ever led, or was involved in, a prosecution, it was never a failure, as the person or company causing the offence had left no other choice.

Unfortunately, prioritising resources also means a high degree of targeting, including what reported pollution incidents are responded to, which leaves parts of the UK under-supported, regardless of best endeavours. Ultimately, like animal welfare and healthy food, we get the environment we pay for. Have a moan, but remember: your voice is much stronger if you do not contribute to the problem through your choice of food.

Most pollution on farms is caused by livestock production, so we can all help make a big difference by changing what we eat. Eat the right balance of healthy whole plant-based foods and we can become healthier, and kinder too.

20. Dutch N Case

A recent court case taken in the Netherlands, known today as the 'Dutch N' case, challenged the country's approach to protecting Natura 2000 (N2K) areas from nitrogen pollution under the European Habitats Directive. N2K is the collective name for Special Areas of Conservation (SAC) and Special Protection Areas (SPAs), marine and terrestrial, protected under European legislation for their important wildlife habitats. That is, cornerstone breeding and nesting sites for rare and threatened species, including some rare natural habitat types. The Court of Justice of the European Union (joint cases C-293/17 and C-294/17) ruled in favour of the plaintiff and pronounced that the Dutch permitting system provided insufficient protection under the directive. The ruling had an immediate impact on thousands of business/housing developments and farming projects already in the planning system, whose nitrogen emissions, or other relevant emissions (e.g. phosphate and ammonia), had the potential to effect areas designated under the directive either directly or indirectly.

The Dutch N case is equally applicable in the UK, creating great uncertainty within the planning system for any planned development which may deteriorate a site designated under the Conservation of Habitats and Species Regulations 2017,[1] such as:

- Special Protection Areas (SPAs).
- Special Areas of Conservation (SACs).
- candidate SACs and Potential SPAs.
- associated Ramsar sites (wetlands of international importance).

Of particular concern is nutrient pollution from farms and municipal waste waters arising from households or industrial premises, underpinned by legislative principles such as the 'precautionary principle', which requires a high degree of scrutiny.

Local planning authorities in the UK have been advised by Natural England, the government's adviser for the natural environment, to take a precautionary approach when making decisions on planning applications in line with existing legislation and case law to address any uncertainties.[2] This includes the principle of achieving 'nutrient neutrality', described by Natural England as:

[...] a means of ensuring that development does not add to existing nutrient burdens and provides certainty that the whole of the scheme is deliverable in line with the requirements of the Conservation of Habitats and Species Regulations 2017 (as amended).

If negative impacts on designated sites cannot be ruled out completely, following a thorough and in-depth scientific examination, planning permission can be withheld.

Special areas affected by nutrient enrichment need to be brought back into a favourable condition too – for example, where phosphorus loss from agricultural practices is contributing to eutrophication. This ties in with an ambition, under the UK 25 Year Environment Plan, to achieve clean and plentiful water by improving at least three-quarters of waters close to their natural state as soon as is practicable.

Agricultural intensification, the expansion of an existing facility or a new 'mega' dairy, pig, poultry or beef farm, can seriously threaten nutrient neutrality or improvement and can be difficult to mitigate against – water and/or air pollution, particularly for farms close to a designated site. Even the construction of a new slurry store can come under scrutiny, if its planned size would, in particular, support/enable livestock intensification.

Expansion would not just be limited to herd size but could be hidden from the planning system if it results from a per-cow increase in milk yield – for instance, from the average 8,000 litres per year (2020) to 11,000 litres per year, adding an extra ten litres per cow per day of excreta. This would only show up if the increased yield involved the need for new infrastructure and associated planning permission. To feed those extra hungry cows would also require a larger area of erosion-ready maize. Any increase in milk yield inevitably increases the phosphate and nitrogen flow into an area, which would need to meet, not exceed, the needs of the soil and crop on the land.

The dairy industry is adept at increasing the amount of slurry it needs to capture and dispose of, and in concentrated areas of the UK, all too often at the expense of the environment. And if ever the livestock industry per se creates more organic manure than can be used without exceeding soil and crop need at any one time, and it's not economical for them to store and/or use/supply it beneficially elsewhere, including to avoid causing pollution, then the law is likely to be blamed, rather than the manure they need to dispose of.

Planning applications for ADs are another concern, not just for the extra nutrients they may bring into a location but also in relation to diffuse pollution arising from feedstock crops grown nearby (erosion-ready maize). AD plants are becoming increasingly common on larger farms, and those that are not may take livestock manures from local farms. Where an AD unit takes livestock manure, this in itself may enable livestock expansion back on a farm. Processing industries may also establish in areas where livestock sectors concentrate/intensify, whose own wastes add to the nutrient load.

Throughout this book I discuss how livestock farms contribute to pollution, and earlier I made reference to the River Axe SAC, which is at particular risk of eutrophication from phosphorus pollution (e.g. bioavailable particulate phosphorus) associated with the intensification of dairy farming. Another at risk of phosphorus pollution and the fallout from the Dutch N case is the River Wye and River Lugg SAC.

Case study

River Wye and River Lugg SAC and phosphate pollution

The River Wye and River Lugg catchment falls within Powys (Wales) and Herefordshire (England). At 25 km long and draining a catchment of around 4,136 km,[2] the River Wye/Afon Gwy is one of England and Wales' longest natural rivers. The habitats supported by the Rivers Wye and Lugg are extremely varied, with a rich, sometimes rare and unique, diversity of flora and fauna, for which the SAC, amongst other outstanding attributes, has been designated. Both rivers' natural nutrient regime is threatened by various activities and developments within the catchment, including phosphorus, nitrate-nitrogen, ammonium-nitrogen and airborne ammonia enrichment from livestock's longer shadow. Over the last ten years particular concern has centred on the exponential growth of intensive poultry farming within the catchment.

Reported in the *Hereford Times* by Carmelo Garcia (2020),[3] there are 'calls to stop the "uncontrolled spread" of poultry farms across Hereford'. A local campaign groups for the CPRE and CPRW (Campaign to Protect Rural England/Wales respectively) estimate that there are around 500 intensive poultry farms, containing 45 million bird places, across Herefordshire, Shropshire and Powys[4] (78 per cent broilers, 19 per cent free-range/fertile eggs and 3 per cent pullets across the three counties in 2019, with the highest percentages of broilers in Herefordshire and Shropshire[5]), which they link, with the exception of those not in the two catchments, to well-reported water quality issues in the River Wye and River Lugg SAC associated with phosphorus and nitrate pollution. According to CPRE branch chairman Bob Widdowson, there is an 'apparent uncontrolled spread' of poultry units in Herefordshire, which 'comes out near the top of the league for the number of [poultry] units.'[6]

The rise of poultry farms has been connected to Cargill's soya crush plant in Seaforth, Liverpool, and their presence and investment in poultry operations in Herefordshire, with associated livestock intensification blamed for the deterioration of water quality in the SAC. However, if the county's poultry farms are causing water pollution, then it will be a part of the problem rather than the single issue (i.e. along with non-livestock sources, such as sewage treatment/septic tank discharges).

The *Hereford Times* published a defensive statement[7] from the NFU, asserting that 'much is often made of the environmental impact of poultry units', whilst highlighting that the 'larger chicken units are regulated by the Environment Agency'. Furthermore, they added, our 'farmers are food processors but also take their environmental responsibilities seriously', and that the expansion of 'fresh poultry meat for the UK markets stave off our reliance on foreign imports and benefit the rural economy'.

It is true that the larger, most intensive poultry farms are regulated under an environmental permit by the Environment Agency (i.e. farms with more than 40,000 places for poultry). However, once manure leaves the premises, pollution interest/control wanes. Furthermore, the fact that up to 66 per cent of water pollution in the River Wye and River Lugg SAC is associated with agriculture (see below) strongly suggests the assertions about farmers, per se, taking their environmental responsibilities seriously in this case, as in many others around the UK, appears weak. This is hardly surprising, as when phosphate fertiliser accumulates on and in the soil the risk of losing it and polluting inland freshwaters increases, especially in areas with high livestock densities and high levels of manure production. What's more, the continued expansion of UK chicken production involves importing extra soya bean meal from foreign shores to feed them, which comes at the expense of further rainforest loss and ecocide (and human health/rights abuses) in South America (see Chapters IV and V).

On top of this is the extra, and very concerning, damage done to marine ecosystems to provide the chickens with fishmeal too. For instance, according to data compiled by Seafish, the recommended inclusion rate of fish meal and oil in diets for broilers is 2 to 5 per cent (2 per cent for layers).[8] A typical broiler eats around 3.2 kg of feed to reach its sale weight of 2.2 kg.[9] Going with 3 per cent, in the UK this would equate to around 100,000 tonnes of fish meal and oil, or around 90 g per bird equivalent. Breaking this down into the number of individual fish consumed per bird, based on a mean weight of 18 to 41 g per fish caught,[10] it works out at two to five fish per broiler. Not only are we unkind to over one billion broilers but to a few billion sentient fish to feed them. For further perspective, a medium Christmas turkey may eat over eight fish.

One large permitted farm in Hereford houses 168,000 chickens in four large sheds[11] (200,000 bird place capacity). Some farms have less and others more, and I use this farm purely as an example. With around 7.7 chicken crops a year, the farm currently supplies around 1.3 million meat chickens for the market (1.54 million at full capacity).

These chickens excrete a lot of nitrogen and phosphate (meat chickens 1.06 grams of nitrogen and 0.72 grams of phosphate a day per animal, and free-range chickens 1.5 grams and 1.1 grams respectively). The 1.3 million birds produce about 55 tonnes of nitrogen and 37.5 tonnes of phosphate per year. From a total of 45 million birds across the three counties, the amount of nitrogen and phosphate produced per year is approximately 16,000 and 11,200 tonnes respectively. My calculations assume 78 per cent are broilers, with a conservative 7.5 x 40-day crops per year (263 million birds x 40); 19 per cent are free-range layers (8.5 million birds); and 3 per cent are pullets (1.35 million birds multiplied by three crops per year). At least 50 per cent of the nitrogen and 60 per cent of phosphate in neat chicken excreta will be in a crop-available form. This makes them excellent organic manures, but if used inappropriately (i.e. in excess of the soil and crop need and/or at the wrong time of year/conditions), they become an excellent source of water and air pollution too.

In some parts of the UK poultry manure is increasingly being used to produce energy in biomass energy plants, in a process which removes the carbon and nitrogen but leaves the phosphate. This is then returned to land as a fertiliser. However, in Herefordshire and Shropshire a significant amount of poultry manure is processed through on- or off-farm ADs and then most likely spread back to land as non-waste – essentially beyond the controls of further direct environmental regulation. Total AD digestate retains the phosphate from poultry manure and contains additional phosphate from other feedstock needed for effective digestion. This may include a variety of food wastes, including dairy processing wastes, and erosion-ready maize grown specifically for effective AD. Erosion-ready maize may then receive the digestate in return. Whatever way poultry manure is treated, it ends up back on the land, normally relatively near to the place of production.

Another addition to the phosphate load is blood. Each 2.2 kg broiler contains about 0.165 kg of blood per bird (about 7.5 per cent of body weight), containing about 1.5 kg phosphate per tonne (15 kg nitrogen per tonne). This amounts to 65,000 kg of phosphate for 263 million broilers. About half of the blood drains out of the carcass when a bird's throat is cut, which is likely to be recovered, rendered and turned into fertiliser; body parts removed for food, or otherwise rendered, will contain blood; and a smaller amount of other blood will end up in waste water/sludges, along with other fluids, cleaning agents and excreta, which will likely be recovered to agricultural land, either directly or following treatment. To this you need to add blood in waste water/sludge from other slaughterhouses, which may also be spread to land in the three counties/SAC catchment.

The total nitrogen content is lower in slaughterhouse waste water/sludge than found in neat blood (c. 0.5 to 0.6 kg per tonne), but the phosphate content per kg can be similar to blood, or higher for treated sludge (e.g. dissolved air flotation – DAF) at between 2–3 kg per tonne. The amount of waste spread to land from the bigger 24/7 plants can be more than 10,000 tonnes per year (maybe up to 30,000 tonnes per year for the largest plants), which at 2–3 kg per tonne adds more phosphate. For the three counties, add around an extra 1,000 tonnes of phosphate for slaughtered poultry. Storage is also often very limited, meaning it is habitually collected and spread on agricultural land at times of high leaching potential, preferential drainage and run-off risk. The amount of phosphate, whilst far less than poultry manure, becomes even much more significant if applied locally.

The majority of soils in the River Wye and River Lugg SAC have a soil phosphorus index of 3 or more,[12] with little or no soil and crop need for phosphate fertiliser – certainly no more phosphate than a crop can use (e.g. 60 kg per year), which would otherwise increase the soil phosphorus index and with it the croplands pollution potential. For example, the land in the lower River Wye and River Lugg is also classified as having a medium erosion risk.[13] A fertile soil, an oversupply of phosphate from chicken manure, and other livestock (see below), combined with an erosion risk, is a recipe for sediment and nutrient loss, and with it water pollution. Free-range chickens are also associated with causing soil compaction and accompanying run-off.

Herefordshire, Shropshire and Powys are also home to a lot of dairy, beef, pigs and sheep, adding thousands more tonnes of nitrogen and phosphate (handled/spread or deposited during grazing). Combined with chickens, the livestock in the area collectively excretes in the order of 35,000 tonnes of nitrogen and 18,800 tonnes of phosphate. My overall estimate, which includes around 100,000 dairy cows from 650 farms,[14] is conservative; it does not include all livestock (e.g. goats and game birds), sewage biosolids or other industrial wastes recovered to agricultural land, which would add significantly to the overall phosphate load. While poultry are not the only source, unless manure is exported out of the catchment then it is a major, and growing, contributor.

Based on 18,800 tonnes of total phosphate, for soil at index 3 or more and using a generous average crop phosphate offtake of 60 kg per hectare, you would need around 313,000 hectares of farmland to spread on. This would retain the nutrient status quo, but in accordance with good agricultural practice a high soil phosphorus index should be reduced, by restricting applications to no more than 30 kg *total* phosphate, which would double the land bank needed (c. 626,000 hectares). Ultimately though, in most circumstances, when a soil is at index 3 or more there is no soil and crop need. You should also not add unneeded phosphate if there is a significant risk of diffuse pollution. Throwing petrol on to a fire is not helpful. For context, 626,000 hectares is over half the size of Herefordshire, Shropshire and Powys combined. Furthermore, about three-quarters of the poultry manure is produced in Herefordshire and Shropshire, with a combined agricultural land bank of around 442,000 hectares, of which around 155,000 hectares is in cereal/general cropping (c. 231,000 hectares in permanent/temporary grass).[15] Most broiler manure would be directed to the smaller cereal/general cropping/horticultural land area, within an economic distance of the poultry farms.

If you add in all the other phosphate likely to be recovered to land in the counties, the landbank required becomes much greater. What's more, distribution/sharing of organic manure around the counties would not be even, but local to the points of production – concentrating the pollution potential.

More specifically, a typical application of poultry manure is around eight tonnes per hectare, which would supply 240 kg of total nitrogen and 200 kg of total phosphate to each hectare spread – over three times more phosphate than an eight-tonne cereal crop yield needs. If spread on the same land each year, the unused phosphate would increase the soil phosphorus index and raise the potential pollution risk. For example, on a medium/heavy soil, with a crop offtake of 60 kg phosphate, applying eight tonnes of chicken manure a year would increase the soil phosphorus index from 3 to 4 within six to seven years (and from index 2 to 3 within four to five years).[16] Furthermore, if spread in the autumn on a clay loam, for example, before a winter cereal, a time when there is also no need for nitrogen, around 28 kg of nitrate-nitrogen would be lost to water per hectare spread. Much more nitrogen would be leached on a lighter soil.

Where spreading exceeds the needs of the soil and crop, or it results in a significant risk of diffuse pollution (significance of risk, not significance of pollution), it would breach The Reduction and Prevention of Agricultural Diffuse Pollution (England) Regulations 2018. Farmers accepting and applying poultry manure need to be aware of these regulations, as do agronomists, farm advisers and local planning authorities. Ultimately though, as the land manager, the burden is on the farmer not to breach the regulations.

Taking all of the above into account, it is of little wonder that between 61 and 66 per cent of the phosphorus pollution source in the River Wye and River Lugg catchment is attributable to agriculture.[17] CPRE Herefordshire[18] report that the 'growth in the industry has coincided with a rise in phosphorus levels in the River Wye SAC and the River Lugg'. The Brecon and Radnor Branch[19] also report that in 2010, 'before [an] explosion in the number of chicken sheds across Powys, the Wye was one of the cleanest rivers in the UK', and now in 2020 'turns a putrid green'. BRB CPRW further describe that since November 2016 alone the number of chicken places in Powys has grown from just over 5 million birds to just under 10 million.[20]

The 66 per cent of pollution attributed to agriculture, as already alluded, will not be down to just chickens, but they are a likely contributing factor. Previously I've already discussed how dairy and erosion-ready maize can contribute to phosphorus pollution, along with other types of livestock farming, and above that there will be other sources of phosphate applied to land here too. An inability to apply these manures at a time when there is a genuine soil and crop need, or without giving rise to diffuse pollution, is a growing problem. The inability for soil to hold on to nutrients due to poor soil management and associated deteriorating soil health on farms (dirt farming) is well documented in Herefordshire and compounds the problem.

For further context, soils with a high clay content have a high buffering capacity and as a result the concentration of phosphorus found in solution is low. Conversely, soils with a low clay content have a lower buffering capacity and a higher phosphorus concentration in solution. The predominant soil types in the catchments are well-drained sandy, silty and medium loamy soils, fitting within the low buffering capacity bracket, and as such are at a higher risk of contributing to diffuse pollution through subsurface phosphorus leakage when it rains sufficiently. Where phosphate-rich manure like poultry litter is spread in excess of crop offtake, phosphorus accumulates, increasing this risk. Generally, though, a better rule of thumb, and fairly widely accepted, is that soils will leak phosphorus at a faster rate once 25 per cent of the soil's maximum phosphorus sorption capacity is saturated.[21] Notwithstanding that, most of the phosphorus stress on water quality comes from particulate phosphorus attached to soil and organic manure particles washed off/from the soil surface during significant rainfall events. If, as appears likely, soil phosphorus fertility is increasing within the catchments, it paints a worrying picture for the future ecology of the SAC unless urgently addressed.

Ultimately, the amount of phosphorus loss is governed by a host of factors. These include, but are not limited to: the soil's clay content, pH, structure and the level of phosphorus present in the soil; rainfall and intensity; slope and slope complexity; the amount, type, frequency and characteristics of the organic manure applied, including the proportion of water-soluble phosphate in the manure (for poultry manure it is high); and how the land has been, and is being, managed. Keeping the soil Olsen phosphorus level close to the agronomic optimum ('critical' crop need) is the best course of action to avoid a significant risk of pollution.

Accruing phosphorus surpluses on farmland, i.e. when there is no soil and crop need, is poor practice, contributes to water pollution and is a waste of a finite resource. The risk is greatest on farms, and/or within areas, where livestock farming concentrates/intensifies. Often these farms do not have enough land to grow all their own feed, and therefore need to import additional feed like soya bean meal. Imported feed contains phosphorus, which livestock uses inefficiently. Unused phosphorus ends up in the livestock's manure, and if the farm has an insufficient land bank to spread on, this results in more being applied than a crop can use, which in turn raises soil phosphorus levels. Farms which produce their own livestock feed often have more land to return manure to, but may still accrue phosphorus over time, just at a slower rate.

Past action to address river water quality has been to provide a 'route for development to be able to proceed in the catchments, even when it may add to the existing phosphorus levels in the river, as increases would [apparently] be mitigated by the River Wye's Nutrient Management Plan (NMP).'[22] The NMP is a partnership initiative to reduce phosphorus levels to below the target level by 2027, in line with Water Framework Directive requirements. However, this way of managing phosphate, or any other nutrient affecting the SAC, or any other SAC for that matter, now looks far from satisfactory given the Dutch N case. Put very simply, there is too much chicken manure and an associated, and growing, phosphorus soil legacy, to achieve nutrient neutrality or to bring the two rivers' water quality into a much more favourable condition. Yet the planning applications for new poultry farms and extra units in the River Wye and River Lugg catchment still come in and are approved. One can only wonder what actual environmental evidence, beyond growing the local economy, the local planning authorities have been looking at when making determinations.

Actions taken to date are particularly challenged when it comes to intensive livestock farming, which shoehorns the environment/sustainability into livestock farming rather than vice versa. Continually growing the livestock problem makes it more difficult to mitigate against pollution, and to a point where solving it can become unfeasible against cost and logistics. It seems this is already the situation in the River Wye and River Lugg catchment unless there is a drastic reversal of the cause – too much livestock and associated maize – or a change of practice (e.g. incineration of poultry manure for energy, or exporting excess nutrients to places where needed). The problem is amplified by an associated increased risk from ammonia-related emissions across the catchment (CPRE&W, 2019).[23]

Phosphorus, and in particular particulate phosphorus, is only one nutrient polluting water and harming the SAC; nitrogen and antibiotic pollution in faeces are major contenders too. This in turn impacts on other developments, including meeting housing needs. Whilst we may need housing, we do not need poultry farms and the pollution which comes with it. There are options, such as pelletising dried poultry manure, reducing its bulk by two-thirds and selling it as a fertiliser product for use on land that needs it. This avoids spreading in the autumn, when the nitrate-N leaching risk is high, or spreading in the spring when soil conditions may be unsuitable and there is a medium risk of causing phosphorus pollution.

The River Wye and River Lugg SAC is far from alone; the soil phosphorus index and nitrogen problem is growing wherever livestock is intensifying and concentrating, in particular cows, pigs and poultry.[24]

In completing this part, the water quality of another Site of Special Scientific Interest, the Somerset Levels and Moors (SLMs), has been found in an 'unfavourable declining' condition, primarily due to high phosphorus levels.[25] The pollution sources stated are the usual suspects: agricultural activities and water industry discharges. Historically, the most notable source was effluent from waste water treatment works. Today the situation has reversed, with nearly half of all phosphorus emissions in SLMs now down to livestock (around 75 percent including arable farming). Whilst the responsible water company, Wessex Water, announced an investment of a further £57 million to reduce pollution from their discharges, the agricultural industry is to be managed through regulatory enforcement and advisory visits to improve compliance. Unless the source of the cause is addressed – largely intensive livestock farming, in particular dairy, and the market failures which limit the adoption of practices which would restrict water pollution – I suggest, from experience, there is limited hope in reversing the situation any time soon without really drastic action.

Agricultural lobbyists have long pointed the finger at water companies, and away from themselves, when it comes to water pollution from nitrogen and phosphorus losses. As the water companies have invested to clean up their act, and despite having someway still to go, in catchments like the River Wye and River Lugg SAC, SLMs, the River Axe SAC and many other N2K areas, agriculture, and in particular livestock farming, is now exposed as the predominant polluter. Livestock farming is also a key cause of pollution by microbial pathogens, such as faecal coliforms, Escherichia coli and enterococci. At first it may seem odd that water quality is declining given water company investment, but not when you consider how soil phosphorus at risk of loss is accruing unsustainably as livestock farming relentlessly concentrates/intensifies.

In 2021 Defra produced soil nutrient balances for the regions of England for the first time.[26] The figures produced are a helpful start for estimating the annual nutrient loading for nitrogen and phosphorus to agricultural soil. However, it would be better to have them at least at a county level, ideally at a catchment level, to be more meaningful. The nitrogen and phosphorus balance ranges from +34.0 kg/ha and -4.8 kg/ha respectively in the arable East of England, and +111.4 kg/ha and +10.2 kg/ha respectively in the North West, where livestock farming predominates. For the West Midlands, which includes Herefordshire, the balance is +107.2 kg/ha for nitrogen and +8.2 kg/ha for phosphorus. Essentially, wherever livestock predominates, livestock farming causes a nutrient surplus, which is bad news for water ecology (nitrate and phosphorus levels), air quality (ammonia emissions) and climate change (nitrous oxide emissions). The surpluses in Herefordshire would be expected to be higher still, as livestock is present in higher numbers than in other parts of the West Midlands, Shropshire excluded, and again where livestock is concentrated with the county. My own estimate for Herefordshire is an average 13 kg/ha phosphorus surplus. I dread to think what phosphorus surpluses may be accruing in parts of the county.

The annual cost of agricultural water pollution has been estimated at between £0.75 and 1.3 billion,[27] and it would cost multi billions to bring water quality to an environmentally optimal level. Due to the market's failure to address the negative externalities behind livestock-related water pollution, and the cost of other polluting emissions such as ammonia and carbon, also associated with livestock farming, the sector and market looks and waits for government intervention. This is an incredible cost given British farmers only produce around £9 billion in gross domestic product (GDP) output per year, of which subsidies are worth around half of that output.[28] Incentives and voluntary incentives have been successive governments' preferred approach rather than regulation, but these have continuously failed to deliver the change needed to tackle pollution in the free market economy. Instead, it is growing. Stronger regulatory intervention is necessary, and individuals can rise to the challenge too by stopping the livestock pollution they are not paying for by adopting a whole-food plant-based diet. Add in the huge cost to health services from eating animal-based foods, and what livestock have to suffer to feed us their flesh and bodily fluids, and it is a no-brainer.

Farmers are slowly waking up to the consequences of The Reduction and Prevention of Agricultural Diffuse Pollution (England) Regulations 2018. When soil phosphorus levels and nitrogen inputs are high, there is no soil and crop need and the nutrients disposed of present a serious environmental risk. Business-as-usual and pollution-as-usual are not an option under the regulations, and nor is continually making the problem worse by further concentrating/intensifying livestock farming. The regulations are not something new; not exceeding soil and crop need has been a part of the codes of good agricultural practice, which livestock farming should have been adhering to since 1985.

Excess nutrients could be exported, albeit at a very high cost which currently isn't commercially viable, to arable areas where there is a soil and crop need. However, livestock sector eyes are more focussed on expanding its negative environmental externalities back into arable areas (e.g. to create a 'modern' version of the mixed farming systems of yesteryear), rather than relocating its excess nutrients, marketing such shortages in arable areas, along with soil organic carbon, as a livestock deficiency problem, which it isn't, to anyone unwise enough to see or who wants to be otherwise intelligent about. If livestock expands into arable areas, even more inefficient land use would follow to feed them, rather than feed people directly.

Citations:

1. Natural England (2020), 'Matters regarding development in relation to the Somerset Levels and Moors Ramsar Site' [Letter], 17 August 2020

2. Natural England (June 2020), *Advice on Achieving Nutrient Neutrality for New Developments in the Solent*

3. Garcia, C., 'Calls to stop "uncontrolled spread" of poultry farms across Herefordshire.' *Hereford Times*, 23 November 2020. https://www.herefordtimes.com/news/18892402.calls-stop-uncontrolled-spread-poultry-units-across-herefordshire/

4. Ibid.

5. Brecon & Radnor Branch of CPRW (2020), *Powys, Poultry and Poo*. https://www.brecon-and-radnor-cprw.wales/?page_id=2026

6. Garcia, C., 'Calls to stop "uncontrolled spread" of poultry farms across Herefordshire.' *Hereford Times*, 23 November 2020. https://www.herefordtimes.com/news/18892402.calls-stop-uncontrolled-spread-poultry-units-across-herefordshire/

7. Ibid.

8. Green, K. (2016), *Fishmeal and fish oil facts and figures December 2016.* Seafish.org

9. Redman, G. (2021), *The John Nix Pocketbook for Farm Management*, 51st Edition. Agro Business Consultants, Melton Mowbray, UK

10. Fishcount.org.uk (2019), *Numbers of fish used for feed in aquaculture.* http://fishcount.org.uk/farmed-fish-welfare/numbers-of-fish-used-for-feed-in-aquaculture

11. Wasley, A. and Davies, M. (2017), 'The Rise of the "Megafarm": How British Meat is Made. *Bureau of Investigative Journalism,* 17 July 2017. https://www.thebureauinvestigates.com/stories/2017-07-17/megafarms-uk-intensive-farming-meat

12. UK Centre for Ecology & Hydrology. SO. SoilTheme.OlsenPhophorus. https://catalogue.ceh.ac.uk/maps#layers/6372b558-ba64-4fbe-8766-019e0 1535b37. Main source: Emmett, B.A. et al. (2016), *Soil physico-chemical properties 2007* [Countryside Survey], NERC Environmental Information Data Centre, UK Centre for Ecology & Hydrology

13. Environment Agency & Natural England (2014), *River Wye SAC Nutrient Management Plan Evidence base and options appraisal*

14. Food Standards Agency (2018), *Registered Dairy Establishments: A list of dairy producers registered with the Food Standards Agency.* https://data.food.gov.uk/catalog/datasets/4025207b-f03b-480b-869f-11860 87dd42a

15. Defra (2016 data), *Structure of the agriculture industry in England and the UK at June (county and unitary authority: 1905 to 2017)* https://www.gov.uk/government/statistical-data-sets/structure-of-the-agricultural-industry-in-england-and-the-uk-at-june

16. PDA, *Phosphate and Potash deficiency correction and nutrient offtake calculator.* https://www.pda.org.uk/calculator/pkcalculator.html

17. Nutrient Management Board Meeting (2020), *Nutrient Management Plan Technical Advisory Group Summary Paper on Updating the NMP Action Plan.* https://www.herefordshire.gov.uk/downloads/file/20953/nutrient-management-plan-board-agenda-and-papers-july-2020

18. CPRE (2020), *Intensive Poultry Units.* https://www.cpreherefordshire.org.uk/issues-were-dealing-with/farming-in-herefordshire/intensive-livestock-units/

19. Brecon & Radnor Branch of CPRW (2020), *Powys, Poultry and Poo.* https://www.brecon-and-radnor-cprw.wales/?page_id=2026

20. Ibid.

21. Withers, P.J.A. (2021), Personal Communication.

22. Nutrient Management Board Meeting (2020), *Nutrient Management Plan Technical Advisory Group Summary Paper on Updating the NMP Action Plan*. https://www.herefordshire.gov.uk/downloads/file/20953/nutrient-management-plan-board-agenda-and-papers-july-2020

23. CPRE/CPRW (2019), *Agriculture Ammonia Emissions: Herefordshire, Shropshire & Powys* [Map]. https://www.brecon-and-radnor-cprw.wales/wp-content/uploads/2019/07/IPU-ALLdataV4-Master-20190707-3-Counties-By-Size-Ammonia-FINAL-1.0-20190711.pdf

24. Defra, *Maps of livestock populations in 2000 and 2010 across England*. https://assets.publishing.service.gov.uk/government/uploads/system/uploads/attachment_data/file/183109/defra-stats-foodfarm-landuselivestock-june-detailedresults-livestockmaps111125.pdf

25. Ends Report (2021), Nutrient crisis: SSSI downgraded to 'unfavourable declining' due to water quality.

26. Defra (2021), Soil Nutrient Balances Regional Estimates for England, 2019 (Provisional)

27. Helm, D. (2020), Net Zero: How we Stop Causing Climate Change. William Collins, London.

28. Ibid.

21. Fleeced

Sheep farming in the UK is nothing new. In 1868, in Great Britain (England, Scotland and Wales), the sheep population stood at a massive 30.7 million. This was due to a boom era for sheep farming between 1837 and 1874, when prices for meat and wool were high.[1] The population stayed high until World War I. The consequence of war and the ensuing recession in the 1930s saw a sharp drop to 19.7 million, before numbers recovered to 26.9 million by 1939. World War II led to another decline, dropping back to 19.7 million by 1945 and to a new low of 16.2 million by 1947 (UK, 16.7 million).[2]

The heady days of the 1868 sheep population was not seen again until 1980, when the population stood at 30.4 million for Great Britain and 31.4 million for the UK. Thereafter the sheep population rose steeply, under the influence of the Common Agricultural Policy (CAP) sheep meat 'headage payment' scheme, which incentivised farmers to stock more sheep per acre of land to maximise income. By 1988 the UK sheep number rose to 41 million,[3] eventually peaking at 44.5 million[4] in 1992. The scheme also encouraged new entrants, perhaps more committed to money than sheep farming.

Most of the growth in sheep farming occurred in England and Wales, and in terms of stocking density most of the increase happened on farms with the largest flock sizes.

Table 20: Growth in sheep farming (England and Wales)

No. of sheep	1975	1986
500–699	4,451	5,251
700–999	3,308	4,638
1000–1499	2,319	3,725
1500–1999	898	1,616
2000 and over	776	1,673

Source: Marks, H. F. (1989), A Hundred Years of British Food & Farming: A Statistical Survey, edited by D. K. Britton. Taylor & Francis, London, UK

2001 saw the last substantial drop in sheep numbers, as a consequence of the foot-and-mouth disease outbreak, dropping from 19.14 million in 2000 to 15.40 million by 2002 in England alone.

The sizeable increase in sheep numbers and associated stocking density, falling out of CAP, was further bad news for wildlife, resulting in considerably more overgrazing and habitat impoverishment.

According to the National Sheep Association (NSA):

[...] sheep support wildlife and plant biodiversity. Without sheep our grasslands, and uplands in particular, would become overtaken by scrub, and coarse vegetation, becoming less valuable to many types of plants, small mammals and ground nesting birds [...] sheep have created and continue to maintain our iconic landscapes.

Around 60% of UK farmland is only suitable to grow grass [...] Land that can only grow grass is rarely (if ever) cultivated, meaning carbon is locked into the ground and not released into the atmosphere to contribute to climate change issues[...] This land also holds and filters water

In reality sheep farming creates an ecologically dire, highly manicured landscape, which resists its full potential. Historically the land would have been adorned in a wild blanket of varied and stratified vegetation, steeped in, rather than fragmented by, the limited biodiversity we have today. While grazing livestock at the right stocking density does help enhance some grassland species diversity, it only does it in relation to the grazing system and maintenance of the economic status quo.[5] The monopoly of sheep grazing creates a landscape of limited biodiversity. According to Hopkins and Tallowin (2002)[6] over 95 per cent of semi-natural grasslands no longer have any significant wildlife conservation interest.

A lack of wildlife and plant biodiversity, in uplands in particular, is not caused by a sheep deficiency.

Sheep damage biodiversity through grazing land very tightly and through having a selective taste for flowering plants, and other 'unwanted' vegetation, which would otherwise 'spoil' the most productive pasture. Overgrazing is particularly hard on plant species diversity, which subsequently causes a further decline in the invertebrate species reliant on them. Biodiversity has to be removed from swaths of land in order to provide sheep with nutritious, young and soft, tasty and palatable vegetation the industry needs them to eat to return a profit.

Soil enrichment through fertilisation, applied in certain situations to fuel sufficient grass growth to feed the extra hungry sheep, has an additional negative impact on plants whose very survival depends on impoverished soil conditions.

Creating and maintaining a suitable landscape and vegetation for sheep in the uplands can also involve land drainage and regular burning. Drainage and burning may also be carried out to create a favourable fragmented habitat needed for red grouse to support recreational shooting.[7] Red grouse require a mixed, fine-scale mosaic of different aged vegetation and density to provide nesting cover and food resource.[8] The two systems may be worked together for mutual benefit in areas of the UK.

Where burning is used to create livestock habitat, it is encouraged on small land strips of around two hectares at a time, ideally over ten- to twenty-five-year rotations,[9] and is often combined with intensive predator control (e.g. crows, ravens, stoats, weasels, foxes and raptors).[10] If the wrong kind of flora and fauna are present, livestock farming has an 'unfortunate' solution!

From Douglas et al. (2015),[11] there is increasing evidence that burning has an environmentally negative impact, including:

- soil erosion, as it creates a bare surface exposed to rainfall until vegetation recolonises.
- alteration of soil processes, including nutrient cycling and soil hydrology.
- deterioration of water quality (e.g. discolouration, lowering pH and contamination of particulate and dissolved organic carbon from peat soils, which in turn can have a negative impact on aquatic life and drinking water quality).
- air pollution.
- habitat condition and biodiversity impacts on vegetation, invertebrates and bird communities.

Whilst burning may sometimes be necessary to control wildfires, it is largely carried out to support sheep grazing, grouse shooting and in places deer stalking. Wildfires can occur naturally in the UK but are rare – once every 200–300 years over peat soils, compared to 10–30 years for managed burning.

Burning established vegetation returns carbon into the atmosphere. It can be particularly harmful in terms of hydrology and peat formation when done over peat bog soils, which causes long-term carbon loss.

Upland sheep farming and the practice of burning is defended by the livestock industry as a way to stop the land 'becoming less valuable to many types of plants, small mammals and ground-nesting birds', and we hardly raise an eyebrow. When we see Brazil's highly diverse Cerrado go up in smoke then be grazed by an army of cattle or turned into a vast monoculture of soya beans, we are alarmed and outraged.

As George Monbiot puts it:[12]

- *Why is it that practices we recognise as destructive when we see them elsewhere in the world are judged 'environmentally friendly' here?*
- *When we see it in Indonesia or Brazil, do we call it conservation, or do we call it destruction?*
- *This burning has sod all to do with protecting the natural world and everything to do with extracting as much grazing from the land as possible.*
- *As for 'overgrown' heathland, 'clearing the ground of dead vegetation' and 'extra growth of plants such as gorse and*

bracken', these are classic examples of the mortal fear of natural processes entertained by conservation bodies in this country [...]. Scrubs grows and then, God help us, trees. Wildlife is returning: quick, fetch the matches!

Land clearance by burning to support sheep grazing is banned between 1 March and 31 August to protect nesting birds, other fauna and flora. However, in Ireland during 2019 there were 409 suspected illegal fires, with conservation support payment withheld in eighty cases.[13] With profits linked to the grazing land area, there is a temptation to maximise earnings through illegal activity, not least as it is notoriously difficult to secure convictions in such cases. Whilst I could not suggest all these fires were deliberate, actual wildfires are not common in the UK. Most fires are connected with authorised, accidental or illegal activity, including arson. In at least eighty of these cases, there was enough evidence to deny payments to the respective landowners.

We interpret our 'iconic' landscape dead zones and what biodiversity they contain in a way that we want to believe, and we convince ourselves, or let ourselves be convinced, that it is ecologically and environmentally friendly.

Sheep are not a simple part of the landscape and its ecology; they are the architects of dysfunctional habitats. When you create such imbalance in an ecosystem, in the name of commerce, the negative consequences derived from it are condoned, and, once money and traditions are at stake, they are defended in spite of our intelligence. Conservation is made to fit around sheep, deerstalking and red grouse hunting, when, at the very least, it should be the other way around. The situation we have created is a delicate one. Unless we are willing or able to acknowledge that a problem exists, the problem will be exceedingly difficult to solve and will persist.

At face value, farming sheep on land on which it would be difficult to grow plant-based foods would appear a good use of the UK's resources, but it has come at a considerable cost to biodiversity, and for very little return in terms of calories and nutrients for the effort and expense involved in sheep farming.

Sheep farming has also contributed to the lowering of the UK's soil carbon stock, through land-use change and land degradation, and in particular in the uplands through overstocking.[14] Sheep are the main cause of organic soil degradation, followed by cattle.[15] Soil with less organic matter holds less water and has less ability to purify water that drains through it.

The relentless trampling of sheep amplifies soil compaction and instability, resulting in an increased risk of surface run-off/erosion and downstream flooding and pollution. Soil compaction not only contributes to flood risk, but also increases the generation of nitrous oxide, a greenhouse gas about 300 times more potent than CO_2.

Another source of pasture-based nitrous oxide emission is sheep urine, which is rich in urinary nitrogen.[16] Whilst I would not want to overexaggerate the amount of extra nitrous oxide released as a consequence of persistent animal treading and urine patches, it nevertheless contributes to the overall carbon footprint of sheep.

Other examples of sheep damage and associated pollution include:

- soil erosion from land used to graze sheep on fodder crops and stubbles.
- soil loss from severely poached stock feed and watering areas, including supplementary feeding areas.
- the erosion of banks of watercourses, where sheep are permitted access and congregate to drink, and where they also defecate in the water.
- run-off from farm trackways.

Sheep are also susceptible to infestation by a number of ectoparasites such as Psorotes ovis and Sarcotes scabiei which cause sheep scab, and for animal health and economic reasons, sheep need to be treated with chemicals, usually using injectables (e.g. Ivermectin and Moxidectin) or by immersion in a diazinon, organophosphate solution (sheep dip). Sheep dip is highly toxic and without great care can pollute groundwater, surface waters and wetlands. The use of sheep dip is also a health hazard to farmers and farmworkers.

Overall, lamb, with beef, is the most environmentally impactful animal production system per unit of volume.[17] This unwanted reputation is a measure of a number of environmental impacts, and includes:

- global warming potential, 100 tCO_2.
- biodiversity ecological footprint per kg consumed.
- eutrophication (nitrate, phosphorus and ammonia).
- land use hectares per tonne.
- waste (volume) tonnes of avoidable meat waste.
- acidification (e.g. urine ammonia acidification).
- abiotic resource depletion (non-replaceable resources).

As a share of overall UK consumption by volume/weighted impact, sheep has the lowest overall impact (2 per cent), with milk/dairy at 55 per cent and eggs at 24 per cent.[18]

Citations:

1. Urquhart, J. (1983), *Animals on the Farm: Their history from the earliest times to the present day*. Macdonald, London, UK
2. Marks, H. F. (1989), *A Hundred Years of British Food & Farming: A Statistical Survey*, edited by D. K. Britton. Taylor & Francis, London, UK
3. Ibid.
4. House of Commons Library (2016), *Agriculture: Historical statistics.* Briefing Paper number 03339, 21 January 2016
5. Tara Garnett (2008), *Cooking up a Storm: Food, greenhouse gas emissions and our changing climate.* Food Climate Research Network, Surrey, UK
6. Hopkins, A. and Tallowin, J. (2002), 'Enhancing the biodiversity and landscape value of grasslands.' *IGER Innovations* 6, pp. 30-33
7. Moorland Association (2006), via https://www.moorlandassociation.org/
8. Douglas, D. J. T. et al. (2015), 'Vegetation burning for game management in the UK uplands is increasing and overlaps spatially with soil carbon and protected areas.' *Biological Conservation* 191, pp. 243-250
9. Defra (2007), *The Heather and Grass Burning Code, 2007 version*
10. Brooks, S. (2014), 'The burning issue.' *John Muir Trust Journal* 57, pp. 28-9
11. Douglas, D. J. T. et al. (2015), 'Vegetation burning for game management in the UK uplands is increasing and overlaps spatially with soil carbon and protected areas.' *Biological Conservation* 191, pp. 243-250
12. Monbiot. G. 'Meet the conservationists who believe that burning is good for wildlife', *The Guardian*, 14 Jan 2016 https://www.theguardian.com/environment/georgemonbiot/2016/jan/14/swaling-is-causing-an-environmental-disaster-on-britains-moors
13. O'Doherty, C. (2020), 'Farmers investigated over 400 suspected illegal fires last year.' *Farming Independent*, 13 April 2020. https://www.independent.ie/business/farming/forestry-enviro/environment/farmers-investigated-over-400-suspected-illegal-fires-last-year-39123397.html
14. Royal Commission on Environmental Pollution (1996), Royal Commission on Environmental Pollution: nineteenth report: sustainable use of soil. HMSO, London, UK
15. Tara Garnett (2008), *Cooking up a Storm: Food, greenhouse gas emissions and our changing climate.* Food Climate Research Network, Surrey, UK
16. Ciganda, V. et al. (2019), 'Soil nitrous oxide emissions from grassland: Potential inhibitor effect of hippuric acid.' *J. Plant Nutr. And Soil Sci.* 82(1), pp. 40-47

17. Kleanthous, A. (2009), *Pigs and the Environment: A report to BPEX.*
https://www.pigprogress.net/PageFiles/24130/001_boerderij-download-PP
5977D01.pdf

18. Ibid.

22. Holy Cow!

Today's cattle descend from the aurochs, a large, fierce bovine which roamed ancient open forest and tall scrubland. It is thought the auroch arrived in Britain long before the loss of the land crossing with Continental Europe, and before the arrival of Neolithic settlers with their domesticated cattle.

In 1866 there were 4.786 million cattle in Great Britain and by the end of the nineteenth century 6.622 million. This level of cattle is not unique. Late-seventeenth-century statistician Gregory King, in his Statistical Accounts of England and Wales, estimated a 1696 population of 4.5 million. What would have been different is the landscape they lived in compared to modern-day farming practices. The late nineteenth century also saw the first introduction of the Friesian, popular for its milk yield and for its offspring's ability to fatten and produce a better carcass than other dairy breeds used at the time. By World War II the Friesian's domination of the milk market had taken root, becoming the principal milking cow. Today the majority of dairy cows are Holstein-Friesians, due to the interdependency of beef raising with dairy cows, aided by hybrid vigour.

By the early 1970s the cattle population in Great Britain had doubled, peaking at 13.582 million in 1974 (UK, 15.202 million), before steadily declining to below 11 million in 1987 (UK, 12.518 million). The number of dairy cows in the UK by 1987 was 3.042 million.[1]

From the mid-1990s the UK dairy herd began to decline. In England it fell from just under 2 million in 1994 to around 1.150 million in 2017, with a respective increase in milk yield from around 5,500 litres to 7,557 (peaking at 7,897 litres in 2014) due to increased productivity. This compares with a mediaeval cow milk yield of around 680 litres and a scrub cow about 1,400 litres.

In terms of beef cattle, modern breeds have been developed through the breeding of selected cattle. Native hardy breeds include Aberdeen Angus, Galloway and Welsh Black, suited to upland situations, and breeds such as Hereford, South Devon and Lincoln Red, which are suited to lowland situations. Imported breeds include Charolais, Limousin, Blonde d'Aquitaine, Simmental and Belgian Blue.

Most land in the UK on which cattle are kept has a medium to heavy soil texture. When wet, the soil becomes plastic and more easily compacted. To limit soil damage, dairy cows and other cattle are often housed during the winter and early spring – for about five to six months. Even so, land invariably becomes poached, particularly on land close to the farmstead, around ring-feeders and accessible areas next to watercourses which cows may have access to cross, drink from and defecate in. Cattle with access to watercourses also cause bankside erosion and disturb sediment on the riverbed through treading. Cattle with direct access to surface water is commonplace in the UK.

Where the land slopes, and sufficient rain falls, water unable to drain through the poached/compacted soil runs across the surface instead. The run-off collects disturbed soil particles and organic matter along the way, which is transferred into surface waters, where present, and contributes to diffuse pollution, which can become chronic in nature.

Other potential hotspots for contaminated run-off are farmyards, livestock tracts, gateways between fields and next to drinking troughs.

These situations are not uncommon, and increasingly occur on farms outside of catchments prioritised for regulation, advisory programmes and grants. Outside of prioritised catchments, there is little incentive for farmers to take mitigation measures to prevent pollution, such as fencing watercourses and bridging field crossing points.

According to research undertaken by Kleanthous (2009), per tonne of animal products consumed milk/dairy has the lowest environmental impact and beef the highest.[2] However, in terms of the share of UK consumption, volume/weighted impact, dairy has the highest impact at 55 per cent, with beef at 5 per cent. Our environment cannot continue to sustain the multiple impacts dairy farming is causing (global warming potential, acidification, eutrophication, biodiversity, water, land use, abiotic resources and waste) or the impact beef has per tonne of product.

Citations:

[1.]   Marks, H. F. (1989), *A Hundred Years of British Food & Farming: A Statistical Survey*, edited by D. K. Britton. Taylor & Francis, London, UK

[2.]   Kleanthous, A. (2009), *Pigs and the Environment: A report to BPEX.* https://www.pigprogress.net/PageFiles/24130/001_boerderij-download-PP 5977D01.pdf

23. This Little Piggy

Domestication of pigs is thought to have occurred after sheep, goats and cows, but still within the Neolithic period. A domesticated version of the European wild boar was possibly introduced into Britain by Neolithic people, with Britain's own abundant wild boar probably domesticated too. For most of the time since domestication, pigs would have had the run of outdoor land and woodland enclosures, where they could scavenge and be fattened naturally.

In more recent centuries, pigs have been progressively kept in courtyards and buildings. Pigs also began to be brought into the domestic setting from around the sixteenth century, with cottagers often keeping a few pigs within the household. However, as people became more urbanised, and due to the odour and the disease risk from keeping pigs in houses, laws were eventually introduced to ban this practice in the nineteenth century.

The modern movement of pigs into buildings started after World War II. With a growing preference to use land to graze cows and sheep, pigs, with their more adaptable diet, including human food waste, increasingly found themselves indoors in large specialist units.[1]

In 1866, Great Britain's pig population stood at 2.5 million, with all but 0.2 million farmed in England and Wales. The pig population dropped to 1.8 million after the World War I and, despite recovering to 3.8 million in 1939, hit a new low of 1.3 million two years later. By 1987 the Great Britain pig population stood at 7.3 million and the UK population at 7.9 million.[2] Now the number of UK pigs is a little under 5 million, with about four-fifths reared in England.

In the UK 60 per cent of the breeding herd are kept indoors and 40 per cent outdoor. For the piglets destined for meat, only a very small percentage spend their whole lives outdoors. However, whilst those living outdoors benefit from fresh air and the ability to show some natural behaviours, rearing year-round outdoors can severely damage soil and lead to groundwater pollution and contaminated run-off and surface water pollution when it rains sufficiently.

Soil becomes bare and damaged through progressive treading/poaching and through the pig's natural nose rooting habit. Run-off hot spots can also develop on rutted/compacted trackways and in heavily used areas around feeding and drinking troughs. The contaminated run-off is a mixture of rainwater, loosened and enriched soil particles, organic matter and nutrients from voided urine and manure.

Sows and growing piglets deposit a lot of manure. If stocked at twenty-five sows per hectare, over a year the total amount of nitrogen and phosphate they deposit can exceed 400 kg per hectare and 300 kg per hectare respectively. Some of the nitrogen is volatilised as ammonia gas, some may be washed away as ammonium-nitrogen and some leached to groundwater as nitrate-nitrogen in the autumn and over winter. Phosphorus also enriches run-off and adds to the harm caused where it enters a watercourse.

High-risk areas for erosion include free-draining, sandy soils, where a lot of outdoor pigs are often sited, moderate to steeply sloping fields and sites with moderate to high rainfall. Many outdoor pig farms fit into some or most of these categories. The stocking rate also has a major influence on soil damage, run-off and soil erosion. A high stocking rate of twenty-five or more sows is particularly impactful and more so on medium to heavy soils in high-rainfall areas. Unfortunately, destocking to help maintain vegetative cover is rarely a mitigation option, as it would likely render the business uneconomical.

Mitigating against pollution, including rotating paddocks and trying to maintain a grass cover, can be extremely challenging. Many farmers fail to varying degrees, in particular where pigs are kept on medium to heavy land not suited to outdoor production and where the land slopes towards a watercourse. Options, including the planting of robust and deep-rooting grass, like Rhizomatous Tall Fescue, are helpful, along with avoiding over-concentrating pigs for any length of time around feeding troughs. However, in my experience, most outdoor pig farms give rise to a significant risk of diffuse water pollution.

It is certainly a better life for a sow and her litter than a life on a slatted floor in a confined, indoor pig unit – not least for the sow cruelly kept in a farrowing crate for several weeks at a time during her productive life, with no room to turn around. However, living a life on an exposed area of land in winter, or on hot sunny days in summer months, is not best for an animal naturally adapted to woodlands. A better life for pigs and the environment combined, other than breeds that can be readily reared in woodlands, may be roomy, deeply strawed free farrowing pens, producing farmyard manure. Easier said than done economically, unless the farm has room and a ready supply of home-produced straw. The best solution of all is to stop eating pigs.

When it comes to overall environmental impact per tonne of product consumed, pig consumption sits in fourth place, beaten only by beef, sheep and milk/dairy.[3] In terms of UK consumption by volume/

weighted impact, the impact of pig production is greater than beef and sheep, though well below milk/dairy and eggs.[4]

Citations:

[1.] Urquhart, J. (1983), *Animals on the Farm: Their history from the earliest times to the present day*. Macdonald, London, UK

[2.] Marks, H. F. (1989), *A Hundred Years of British Food & Farming: A Statistical Survey*, edited by D. K. Britton. Taylor & Francis, London, UK

[3.] Kleanthous, A. (2009), *Pigs and the Environment: A report to BPEX*. https://www.pigprogress.net/PageFiles/24130/001_boerderij-download-PP 5977D01.pdf

[4.] Ibid.

24. Fowl Play

For much of its history, at least into the nineteenth century, farming hens for meat and eggs was a supplementation to cattle and sheep farming. However, by the end of the nineteenth century breeders began showing interest in table hen strains being bred in America.

To boost egg production, breeders also took an interest in the Australorp, a super-laying strain from Australia, and the Barnevelder hen from Holland. However, by World War I hens primarily remained as farmyard birds.

Food shortages and a resulting high egg price eventually captured the interest of UK farmers. By the end of the war, farms of all sizes began investing in poultry farming, though the practice was short-lived due to the 1920s and 1930s depressions, which hit the fledgling industry hard. The uncontrolled way poultry were being farmed in housed units also proved inadequate. Many hens died from being fed a poor diet, through stress and through disease from living together in higher numbers in cramped conditions.

By the time of World War II, having learned from and fixed many of their failings, the industry began to prosper, only to be set back again as war struck and feed became increasingly scarce. As a consequence, poultry farming reverted to more traditional free-range practices, with birds being fed on household scraps. The industry took off again after the war, with the majority of hens kept in portable ranges. This enabled farmers to regularly move them around and manure the whole field.

However, the cost and scarcity of labour from the 1950s saw a movement back to large, confined strawed yards and houses and battery houses for egg-laying hens.

Poultry farming has come a long way since the nineteenth century, when in 1884 the number of birds stood at 12.7 million (Great Britain). By the start of World War II, the population was 60.9, million before dropping to 41.3 million in 1945. Since, the number of birds has largely been on an upwards spiral, with a dip for a few years after 1975[1] before recovering and rising to 170 million today as a UK total, with about 130 million birds in England.[2] In England, the majority of poultry farmed are table chicken (broilers), about 85 million, with around 33 million breeding and laying hens. There are a further 10 million ducks, geese, turkey and other birds.[3]

Whilst chickens are not the most obvious of farm animals to cause soil compaction and run-off, they certainly do. Intensively stocked free-range poultry routinely wear away grass cover outside of poultry houses, and their constant trafficking compacts and caps the soil. Where land slopes and sufficient rain falls, water unable to infiltrate the soil accumulates and runs across the surface, carrying with it detached soil and excreta. In storm conditions the volume of run-off can be considerable and difficult to mitigate against both practically and cost-effectively.

The risk is greatly reduced on farms with small flock sizes, which can be provided with good, maintainable vegetative cover.

Another notable run-off problem arises from the litter-free manure produced in caged egg production. When removed from the sheds, placed in field heaps and left uncovered, subsequent rainfall super-saturates the manure until it slumps and begins to flow. Even if covered, in the UK's windy weather the cover can become detached and/or torn, allowing water in. Super-saturated poultry manure is problematic to handle, causes nutrient hotspots, and the run-off from it can pollute water.

When it comes to environmental impact, egg production is in sixth place in terms of its share of UK consumption (per tonne consumed), and by volume/weighted impact as consumed its impact is second only to milk/dairy at 24 per cent. Poultry is in third place for both impacts.[4]

Citations:

1. Marks, H. F. (1989), *A Hundred Years of British Food & Farming: A Statistical Survey*, edited by D. K. Britton. Taylor & Francis, London, UK
2. Defra, Livestock numbers in England and the UK. *Annual statistics on the number of livestock in England and UK in June and December*
3. Ibid.
4. Kleanthous, A. (2009), *Pigs and the Environment: A report to BPEX.* https://www.pigprogress.net/PageFiles/24130/001_boerderij-download-PP 5977D01.pdf

25. You're Kidding

Goat farming is small compared to other livestock sectors within the UK. Nevertheless, goat farming has its place, providing milk, cheese, meat and other commodities. Historically, goats tended to be kept in areas less suited to sheep farming. They were commonly reared by cottagers for meat and as a cheap milk alternative to cows, given their versatility to graze poor herbage.

Goats were particularly commonplace throughout Scotland and Wales until the end of the eighteenth century, with goats more common in Scotland than sheep. In England, the spread was more sporadic, with most of the goat population located in Cornwall. As UK farming advanced, land was increasingly used for more profitable sheep farming and the goat population dramatically declined. Whilst goat farming waned, it allowed 'unfavoured gorse' and other 'nuisance' plants, which sheep cannot eat, to thrive across the mountains and moorlands.

In a very short space of time goats largely disappeared from the UK landscape. Today goat farming is becoming commercially popular again, due to improved returns and mooted health benefits over cow's milk (e.g. lower cholesterol). The larger UK farms can be home to 400–500 or more goats, and the largest has over 3,000.

Goats contribute to pollution in a similar capacity to sheep when allowed to overgraze. As goats aren't fussy eaters, they can create particularly barren land, stripping most of its biodiversity interest.

26. Something's Fishy

The two major farmed fish in the UK are intensively reared rainbow trout and salmon, with trout having the longer history. In terms of intensification, in Europe the practice of rainbow trout farming started in the 1960s, with Atlantic salmon following suit towards the end of that decade.[1]

In the UK rainbow trout are reared inland, connected to rivers, or in deep Scottish freshwater lochs. In trout farms up to 27 trout, each around 30 cm long, can be crammed into the volume of a bathtub.[2]

Atlantic salmon are mostly reared in large floating cages, in sheltered coastal marine sites in the west of Scotland. Scottish salmon farms are also concentrated along the length of the eastern coast of the Outer Hebrides and around the coasts of Orkney and Shetlands.

A modern-day Scottish fish farm consists of a series of 120 m round (38.2 m diameter) plastic, cylinder-net cages, mostly hidden beneath the sea line. A central turret supports a net aloft of the cage to keep birds out and the fish in. Salmon cages, depending on size, can be home to between 50,000 and 90,000 fish.[3] As aptly described by James Merryweather (2014),[4] the problem with these cages are the millions of holes that keep fish in and let all the fish faeces and other wastes generated out – produced by about a hundred million salmon, counted across all Scotland's farms.

Atlantic salmon are reared for up to two years, in around 200[5] to 250[6] loch farms. The density they are kept at is high and as the salmon increase in size the density progressively leads to considerable stress, injury, deformity, disease, sea lice infestation and mortality. According to a 2019 *Panorama* investigation, up to 20 per cent (around 9.5 million) salmon die in the cages each year. Most deaths are the result of infestations of sea lice, which first attach to the salmon's gills before spreading more widely across their bodies. Basically, the sea lice eat the salmon alive – and slowly! Chemical treatments are applied to control infestations, which in turn pollute and damage the surrounding ecology.

Dead salmon are pumped out from the bottom of the cages, unlike the huge amount of faecal matter they excrete daily along with any uneaten fish food. It is impossible for salmon not to ingest a proportion of each other's faeces.

In 2000 Scottish salmon production was projected to be around 125,000 tonnes, with the salmon excreting an estimated 7,500 tonnes of nitrogen and 1,240 tonnes of phosphorus into the lochs.[7] This is comparable to the sewage input, for nitrogen, of 3.2 million people, and for phosphorus 9.4 million people.[8]

In 2017 the level of salmon production was estimated to be 189,707 tonnes.[9] Analogous to the 2000 calculation, this means the population was excreting around 11,250 tonnes of nitrogen and 1,860 tonnes of phosphorus. This is equivalent to a sewage input of 4.8 million and 14.1 million people respectively. For some perspective, Scotland's current population is just shy of 5.5 million people!

From Merryweather (2018), the annual volume of untreated excrement discharged from Scottish salmon is around 250,000 tonnes,[10] from around 250 farms each on average producing 1,000 tonnes of excrement.[11]

From this, a further comparison can be made with other livestock.

The volume of excreta a single beef cow (from twelve months to less than twenty-four months old) produces in a year is about 9.36 m^3, containing around 50 kg total nitrogen. A dairy cow yielding between 6,000 and 9,000 litres of milk produces about 19.08 m^3/year, containing around 101 kg total nitrogen. A pig (one dry fed finisher, 66 kg and over), 1.56 m^3/year, containing 10.6 kg of total nitrogen.[12]

Based on 250 salmon farms discharging on average 1,000 tonnes of untreated effluent, each one would release 45 tonnes of nitrogen and 7.44 tonnes of phosphorus a year (17 tonnes of phosphate equivalent). In terms of nitrogen, this equates to the excreta produced from 900 beef cows, 446 dairy cows and 4,248 pigs. For phosphate, 1,022 beef cows, 385 dairy cows and 1,702 pigs.

From this, if I were to apply for a permit to build a floating, slated-floored livestock farm on top of a Scottish sea loch to externalised all my animals' excreta, I should in theory have no trouble getting one. Provided, of course, I kept my livestock's nitrogen pollution within similar bounds to the salmon farms.

My chances of getting a permit, though, would most likely be nil, but could be better if I dealt with the aesthetics, given a salmon farm's sunken pens would be far less visually intrusive than a flotilla of cows and pigs. So, I could, as an alternative, keep them below the waterline. Perhaps with a mirrored cover to mimic the loch water. Problem solved? Perhaps not.

Not one to easily give up, I could promote my farm on the basis that: I do not need to shoot seals to protect my slats from damage and to prevent my livestock being eaten; my cows and pigs will not suffer from or spread sea lice to the loch's wild Atlantic salmon; and if any of my cows and pigs escaped, they would not breed and dilute the genetics of the wild salmon. I fear, though, whatever I do would not be good enough.

With the growth of Scotland's salmon farms expected to double in the future, I could simply give up on farming my flotilla of cows and pigs and do the same as the rest with impunity and pollute the water with farmed salmon.

But maybe I do have one last hope, if I could convince the middle class and elite consumers – who home and abroad buy Scottish premium salmon, off the back of the pure landscape and 'pristine' water in which they are raised – that my loch-farmed cows and pigs are equally as pure.

Whatever way I choose to go, I would be left with another topical problem: seagrass meadows (Zostera marina) – the sea's rainforests – some of which could be threatened by loch-farming salmon and other livestock. All that nitrogen released, along with oodles of phosphorus too, causes eutrophic water conditions that can harm seagrass. To quote from Burkholder et al.: 'chronic exposure to nitrate enriched waters is directly lethal to *Zostera marina* even at low enrichment levels, and likely represents an important causative agent in the disappearance of eelgrass meadows from many quiet embayments and coastal lagoons throughout the world'.[13]

Let's not also neglect the direct impact on the benthic environment beneath, and within, the vicinity of each net, where a lot of the excrement and waste feed settles to form a rich organic sludge. This sludge changes the nature of the sea bed's ecology, its chemistry and the life which is able to exist.

Salmon farming in Scotland is big business, worth around £1 billion. A lot of salmon is exported as a value commodity and is one of the UK's biggest export earners. Given the money involved, positive changes will be slow to see, not least given the industry's current ability to externalise their huge impact on the environment. The products are cleverly marketed, with images of breathtaking scenery and water purity. The packaging exudes a perfect picture of health. Should the images portray actual fish life below sea level – the sea lice, pollution, the mortality and the seabed desert beneath and in proximity to the cages – customers would not be so quick to purchase. We may also be led to believe that the animals' welfare and best interests are beyond question, with the UK being a leading example for the rest of the world to follow. The reality is somewhat different, with the UK having a lower standard of regulation and regulatory scrutiny compared to fish farms in Norway.[14]

I'm reminded, looking into Scottish salmon farming, of tales of how England and Wales's soil would be capable of treating diffuse nutrients from livestock back in the early 1990s. I sense a similar belief for the pollutants from caged fish being swept away by the tide.

When the industry started some fifty years ago, the marine environment may well have been able to handle those early small farms. But as cage after cage has been added, and with an estimated doubling of fish farms in coming years, which I can well believe from listening to retailer conversations and Scottish politicians, the situation is more likely to end very badly for the localised ecology.

My career has been spent, often frustratingly, addressing the excessive loss and cost of livestock production for the environment. I fear more of the same from Scottish fish farms, which are already upsetting the ecology of the sea lochs through the rain of faecal matter and associated nutrient enrichment. Excrement settles and builds up along with any uneaten fish food as a sediment on the sea bed, and both substances are likely to be contaminated with toxic drugs used to control sea lice.

A solution being looked at in Norway, the biggest producer of Atlantic salmon and therefore a bigger polluter overall, is the use of closed pens that retain all pollutants, which, along with the dead fish, can be regularly pumped out and waste disposed of more appropriately. Closed-cage systems also prevent sea lice infestation and prevent escapees breeding and genetically diluting the natural Atlantic salmon population. Such systems would be an improvement, but the fish would still be intensively reared and have far from a life worth living and end up being slaughtered against their wishes in unkind ways. The closed cages would be much more expensive to build and run, given the infrastructure required and the need to responsibly manage fish excrement.

Salmon could also be reared in land in a controlled environment, not that I would promote that myself either, with the industry actually taking responsibility for its wastes. This is unlikely to happen in any major way, if at all, as to do so would take away the very basis for the industry's home and export premium it so much enjoys. Land-based farms would also be very expensive in comparison, when the drive is for ever cheaper animal-based food. Why spend money on inland solutions or on investing in closed-cage systems when the Scottish government allows Scottish salmon farms to externalise all their pollutants, leaving an almost impossible job for its resource-depleted regulator, the Scottish Environment Protection Agency, to protect the environment. A familiar tale in all walks of the livestock industry life.

Another area of concern with Atlantic salmon is the food we feed them, a diet loaded with fish oil and fishmeal processed from small, wild-caught oily forage fish, such as anchoveta, mackerel and sardines. The ground-up forage fish fed to salmon enrich their flesh with omega-3, an essential nutrient for the human diet, which Scottish Atlantic salmon are also heavily marketed on. Prey fish fed to the salmon are primarily captured within European, South American and West African waters.[15] The fish caught may be sourced under marine 'sustainability' schemes, like other fish caught for direct human consumption, such as the Marine Stewardship Council (MSC) certification. These schemes may provide comfort to those buying reared salmon, and other ocean-derived products, that no more fish are being taken than can be sustained – if you like, living off the interest rather than the balance. However, the reality is not proven, and what is taken can severely damage marine habitats and disrupt the natural food web, depriving higher-trophic-level predatory fish, marine mammals and seabirds of the food they need, causing their populations to crash. Given that fishing kills, including indiscriminately, many predator species who feed on the fish we take from the oceans, it is perhaps one way of achieving sustainability throughout the ocean food web.

Being sustainable is more than single things and species; it involves the whole story, including ghost fishing gear which is discarded/lost in the sea and goes on to entangle and kill many forms of ocean life for tens to hundreds of years to come; bycatch, including vast numbers of sharks, dolphins, turtles and other species; seabirds; non-target/non-commercial fish numbers, which are part of the ocean food chain/life-support system; ocean deforestation, from bottom trawling; destruction of coral reefs, and other forms of exploitation and damage done to marine ecosystems; ocean-related climate change; and the inefficient use of millions of tonnes of fish in aquaculture and livestock farming, fish and land-based livestock, which in turn are a very inefficient and polluting foods in comparison to plants.

Ocean life is a shadow of what existed before human exploitation, and feeding fish to feed livestock to feed us cannot be considered a sustainable activity.

If you are not ready to stop eating animals, then sourcing sustainability 'ticked' fish is a better option than not, but do not kid yourself that sea fishing is sustainable. It is defined as sustainable within its own context, by those with a vested interest in fishing. Keeping the conversation on sustainability also glosses over the immense pain and distress each fish caught suffers to land on our plates, to the bycatch and to those dying slowly whilst entangled in ghost gear.

As well as fishmeal and fish-oil-based diet, Salmon are fed, increasingly, human-edible soya bean meal protein, vegetable oils and cereals (see Table 21). Other alternatives not listed in Table 21 may include algae and insect meal.

Table 21: Composition of Scottish salmon feed[16]

Ingredient	Estimated percentage of total diet
Fishmeal	14.7 – 25
Fish oil	10.3 – 15
Vegetable protein (soya bean meal, sunflower expeller, wheat gluten, peas, fava beans, etc) and guar gum	23 – 53
Vegetable oils (often rapeseed)	18 – 22
Cereal crops (wheat, corn)	19.5 – 25.5
Supplements	4 – 5

Source: Reproduced from Fishy Business: The Scottish salmon industry's hidden appetite for wild-caught fish. Feedback. 2019.

Fish meal, fish oil and vegetable proteins like soya cause harm to oceanic and land-based ecosystems and biodiversity. Soya bean meal contributes to deforestation and climate change.

Current production of around 190,000 tonnes of salmon is expected to increase to 300,000–400,000 tonnes by 2030.[17] According to Feedback,[18] the estimated number of wild-caught fish used by the industry is currently around 460,000 tonnes, a figure likely to increase by a further 300,000 to 760,00 tonnes by 2030. Feedback based this assumption on forage fish dependency ratios for oil not improving.

UK trout farming is also big business, particularly in central and southern Scotland, south England and North Yorkshire.[19] According to the British Trout Association there are around 360 trout farms, producing in the order of 16,000 tonnes of rainbow trout. Trout farms require a supply of fresh, clean, cool, well-oxygenated river water, diverted into nearby ponds, to be successful, not least to maximise the number of trout that can be farmed. A constant water flow is necessary to remove fish excrement from the ponds. Or in Scotland, you can also keep the trout in submerged nets in deep freshwater lochs, and thereby externalise your pollutants.[20] At least in river water farming the trout industry is required to treat the effluent leaving the ponds before it is returned into the river, and in accordance with the water quality standards set out in the farm's discharge consent.

Citations:

1. European Commission. *Aquaculture.* https://ec.europa.eu/oceans-and-fisheries/ocean/blue-economy/aquacultur e_enhttps://ec.europa.eu/fisheries/cfp/aquaculture/aquaculture_methods/ history_en

2. Lymbery, P. (2002), *In Too Deep, The Welfare of Intensively Farm Fish: A Report for Compassion in World Farming Trust.* https://citeseerx.ist.psu.edu/viewdoc/download?doi=10.1.1.180.6550&re p=rep1&type=pdf

3. James Merryweather (2014), *Holes: Scotland's Salmon Sewage Scandal.* Blue-Skye Books, Auchtertyre, UK

4. Ibid.

5. BBC, *Panorama,* 'Salmon Farming Exposed', first broadcast 25 May 2019

6. James Merryweather (2014), *Holes: Scotland's Salmon Sewage Scandal.* Blue-Skye Books, Auchtertyre, UK

7. MacGarvin, M. (2000), *Scotland's Secret? Aquaculture, nutrient pollution eutrophication and toxic blooms.* Report by modus vivendi for WWF Scotland. https://www.wwf.org.uk/sites/default/files/2000-01/secret.pdf

8. Ibid.

9. Marine Scotland Science. *Scottish Fish Farm Production Survey (2017).* https://www.gov.scot/publications/scottish-fish-farm-production-survey-2017/

10. James Merryweather (2018), *Why Salmon Farms Pollute.* https://salmonaquaculturescotland.wordpress.com/2018/03/11/w hy-salmon-farms-pollute/

11. Munro, L.A. and Wallace, I. S. (2016), *Scottish Fish Farm Production Survey 2015.* Report prepared by Marine Scotland Science

12. Defra (2013), *Guidance on complying with the rules for Nitrate Vulnerable Zones in England for 2013 to 2016*

13. Burkholder, J, et al. (1992), 'Water-column nitrate enrichment promotes decline of eelgrass *Zostera marina*: evidence from seasonal mesocosm experiments.' *Marine Ecology Progress Series* 81, pp. 163-178

14. BBC, *Panorama,* 'Salmon Farming Exposed', first broadcast 25 May 2019

15. Feedback (2019), *Fishy Business: The Scottish salmon industry's hidden appetite for wild-caught fish.* https://feedbackglobal.org/wp-content/ uploads/2019/06/Fishy-business-the-Scottish-salmon-industrys-hidden-appe tite-for-wild-fish-and-land.pdf

16. Ibid. Percentages estimated from Shepherd, Monroig and Tocher 2017, MOWI 2018a, Ytretøyl et al. 2011.

17. Scottish Parliament (2016), *Food and Drink*
18. Feedback (2019), *Fishy Business: The Scottish salmon industry's hidden appetite for wild-caught fish.*
     https://feedbackglobal.org/wp-content/uploads/2019/06/Fishy-business-the-Scottish-salmon-industrys-hidden-appetite-for-wild-fish-and-land.pdf
19. British Trout Association.
     https://britishtrout.co.uk/about-trout/trout-farming/
20. Ibid.

27. BEE-have

Commercial beekeeping is a less obvious animal welfare and environment concern, but the UK's 400-odd professional bee farms, some managing around 3,000 hives housing up to 80,000 bees,[1] are not quite so harmless or beneficial as they may seem.

Years of industrialised and chemical farming has had a devasting impact on the UK's 270 bee species, of which 250 are solitary. Today, there are simply too few wild bees, and other pollinators, to pollinate all our flora and croplands. Commercial honeybees are becoming a happy solution to help farmers maintain fruit and crop yields, such as oilseed rape. However, under-pollination of wild and food crops is not a honeybee deficiency disease. It is an industrial and chemical farming disease, which has been allowed to spread unchecked across our landscape for far too long.

The commercial honeybee has become a carrier of the Varroa destructor mite and the chronic bee paralysis virus, which they can pass on when they visit the same flowers as wild bees. The commercial western honeybee can also outcompete wild bees, and other pollinators, leaving them short of pollen and nectar. According to Gonzalez-Varo and Dr. Jonas Geldmann, of the Cambridge University Zoology Department:[2]

Honeybees are artificially-bred agricultural animals similar to livestock such as pigs and cows. But this livestock can roam beyond any enclosures to disrupt local ecosystems through competition and disease.

The crisis in global pollinator decline has been associated with one species above all, the Western honeybee. Yet this is one of the few pollinator species that is continually replenished through breeding and agriculture.

Saving the honeybee does not help wildlife. Western honeybees are a commercially managed species that can actually have negative effects on their immediate environment through the massive numbers in which they are introduced.

From this it would be a mistake to over-play the role of commercial, and even domestic, beekeeping in conservation. The problem, including the spread of mites and viruses, may be less about hives temporarily brought onto farms to pollinate crops, and more about where the bees are kept in between, where they are more likely to come into contact with or outcompete vulnerable pollinator species in more natural settings.

In parts of the world, and more notably in the USA, some commercial beekeepers also kill their bees at the end of the season to avoid the cost of keeping them alive during the winter. Bees can be replaced with new stock from companies who supply artificially inseminated bees.

An investigation into honey production, reported by the Ethical Consumer in 2020, found little evidence of culling associated with twenty-eight commercial honey brands sold in the UK. However, they reported mentions of the practice occurring on some beekeeping forums.

The Ethical Consumer also looked into bee welfare policies, including wing clipping, artificial insemination (AI) and the use of harsh chemicals in hives. Only three companies were found to have a comprehensive policy and to prohibit harmful practices. A few others were determined to have adequate policies, but they also reported that the absence of a policy does not necessarily mean bad practice is taking place. Asda, Sainsbury's, Tesco and Morrisons were singled out as having no bee welfare policy.

In bee AI, semen is collected from mature drones to artificially inseminate queens. The thorax of a mature drone bee is first crushed in the fingers to kill it. Pressure is then applied from the base of the abdomen towards the tip, by rolling the fingers together, to squeeze out the bee's semen. This is known as the eversion process. A drop of saline on a syringe is then applied to make contact with the semen on the bee's endophallus (the inner wall of the bees aedeagus – intromittent organ or penis) and using capillary action the semen is then drawn up into the syringe, avoiding contamination with other bodily fluids/mucus and air bubbles.

Each drone yields about one microlitre (µg) of semen, with 8–10 µg needed to fertilise each queen. Not all drones yield semen and eversion is not always successful, so many drones may be killed to collect enough semen.

In natural settings, lots of drones perform their solitary mating function with a new queen and die immediately after. Those that do not mate eventually find themselves surplus to requirements and driven out of the hive, and as they can't fend for themselves they either die from the cold or through starvation. Nature can be harsh, but that is not a justification for interfering in natural processes.

The next stage is to anaesthetise the queen with CO_2. She is then placed into a tube, and gently maneuverered to expose her abdomen through a narrow opening at one end. The sting hook is then surgically lifted out of the way to expose her valve fold (flap of tissue), which needs to be bypassed to enable semen to be placed into her median oviduct for insemination, using a syringe. The queen is then removed, the tips of her wings clipped, and marked.[3] Not even bees can escape the growing livestock AI phenomenon.

AI has become common within livestock farming, and the trend is now starting to gain some traction in the UK for bees. The process on bees is also known as instrumental insemination. Whenever we interfere artificially in animal breeding, it is not done for animal benefit, and they suffer. With AI also comes an increased risk of commercial UK beekeepers turning to culling bees to save on winter care and cost.

A more sustainable practice would be to provide large inter-connected areas of varied pollinator ecology to boost pollinator numbers, and to phase out the use of farm practices harmful to them. Whilst the right mix of animals at the right density and in the right place can serve conservation, livestock farming is a poor substitute for natural processes.

Citations:

1. Allison, R., 'How Britain's bee farmers can help raise crop yields', *Farmers Weekly* 24 April 2018.
   https://www.fwi.co.uk/arable/britains-bee-farmers-can-help-raise-crop-yields

2. Knapton, S., 'Urban beekeeping is harming wild bees, says Cambridge University', *The Telegraph* 25 January 2018.
   https://www.telegraph.co.uk/science/2018/01/25/urban-beekeeping-harming-wild-bees-says-cambridge-university/

3. Cobey S., Tarpy, D. and Woyke, J. (2013). 'Standard methods for instrumental insemination of Apis mellifera queens.' *Journal of Apiculture Research* 52(4), pp. 1-18

28. Biodiversity & Livestock Farming

Read literature from the UK livestock sector and you will be told that livestock farming is beneficial to biodiversity, and the removal of livestock would be detrimental to biodiversity. An example of that portrayal is set out in the AHDB's 2018 report 'Landscapes without Livestock, Visualising the impacts of a reduction in beef and sheep farming on some of England's most cherished landscapes'.[1] Given the AHDB exists to enhance the profitability and sustainability of livestock farming, then it is unrealistic to expect anything from this publication other than a sound defence of the sector's place in our landscape.

The AHDB report largely focuses on preserving landscape value, which over generations people have become accustomed to, rather than the land's biodiversity potential, and chooses to concentrate on how much 'poorer' landscape would be without livestock.

However, the biodiversity which the report values, and what we have come to see as a breathtaking landscape, is a shadow of the land's potential. What we actually have today is a species-poor landscape, dotted with sheep, cattle and their droppings. A landscape we have become accustomed or led to believe is natural, which it isn't at all.

Don't take my word for it; you can see it for yourself. Take a walk in the uplands, like the Brecon Beacons, and take time to look for the biodiversity (flora and fauna) the livestock industry promotes as worthy of preserving over what can be achieved through natural regeneration/ rewilding. Then, if you are not able to visit the Knepp Estate rewilding initiative, buy or borrow Isabella Tree's telling book *Wilding*[2] from your local library and read it cover to cover. The dairy farm that preceded the estate's 'return to nature' is an example of the kind of landscape and its biodiversity verdantly defended by the livestock industry and within 'Landscapes without Livestock'.

The Knepp Estate, like other rewilding initiatives, has allowed nature to flourish through thoughtful stewardship and allowing the animals to rule. The approach taken allows for natural regeneration, with careful consideration and introduction of animals such as Old English longhorns (substitute for aurochs), Exmoor ponies, Tamworth pigs (substitute for wild boar) and deer needed to provide a functioning habitat with the minimum of barriers.

Although the AHDB report does not portray in a favourable way the transformative change needed to maximise biodiversity in the UK, it usefully reminds us that we cannot simply vacate land and abandon farmers. Farmers whose interests and fortunes rest in the land need to be given the means, incentives and secured income to manage and share land for the betterment of themselves, wildlife, our future climate and to reconnect people with nature. There is room for open landscape within a much broader mosaic of habitat, hospitable to the widest possible biodiversity. A landscape within which we can reengineer our relationship and kinship with nature.

Herbivores and omnivores – and carnivores too, if they could ever be reintroduced – are a necessary part of a healthy, functioning ecosystem. With no top predators, some form of control would be inevitable; however, this does not need to involve farming or hunting animals for meat.

We cannot systematically continue farming livestock in the way we do in wells of silence. We need to aspire to a greater good for the sake of our environment, our health and well-being, for the animals and for future generations.

The AHDB project, in their words, is described as follows:

For this project, AHDB Beef & Lamb identified five different environments in England typically grazed by beef cattle and sheep and in which these livestock play a major role in maintaining the distinctive landscape character. At the start of this project, these landscapes were identified as:

- *Less-favoured area (LFA) Upland*
- *Less-favoured area (LFA) Hillsides*
- *Rotational Pasture*
- *Permanent Pasture*
- *Moorland*

During the study, specific sites were identified by AHDB Beef & Lamb.

At the beginning of the project panoramic photographs were taken 'representing the quintessential characteristics of the current landscapes as "control" images'. From these photographs, AHDB made nature-based assumptions on what the landscapes would look like as 'photomontages at Year three, Year 10 and Year 30' timeframes 'to illustrate the visual impacts of [*deleted word*] changes at each location.' In doing so, the aim was to approach the project in a way that 'would stand up to scrutiny, prepared with input from ecologists, farmers and landscape specialists'.

In one account landscape abandonment would result in naturally colonising shrub; planted woodlands; hedgerow removal; bigger fields; industrialised vegetable pack houses; creeping thistles and ragwort; drainage of wet pastures; nutrient enrichment and pollution from run-off of agricultural pesticides, herbicides and fertilisers; solar panel fields and energy crops; horses; decimation of farmland birds; and a whole load more wows.

The demise begins in the uplands with the encroachment of dense bracken and accompanying areas of gorse and small hawthorn trees. By Year 30, 'the patterns of fields have become very largely obscured by extensive growth of mature gorse scrub and young self-sown oak and willow woodland'. Gone is the '2018 landscape – with its smooth pastoral fields of varying shades of green ... favoured by birds like merlin, red grouse, meadow pipit and upland waders'. The report goes on to describe the decline of several rare butterflies and moths. Red grouse is a particularly interesting mention, as without them, those interested in game hunting would have less of these preserved birds to shoot!

In another account mixed grassland landscapes are replaced with continuous arable and energy crops, tightly trimmed and less prominent hedgerows and a caravan and campsite (I'm not sure how that would attract a caravan and campsite). By Year 30 we have large blocks of land, as more field boundaries have been removed to increase some field sizes, the removal and non-replacement of many hedgerows and field trees. Isolated 'stag-headed' trees have also taken shape due to regular ploughing and associated tree root damage under their canopies. The growth of a caravan site has also further ruined the landscape, bringing with it a 'network of concrete tracks dividing up the mown grass fields'. The decline in farmland birds, like the grey partridge, linnet, corn bunting, reed bunting and tree sparrow, is also highlighted.

The accounts go on, and the reader of the AHDB document is left in little doubt of the despair removing beef and sheep would have on the UK's landscape/biodiversity, how it would diminish the public's enjoyment of the land and how the removal of livestock could also be bad news for water quality and our heritage.

Anyone persuaded by this kind of rhetoric should not only read *Wilding*[3] by Isabella Tree, as already mentioned, but visit the Rewilding Britain website too, to take a look at other similar initiatives, to get a rewilding reality check.[4]

The AHDB's account provides no balance and neglects how much richer in biodiversity these landscapes can become through systematic and sympathetic regenerative changes to existing land-use practices. It is a failed opportunity to acknowledge the problems facing our place in nature, livestock farming's longer shadow and our future food security and health; and to acknowledge what intensification of livestock farming has taken away and has been replaced with, and how things can become so much more than they are at present.

We may not be living in America, with its huge feedlots of hormone-fed beef cows, yet (although we are on our way), but of the many beef cows and other livestock we farm, only a very small percentage can be claimed to be part of regenerative agriculture, the name given to unconventional livestock systems which aim to make meat and dairy a by-product of nature and to prioritise soil health over dirt. The majority of the UK population cannot afford to eat livestock 'by-products of nature' and instead source what they can pay for from subsidised and pollution-externalising conventional and industrialised farming enterprises.

Land-use change is an absolute necessity to meet the UK's greenhouse gas (GHG) emissions commitments, and it best happens on land which provides the least nutrition. Reduced animal consumption and increased plant-based nutrition is also recognised by all but the most hardened carnivores and livestock lobbyists to be an absolute necessity to improve our nation's health.

AHDB is not alone in providing an exaggerated picture of the actual situation. *The UK Dairy Roadmap, Showcasing 10 years of Environment Commitment* points out three key biodiversity achievements and commitments since 2008:[5]

- At its peak, 70 per cent of agricultural land was in agri-environmental stewardship schemes.
- Under the Campaign for the Farmed Environment, 30,000 km of hedgerows and 37,000 km of grass margins have been planted.
- By 2025 – dairy farmers to enhance and promote action being taken to improve biodiversity.

The Campaign for the Farmed Environment, mentioned in the second bullet point, was launched in 2009 as an industry-led alternative to regulation, focusing on farmland birds (not all wildlife birds), resource protection (primarily soil conservation and avoiding diffuse water pollution) and biodiversity provision.[6]

According to a Defra Campaign for a Farmed Environment (CFE) survey, land managed voluntarily (unpaid, i.e. without farmers receiving a public subsidy) in the 2014/15 farming year for England amounted to almost 269,000 hectares. Of this, 250,000 hectares was land managed voluntarily under CFE with 18,000 hectares of land classed as enhancements to basic Ecological Focus Area (EFA) fallow land requirements. The 269,000-hectare total represented a *drop* of *41 per cent* on the 2013/14 CFE area. This drop was not mentioned in the *Dairy Roadmap*!

Of particular note, grass buffer strips next to watercourses or ponds totalled 17,197 hectares in 2013, 17,000 hectares in 2014 and 12,627 hectares in 2015. Infield grass strips were 3,752 hectares in 2014, 4,562 in 2014 and 1,776 hectares in 2015. Unpaid environment management also dropped for management of maize fields to avoid erosion (-2,663 hectares) and fencing next to watercourse too (-2,907 hectares). Only two positive changes were recorded: pollen and nectar mixes (+2,125 hectares) and legume and herb-rich temporary grass (+589 hectares).

Whilst land under CFE is greater than before the scheme started, these recorded drops from 2013 to 2015 represent a marked regression against previous achievements. Not exactly an achievement. Notwithstanding that, most of the land attributable to CFE is arable and not dairy.

Defra stopped collecting CFE statistics after 2014/15, after which responsibility for recording etc. was passed on to the industry. I'm not aware of any updated corresponding statistics provided by the industry since. There may be a good reason for that!

But how does this relate to the achievements of 30,000 km of hedgerows and 37,000 km grass margins? Very little. These statistics are from publicly funded agri-environment schemes and not from unfunded CFE. CFE would have encouraged farms to take up agri-environment schemes, but the dairy achievement is misleading.

In a paragraph later in the document a distinction from CFE is somewhat made, but you would still be forgiven for believing this was *all* achieved by dairy farmers. It was not. The 30,000 km of hedgerows and 37,000 km grass margin statistics relate to the whole of agriculture and not just dairy, of which dairy would have only, like the rest of the CFE statistics, been a part.

Since 1947 around 200,000 miles (320,000 km) of hedgerows[7] have vanished from the nation's countryside due to increased farm mechanisation and intensification, and through other development. What's left, including trees and other vegetation, some in the livestock industry want to count towards their on-farm Net Zero carbon ambition. Much of this wanton destruction has been encouraged by generous, targeted government handouts supported by farmers and agricultural lobbyists alike. Hedgerows that remain are commonly damaged or neglected (including from the drift of chemicals and fertilisers), along with the life in and around them.

Many bird species are associated with species-rich hedgerows and hay meadows; many invertebrates and mammals too, with older hedgerows in particular offering an important breeding site, habitat, food and cover. Hedgerows should be part of an ecological system, not outliers within monocultures.

Being near the top of the food chain, bird populations are a good indicator of vibrant, healthy ecosystems. Defra's 'Agriculture in the United Kingdom' statistics for 2018[8] depressingly reports of a 70 per cent and 12 per cent decline respectively of the twelve farmland specialist and seven generalist bird populations between 1970 and 2017, with the steepest decline of specialist species occurring between 1978 and 1988 with a steadier decline thereafter. A similar trend is shown for all the nineteen farmland species monitored. The cause is mainly the negative impact of rapid changes in farmland management.[9] In more recent times (2011 to 2016) 21 per cent of the species population increased, 47 per cent showed no change and 32 per cent were still in decline.[10]

Given 71 per cent (17.4 million hectares) of the UK's land is used for agriculture,[11] the loss of farmland birds adds further to the nation's silent footprint. 10.072 million hectares of this is under permanent grassland and 6.084 million hectares is arable land.[12] According to Henri de Ruiter et al. (2017),[13] 44 per cent of the UK wheat crop and 87 per cent of all barley was used for animal feed in 2011. Another 1.759 million hectares in modern crop currency used to grow animal feed.[14] The cause of most of our countryside's silence laid bare, over seventy years in the making.

Features like hedgerows are being replanted, following years of removal to support agricultural production. This is welcomed, but it is somewhat wilful for the dairy industry to boast about replacing something it has, albeit directed by government policy and previous profit, ripped up in the first place. Replanting of hedgerows is also largely being accomplished through public subsidy payments, with only a part directly attributable to dairy.

Furthermore, UK dairy farming's offshored share of deforested rainforest used to grow imported soya bean meal for feed is 150,000 hectares (see chapter IV). With a dairy cow population of 1.9 million, that's 790 m^2 per dairy cow, or 12 hectares for an average herd size of

151 cows. Assuming a hedge is 2 m wide, a 1 km length of a hedgerow would occupy about one-fifth of a hectare (2,000 m²), or a 70 km length of hedgerow in 12 hectares of land (5 x 12). For a dairy factory farm with 1,200 milking cows that's 948,000 m² or 95 hectares, or a 474 km length of hedgerow. This amount of hedgerow contains a lower level of the biodiversity and carbon storage than an equivalent-sized rainforest too.

Let us not forget either the knee-high hay meadows foregone, 97 per cent of which have fallen to the plough since World War II. Meadows with throngs of wild flowers (e.g. yellow rattle, meadow buttercup, common bird's-foot-trefoil and lady's mantle, to name but a few) and grasses nodding rhythmically in the summer breeze, dancing butterflies, velvety bees gathering brimfuls of nectar, quivering hoverflies, the songs of chirping crickets and the sweet sound of skylarks.

The soil beneath provides an even greater reservoir of biodiversity, including microfauna (nematode, protozoa and rotifers), meso fauna (mites, springtails and enchytraeid worms), macrofauna (earthworms, ants, beetles, millipedes and centipedes) along with a network of arbuscular mycorrhizal fungi, which forms symbiotic relationships with many plants, helps develop soil structure and is an extremely important carbon sink.

The plough, copious chemical and organic fertilisers, pesticides, herbicides, fungicides, overgrazing and compaction over time has laid waste to much of this life, leaving us instead with green, lifeless deserts of Italian rye-grass.

Ironically, hay meadows, although species-rich, are termed 'unimproved grasslands'. In contrast, converted grasslands/pastures managed for livestock are perversely termed 'improved grasslands', based on the sward's quality, nutritional quality and its ruminant health-giving properties. The latter, ruminant health, is equally ironic given how traditional hay meadows were once used by farmers for their *grazing* medicinal/curative properties for sickened livestock.

Farmers largely put land in voluntary schemes like CFE whenever the threat of future legislation looms, though some do go above and beyond normal industry expectations. They are only able to go into such schemes when they can afford to. Voluntary initiatives often prove ineffective over time too. Good will/intention at the beginning somewhat withers and dies when profit erodes over time, or to avoid financial loss.

As a replacement to countryside stewardship schemes under the EU's Common Agriculture Policy (CAP), the UK is transitioning towards a new environmental land management (ELM) scheme. This could be a positive outcome from Brexit and is intended to be fully rolled out by the end of 2024, replacing payments to farmers based on how many eligible hectares they own to one where public money is only paid out for the delivery of public goods.

As described by Defra:

Under the ELM scheme, farmers will be paid for work that enhances the environment, such as tree or hedge planting, river management to mitigate flooding, or creating or restoring habitats for wildlife. Farmers will therefore be at the forefront of reversing environmental declines and tackling climate change as they reshape the future of farming in the 21st century.[15]

These are positive words, and ELM could increase the amount of land turned over to wildlife from around 4 per cent today to over 8 per cent in the future. However, whilst ELM has the potential to make a positive and welcomed contribution to biodiversity, they will not be anywhere near enough to achieve what is needed. The ELM schemes, like previous CAP schemes, is hindered by having to fit biodiversity around farming and in a way more acceptable to us than to nature. Furthermore, as most farmers will tell you, delivering for wildlife and the environment is expensive, and the funds to be made available are likely to fall way short of what is needed to enable the level of change we need to see (and to ensure take-up). As long as that is the case, what will be achieved will fail to function in a complete manner. Instead, we need to build on ELM with a transformative approach which makes farming a true part of functioning ecosystems. This can't be achieved without making space for wildlife and the most sustainable and climate friendly way to do that is to remove livestock's longer shadow from our landscape and allow nature to move in and properly flourish. Nevertheless, ELM is an important step in the right direction.

According to the Food and Agriculture Organisation of the United Nations (FAO), by 2050 meat consumption is projected to rise by 73 per cent and dairy consumption by 53 per cent.[16] This assumes we continue eating our current Western diet, combined with a spiralling increase in per capita animal protein consumption in developing countries as their affluence and population grows. Should that come to pass, we can reasonably expect further livestock protein growth through intensive scale production within and outside of the UK. This is unhelpful in our quest for a more sustainable and healthy food production system, and risks countering biodiversity gains and the livestock industry's own Net Zero GHG emission ambition – a GHG emission ambition which would already require considerable land use change, a mammoth trans- formation of farming practices and an eye-watering amount of money to actually achieve.

This is not in harmony with the planetary health diet created by thirty-seven world-leading scientists, commissioned by the EAT-Lancet Commission on Food, Planet, Health, which concluded, based on hard epidemiological evidence, that the UK and our fellow Europeans need to eat at least 77 per cent less red meat, a lot less eggs and dairy, and make the majority of our diet plant-based. The livestock industry in the UK and the rest of the world are leading us in the wrong direction.

On a more positive note, in researching this section, I was somewhat encouraged to see the Wild East initiative, founded by three farmers, Lord Hugh Somerleyton, Oliver Birkbeck and Argus Hardy, to 'restore space for nature and to restore our connection to nature too.'[17] The aim of this initiative is to rewild 250,000 hectares of East Anglia, about the size of Dorset, which they hope to achieve with the support of fellow farmers, and other landowners, by returning at least 20 per cent of their land back to nature. When it comes to restoring wildlife, the landscape's size, scale and interconnection is vital, which these farmers, and those supporting them, are recognising.

In farming literature, it is not helpful to give an unjust representation of how well livestock farming is doing. Many livestock farmers do what they can within current government policy and their own means, and we cannot realistically expect them to do what's needed without the right support. Whilst most farms are land-rich, much land is held as collateral against loans/debt. Much of our farmed land is also under tenancy, restricting some farmers' better intentions. Average farm income for many farmers falls well below the national average, with farms often having to increase livestock numbers and/or make farm animals work harder to offset ever-diminishing returns. Society can only start to help farmers when we are all honest about the situation. Any betrayal of the actual situation serves to undermine any case for support.

In my book *Carnivorous Plants of Britain and Ireland* (2016)[18] I discuss habitat loss of peatlands, which many moons ago were regarded as unproductive wastelands similar in value to 'unimproved grasslands'. Next to a former peatland in Wareham, Dorset, England, is a commemorative plaque which states:

PARSONS PLEASURE

This experiment was designed by Frank Parsons M.B.E.,
Chief Forester at Wareham from 1950 to 1968.
The waterlogged heath in this area is one of the worst sites
in Europe. The experiment demonstrates a practical way of
making it plantable by mechanical drainage and fertilising.

Like pre-existing landscapes turned into livestock farming pastural monocultures, unless tamed to provide a commodity which can be sold, peatland was once seen as a worthless piece of our landscape which needed to be improved by mechanical and fertilising interventions. Today in Wareham, logged trees are not being routinely replanted and great effort is being made to reverse the decades of damage done and to return this once highly valuable habitat to its former glory and nature. Lessons have been learned, and many other areas of damaged peatland habitat across the UK are being restored, breathing life back into the landscape, returning carbon to the soil and helping to reduce flood risk by capturing and storing water.

A commemorative plaque like Parsons' may not feature on the landscapes we have relinquished to livestock farming, but it nevertheless exists in kind in what the agricultural industry refers to as 'improved pasture': vast areas of previous breathtaking biodiversity and functioning ecological food webs turned into barren, relatively lifeless green deserts.

We need a new sign, and for this I turn to the enlightening words of Rewilding Britain:

LEGACY'S PLEASURE

[The pasture which once dominated this landscape,
and once fed animals to feed us, was one of the worst sites in Europe.]
This rewilding initiative demonstrates a practical way to enable
the large-scale restoration of ecosystems where nature can take
care of itself. It has reinstated natural processes and has allowed
missing species to reshape the landscape and the habitats within. It
has encouraged a balance between people and the rest of nature,
where each can thrive. It has provided opportunities for
communities to diversify and create nature-based economies; for
living systems to provide ecological functions on which we all
depend; and for people to re-connect with nature.[19]

Citations:

1. AHDB (2018), *Landscapes without Livestock, Visualising the impacts of a reduction in beef and sheep farming on some of England's most cherished landscapes.*

2. Isabella Tree (2018), *Wilding: The Return of Nature to a British Farm.* Pan Macmillan, London, UK

3. Ibid.

4. Rewilding Britain. https://www.rewildingbritain.org.uk/ and https://www.rewildingbritain.org.uk/rewilding/what-does-rewilding-look-like.

5. Dairy UK (2018), *The UK Dairy Roadmap: Showcasing 10 Years of Environment Commitment*

6. Defra (2013), *Campaign for the Farmed Environment: Summary of Evidence.* Defra Agricultural Change and Environment Observatory Research Report No. 33. https://assets.publishing.service.gov.uk/government/uploads/system/uploads/attachment_data/file/183824/defra-stats-foodfarm-environ-obs-research-setaside-farmenviroment-cfeevidencesummfeb2013-130214.pdf

7. RSA Food, Farming & Countryside Commission (2019), *Our Future in the Land*

8. Defra (2018), *Agriculture in the United Kingdom.* National Statistics at Gov.uk

9. Defra (2018), *Wild Bird Populations in the UK, 1970 to 2017*

10. Ibid.

11. Defra (2018), *Agriculture in the United Kingdom.* National Statistics at Gov.uk

12. Ibid.

13. de Ruiter, H. et al. (2017), 'Total global agricultural land footprint associated with UK food supply 1986-2011.' *Global Environmental Change* 43, pp. 72-81

14. Defra (2018), *Agriculture in the United Kingdom.* National Statistics at Gov.uk

15. Defra (25 February 2020), *New details of the flagship Environmental Land Management scheme unveiled by Environment Secretary.* https://deframedia.blog.gov.uk/2020/02/25/new-details-of-the-flagship-environmental-land-management-scheme-unveiled-by-environment-secretary/

16. Food and Agriculture Organisation of the United Nations (2011), *Major gains in efficiency of livestock systems needed* http://www.fao.org/news/story/en/item/116937/icode/.

17. Wild East. https://wildeast.co.uk/

18. Bailey, T. and McPherson, S. (2016), *Carnivorous Plants of Britain and Ireland.* Redfern Natural History, Poole, UK

19. Rewilding Britain. https://www.rewildingbritain.org.uk/rewilding/.

29. Water Quality and Water Resources

Water is an essential resource. Maintaining its abundance and securing its quality is of utmost importance.

The UK's complex weather patterns provide plenty of rain, but not always in the right place, at the right time and in the right way to replenish all water resources. Some parts of the UK could do with a bit more rain, in some places a bit less, and sometimes we need to take water from one place to make good a shortfall in another.

In recent years, as the effects of climate change take their course, we are seeing more extreme weather events – so much so that it is starting to become normal. In the autumn farmland can be inundated with rain and in the spring be bone dry. Both can make planting and getting a crop away extremely challenging, no less so in autumn 2019 when so much rain fell farmers struggled to get their crops in the ground, some that could were spoiled and others like erosion-ready forage maize had to be harvested late in very poor soil conditions. Then in spring 2020, coinciding with the first Covid-19 pandemic lockdown, hardly a drop of rain fell in many areas, resulting in poor establishment conditions and the need for some farmers to dip early into their allotted water abstraction allowance.

Generally, in the UK, the majority of dairy and other livestock farms are located in areas with high rainfall, where grass and other forage crops grow well without the need for irrigation. Rain, in footprint terms referred to as 'green' water, provides most of the water needed to grow grass and many of the other crops grown in the UK to feed livestock. Only a relatively small amount of 'blue' water (i.e. water abstracted directly from surface water or groundwater or taken from mains supply, and whose use can contribute to lowering local water tables, drying wetlands, lowering of river flows and reducing water quality) out the total water footprint (the sum of green, blue and grey water) is used in the UK to support livestock farming compared to very dry, hot, sunny places like California.

Total on-farm water use in England and Wales has been estimated at above 300 million m^3 per year,[1] with irrigation accounting for 50–120 million m^3 depending on rainfall in any given year.[2] Irrigation is mainly used to grow field crops, in particular potato, vegetable and soft fruit, which can use around 85 per cent of total water.[3]

In terms of livestock use, a study by King et al. (2006)[4] estimated that livestock rearing in England accounted for 119 million m^3 per year of blue water use:[5]

Table 22: Livestock water use

Livestock type	Blue water use (million m³/year)	Of which, drinking water (%)
Cattle	82	79
Sheep	17	>99
Poultry	12	96–99
Pigs	8	87–99

The vast majority of water use is, as shown in Table 22, used for drinking, with dairy consuming the most other water (e.g. washing and refrigeration) at around 21 per cent of the total.[6]

Whilst blue water use is a sustainability issue, caution has to be applied when comparing global water footprint figures to UK livestock water use without adding context, including breaking down the figures into their constituent parts of blue, green and grey water. For instance, the England and Wales 300 million m³ per year water use represents around 1.5 per cent of total freshwater withdrawals,[7] and green water, from rainfall, makes up most of the UK's livestock water footprint. However, grey water cannot be so readily ignored while the UK livestock industry continues, unabated, to externalise their water pollution costs.

Grey water refers to pollution and can be defined as the volume of freshwater that is required to assimilate the load of pollutants, given natural background concentrations and existing ambient water quality standards. Grey water has the potential to:

- lower the chemical quality and ecological status of fresh and coastal water (from inputs of nutrients, pesticides/herbicides/fungicides, FIOs, sediment etc.).
- cause detriment to water's biological community, and use as a water resource in its broadest possible sense.
- Lower the chemical quality of groundwater and increase FIOs.

Grey water is what we have relied on for many years to dilute and disperse pollutants from livestock agriculture in the vain hope those pollutants would be retained in soil, and be broken down or released in a controlled manner within the capacity of the environment. This is often largely dependent on a myriad of voluntary initiatives, codes of practice and guidelines, which without sufficient regulatory scrutiny and investment in pollution prevention measures have failed to deliver adequate environmental protection. This is far from satisfactory, with the farm-gate price of livestock products not reflecting the true cost of mitigating the environmental consequences associated with them.

In dairy, the volume of drinking water needed is associated with the amount of dry matter fed to the cows, their expected milk yield and the temperature at different times of the year. Grazing cows have the lowest impact.

Retailed British dairy milk has, on average, a blue water footprint of around 10 litres per kg. The majority is used up to the farm-gate, at around 8 litres per kg of fat-corrected milk (FCM).[8] FCM is a method used to standardise milk production for comparison between cows.

Table 23: Average blue water used in British dairy systems (FCM) at the farm gate

Production system	Blue water, litres/kg FCM
Spring calving	7.4
Autumn calving	7.5
All-year calving	7.5
Zero grazing	7.6
Organic	8.1

Source: Table 1. The Volumetric Water Consumption of British Milk. Cranfield University. November 2012.

Post farm milk processing adds a further 1.3–2.5 litres of water per kg of milk.[9]

Typical freshwater sources used in dairy production is shown in Table 24.[10]

Table 24: Freshwater sources for dairy production

Water source	% of farmers using source[1]
Collection of roof water	15
Spring	32
Borehole	36
Mains – non-metered	22
Mains – metered	75

Table notes:

[1] Percentages do not add up to 100 per cent because farmers use more than one source of freshwater.

Typical daily water use by livestock is shown in Table 25.

Table 25: Overview of typical water use requirements by all livestock

Livestock type	Litres/day
Dairy cow	104–122 (lactating)
	20–59 (dry period)
Beef cow	25–45
Calf	5–25
Sheep	3.3–7.3
Pig	6–10 (dry sow)
	15–30 (farrowing sow)
	3–6 (grower/finisher)
Poultry	0.09 (pullet)
	0.20 (broiler/caged layer)
	0.22 (non-caged layer)

In 2010, Cranfield University calculated the blue water footprint of beef and lamb systems for England.[11] The national baseline for beef came out at 67 litres per kg of carcass weight (including green and grey water, 17,657 litres per kg) and for sheep 49 litres per kg of carcass weight (including green and grey water, 57,759 litres per kg). The total figures are largely made up from green water: 14,900 litres per kg and 55,800 litres per kg respectively. For beef this is higher than the global average commonly used figure of 15,500 litres per kg, though as UK livestock is often located in areas where the rainfall and rainfall surplus is high, it would be wrong to draw direct comparisons with other countries.

Cranfield University's English beef and lamb production report concluded:

> [...] that the total water footprint of English lamb and beef (m^3/kg) is of a similar order to estimates from other countries, however, the overwhelming dominance of green water in the total figure demonstrates that, compared to livestock production systems that rely on irrigated feed, the hydrological impact of English meat production is very small.

Because of the variability of blue, green and grey water footprints across countries, they cannot be compared like for like. The use of green water in figures in this respect is particularly hotly debated, given it would exist with or without livestock. I would express caution in citing global and UK water footprints as the same, as they are not. However, for me, there is a sound justification for combining the UK blue and grey water footprints.

For comparison, livestock's 119 million m^3 per year blue water footprint is enough water to supply over 700,000 average four-people households (165 m^3 per person per year). With many farms intensifying, the drain on water resources has also become increasingly localised, which could prove harmful if the shortfall was abstracted from springs, rivers or groundwater. For example, an average cow would use around 37 m^3 per year and an average herd of 150 dairy cows 5,500 m^3 per year. A super-dairy, assuming the same water use and with ten times the number of cows, would use 55,000 m^3 per year (333 four-person households).

Many farmers still allow their livestock to access watercourses too, which they drink from and defecate in, and so that needs to be factored in. And, as I repeat elsewhere, water resource use is just one part of livestock's longer shadow. All footprints need to be considered as a whole.

Before leaving this account, it would be remiss not to briefly discuss how the UK's livestock's longer shadow otherwise impacts green water. Livestock farming causes a lot of soil damage, primarily through the sowing of rye-grass monocultures and maize, heavy grazing and poaching which contribute to run-off, erosion, flooding and increases nitrous oxide emissions. This is because dysfunctional soils are unable to capture, store and filter rainwater, disrupting the natural soil hydrological cycle.[12]

Natural soils buffer the flow of water into surface water and groundwater, but once damaged, they retard air and water movement and create anaerobic soil conditions. The lack of diverse vegetation compounds the problem, increasing the risk of run-off, which collects sediment and nutrients from fertilisers and manures along the way, and which in turn harms the quality of aquatic ecosystems or terrestrial ecosystems dependant on them. The hydrological impact of English meat production is not quite so small after all.

Reducing risk at source has to be an essential part of flood prevention strategies, and farmers need to be sufficiently rewarded for delivering this public good. We will fail those affected by flooding if we do not do that, as a failure to capture and store water results in millions of cubic metres of *avoidable* rainwater pouring off land, into watercourses and into properties lower down in catchments. This is particularly important with increased extreme weather events, which can result in a month's rainfall or more falling within a very short time. To do this, we need to rewild huge swathes of grassland and reduce the effectiveness of artificial land drainage, including drainage on peat bogs. Capturing, filtering and retarding rainwater water releases it slowly to watercourses, helping to maintain healthy river levels, recharge vital groundwater resources and rebuild peatland habitats.

Addressing flooding at source has multiple benefits, such as biodiversity, carbon sequestration, cleaner water and resources. Rewilding also allows us to reintroduce species such as beavers, which contribute to flood prevention and improved water quality.

Citations:

1. Global Food Security, Sub Report, *Farming and Water 2, Agriculture's Impact on Water Availability*
2. Defra, ENV15 – Water abstraction tables for England. https://www.gov.uk/government/statistical-data-sets/env15-water-abstraction-tables
3. Global Food Security, Sub Report, *Farming and Water 2, Agriculture's Impact on Water Availability*
4. King, J., Tiffin, D., Drakes, D. and Smith, K. (2006), *Water Use in Agriculture: Establishing a Baseline*. Defra Project WU0102
5. Defra (2012), *Sustainable water for livestock*. Defra Project WU0132
6. Ibid.
7. Global Food Security, Sub Report, *Farming and Water 2, Agriculture's Impact on Water Availability*
8. Cranfield University (2012), *The Volumetric Water Consumption of British Milk*. Report for DairyCo
9. United Nations Environment Programme Division of Technology, Industry and Economics and Danish Environment Protection Agency / Danish Ministry of Environment and Energy. Cleaner Production Assessment in Dairy Processing (2000), *Cleaner Production Assessment in Dairy Processing*
10. AHDB (2009), *Dairy Co Farmer Water Survey Report*
11. Chatterton, J. et al. (2010), *The Water Footprint of English Beef and Lamb Production: A Report of EBLEX*. Cranfield University, Cranfield, UK
12. Stika, J. (2016), *A Soil Owner's Manual: How to Restore and Maintain Soil Health* (independently published)

30. Pandemics and Foodborne Illnesses

Unsustainable exploitation of the environment due to land-use change, agriculture expansion and intensification, wildlife trade and consumption, and other drivers, disrupts natural interactions among wildlife, livestock, people and their pathogens and has led to almost all pandemics.[1]

This is a summary statement of the Intergovernmental Platform on Biodiversity and Ecosystem Services (IPBES), a multidisciplinary expert panel consisting of twenty-two experts from all regions of the world, of whom seventeen were experts nominated by governments and organisations.

According to IPBES,[2] 70 per cent of emerging diseases (e.g. Ebola, Zika, Nipah encephalitis), and nearly all recognised pandemics (e.g. influenza, HIV/AIDS, Covid-19) are zoonoses (i.e. stemming from microbes of animal origin, which 'spill over due to contact among wildlife, livestock, and people'). For example, the emergence of the highly infectious and pathogenic avian influenza (bird flu) from the intensification of poultry production and the enhanced spillover of the lethal zoonotic Nipah virus ('barking pig syndrome') transmitted from fruit bats to intensively farmed pigs and then from the pigs to us (and to a host of other farm animals and domestic pets).

Another harmful zoonotic pig pathogen which can infect people is the highly virulent, meningitis-causing Steptococccus suis virus, which has shown up in China, North America and in Europe.

Livestock farming is not the only vector, with human settlement and urbanisation, and wildlife farming, also significant drivers. Notwithstanding that, livestock farming uses around 80 per cent of the world's farmed land and as such is central to most land-use change (e.g. clearance of biodiverse habitats for the keeping of livestock and the growing of crops to feed them, livestock intensification and the close confinement of millions of animals in buildings and other unnatural spaces).

The IPBES[3] makes it clear that pandemics are due to anthropogenic changes, and that 'blaming wildlife for the emergence of diseases is thus erroneous, because emergence is caused by human activities and the impacts of these activities on the environment'.

The solution to lowering our exposure to pandemics is significantly linked to addressing patterns of unsustainable consumption, including 'reducing excessive consumption of meat from livestock production' and not through our current business-as-usual approach of containing and controlling disease, but rather 'reducing the drivers of pandemic risk to prevent them before they emerge'.[4]

Unfortunately, the livestock industry's response, which remains resolute even within the grip of the Covid-19 pandemic and the climate crisis, is to further intensify and confine farmed livestock and to expand the trade of meat, dairy and eggs globally. At a time when us humans are told to live in bubbles, wear face masks and reduce our personal contact with each other, the livestock industry has opposing ideas for their unfortunate residents.

In the face of questions about the risk of rearing and spreading the next pandemic in factory farms, systems primarily driven to stock the shelves of supermarkets and other retailers with ever cheaper intensively farmed, poorer quality animal flesh for the mass market, the public relations and communication teams go into overdrive, gaslighting their customers with claims of rearing to Red Tractor or RSPCA Assured farm standards.

Keeping livestock like chickens a few more inches apart than the law would allow is not something that would fill me with much confidence in terms of avoiding disease transmission, not least when bird flu outbreaks are regularly confirmed on farms and we now have human cases of Covid-19 in Denmark with the SAR-CoV-2 variant (the virus which causes Covid-19) associated with the vanity farmed mink trade. So far outbreaks of bird flu have not been associated with the deadly H7N9 strain, but it's worth noting that the devastating 1918 flu pandemic, which killed 50 million people, was thought to have originated in birds.

As widely reported in January 2021, Russia registered the first case of the bird flu strain H5N8 being passed from poultry to humans. The infection happened in December 2020, when seven workers were infected. All seven recovered, fortunately. No human-to-human transmission was found, though the potential for a future mutation of the virus cannot be ruled out. H5N8 joins the list of other virus strains that have jumped the species barrier to people, some of which have led to deaths. Only time will tell if a future mutation enables the virus to spread amongst humans. What is certain, is the more poultry factory farms there are the greater the risk. The UK is not exempt from H5N8, with five cases reported in the UK in November 2020 (Gloucestershire, Dorset, Devon, Cheshire and Kent). All infections were contained, with all birds being slaughtered to stop the spread.

We do not need to live with these risks, or the animal unkindness that goes hand in hand in factory farms. Whilst farmers take their biosecurity seriously – their livelihoods depend on it – they are not leakproof. Furthermore, interest in biosecurity stops once livestock reach its slaughter weight and is crated, transported and slaughtered in open systems.

Whilst farmers employ some biosecurity measures to avoid the next pandemic, they, and slaughterhouses, are much less successful in containing foodborne illnesses such as campylobacter, E. coli 0157, salmonella and listeria too. Campylobacter is one of the most common causes of food poisoning in the UK, courtesy of raw chicken, turkey and other poultry (299,392 cases costing £712.6 million); E. coli 0157 is found in undercooked beefburgers, minced beef, contaminated cooked meats and unpasteurised milk (468 cases costing £3.9 million); salmonella poisoning is found in undercooked meat and poultry, untreated milk and raw or undercooked eggs (31,601 cases costing £212 million); and listeria which can be present in unpasteurised soft cheeses (e.g. brie and camembert) and meat pâtés (162 cases costing £37.4 million).[5] Figures in brackets are the median numbers of recorded UK foodborne illness-related cases per year and the total cost burden associated with them (2018).[6] Many more unattributed foodborne illnesses (UFI) occur, estimated to treble the number of cases and associated costs (to about £3 billion) for these four pathogens. Add in a good dose of Clostridium perfringens, found in beef, poultry and even dried or precooked foods, and the annual median count increases by nearly 85,000 cases per year with a total cost of £101.5 million (>£300 million, including UFI).[7]

By properly preparing, storing, cooking and handling raw animal-based foods, and avoiding cross contamination, we can avoid getting diarrhoea, stomach cramps, vomiting, fever, nausea, headache and dizziness. One could be forgiven for wondering why these common foodborne poisons could make us so ill if we evolved so successfully to eat animals. No other animal-eating species seem to have a problem, just us. While the quantity of contamination is unnaturally increased by intensively confining and unkindly treating animals (in amongst each other's excrements and other bodily fluids), and by the ways we slaughter animals, humans are incapable without preserving or cooking, of eating raw animal-based foods without risking illness from these common pathogens. Foodborne illnesses are much less common in plant-based foods, unless they have been cross-contaminated with animal foods and animal excreta which may have been applied to grow them.

The total cost of animal agriculture on our health (and environment) runs into eye-watering £billions, which we pay for in our health and taxes.

This topic is so huge I cannot possibly do it the justice it deserves in this book. For those who would like to understand more I would thoroughly recommend reading Dr. Michael Greger's book *How to Survive a Pandemic* (2020).

Quoting from Dr. Greger's book, 'If you want to create catastrophic pandemics, then build factory farms.'

Citations:

1. Intergovernmental Science-Policy Platform on Biodiversity and Ecosystem Services (2020), *IPBES Workshop on Biodiversity and Pandemics: Workshop Report*.
2. Ibid.
3. Ibid.
4. Ibid.
5. Daniel, N. et al. (2020), *The Burden of Foodborne Disease in the UK 2018*. Food Standards Agency
6. Ibid.
7. Ibid.

CHAPTER IV

Climate Crisis

If global animal agriculture was a country it would be the second-highest emitter of greenhouse gases. [...] Factory farms are undermining both the climate ambitions of high-street brands and the viability of the Paris Agreement.

Jeremy Coller, FAIRR network

31. Greenhouse Gases

The process of measuring and reporting greenhouse gas (GHG) emissions started in the UK in 1988, before being amended to fulfil the UK's obligations under the United Nations Framework Convention on Climate Change (UNFCCC), which came into force on 21 March 1994.

Following on, in December 1997 the Kyoto Protocol, the international treaty on global warming, was agreed and came into force on 16 February 2005.

The protocol applied to 192 countries ('Parties'), each committing to cutting GHG emissions in the atmosphere to 'a level that would prevent dangerous anthropogenic interference with the climate system'.[1]

On an annual basis Parties are required to submit information on their net national greenhouse gas inventories to the UNFCCC secretariat, in compliance with UNFCCC requirements. The requirements include the use of source data and methods consistent with Intergovernmental Panel on Climate Change (IPCC) inventory reporting and good practice guidelines.

The original EU commitment under the Kyoto Protocol was a greenhouse gas emission reduction of 8 per cent below 1990 levels by 2012. Due to growing concerns, in December 2008 the EU agreed to a more stringent target of 20 per cent below 1990 levels by 2020, or 30 per cent subject to a new global agreement.

Further along, the UK established new legal requirements to monitor and report on greenhouse gas emission reductions under the Climate Change Act 2008. The act raised the greenhouse gas emission reduction bar to a more challenging 80 per cent by 2050 compared to the 1990 baseline (for carbon dioxide, methane and nitrous oxide). For hydrofluorocarbons and perfluorocarbons and sulphur hexafluoride the baseline year is 1995.An amendment to the act in 2009 sets a revised 34 per cent interim target reduction by 2020 (rather than 26 per cent under the previous 2008 act).

Powers to implement greenhouse gas emission reductions are devolved to the Scottish government, Welsh government and the Northern Ireland Executive. As such, the devolved administrations' emission reduction targets vary somewhat from England's.

Agriculture is a major source of greenhouse gases, with methane (CH_4) largely arising through enteric fermentation by ruminants (cow burps) and from animal manures; nitrous oxides (N_2O) losses from manufactured fertiliser and organic manure applications; and carbon dioxide (CO_2) from fertiliser manufacture and other fossil fuel use. Disturbance of soil and land-use change (e.g. deforestation and grassland conversion to arable) are also contributing factors.

In 2017, UK greenhouse gas emission were estimated at 460.2 million tonnes of carbon dioxide equivalent (MT CO_2e).[2] The estimate is the sum of seven greenhouse gases covered by the Kyoto Protocol. Of the seven, three are of importance to agriculture: carbon dioxide, methane and nitrous oxide. Together they contribute 45.6 MT CO_2e, approximately 10 per cent of all UK greenhouse gas emissions (CO_2, CH_4 and N_2O: 5.6, 25.7 and 14.3 MT CO_2e respectively) with about 72 per cent (33.03 MT CO_2e) of that associated with the livestock sector (7.2 per cent). 10 per cent may seem small compared to other sectors, but food production is only 0.5 per cent of UK GDP, making it a very large contributor to emissions and climate change, relative to the rest of the economy. Furthermore, the 10 per cent excludes greenhouse gas emissions released from soils and peatland as a result of agriculture (the total includes livestock and manure, synthetic fertiliser, fuel and machinery).

Agriculture is unusual compared to other sectors, with CH_4 and N_2O making up most of the total. For the other sectors CO_2 is by far the most dominant greenhouse gas.

Table 26: UK greenhouse gas emissions[3]

	CO_2	CH_4	N_2O	Other gases	Total (MT CO_2e)
Transport	124.6	0.1	1.2		125.9
Energy supply	106.0	5.8	0.8		112.6
Business	66.1	0.2	0.9	13.0	80.1
Residential	64.1	1.0	0.2	1.6	66.9
Agriculture	5.6	25.7	14.3		45.6

Source: From Tables 3 to 7, Final UK greenhouse gas emissions national statistics 1990–2017 Excel data tables.

Overall emissions in the agricultural sector have increased since 2010, from 44.6 MT CO_2e to 45.6 MT CO_2e in 2017, though they still remain lower than the 1990 baseline of 54.0 MT CO_2e.

Care must be taken when quoting greenhouse gas emission figures as the percentages are proportionate to the level of each sector activity that occurs in England and each of the devolved administrations. For example, England is much more industrialised than Northern Ireland (NI), and therefore it should not be read that NI's agriculture/livestock sector is less efficient and therefore more polluting per unit of animal-based production.

Table 27: Percentage agriculture and livestock emissions[4]

	Agriculture sector		Livestock sector	
	MT CO_2e	% of country's emissions	MT CO_2e	% of country's emissions
England	27,032,000	7.85%	18,932,000	5.5%
Scotland	7,565,000	19.57%	5,416,000	14%
Wales	5,613,000	13.44%	4,314,000	10.33%
Northern Ireland	5,385,000	26.09%	4,373,000	21.9
	45,595,000	10%	33,035,000	7.2%

Source: Based on National Atmospheric Emissions Inventory.

The same is true when drawing comparisons between countries, for example with the US. On paper, US agriculture contributes about 8.4 per cent to their greenhouse gases inventory compared to the UK's 10 per cent,[5] but that does not mean they are better. One way around this is to look at greenhouse gas emissions on a per capita basis.

When this is done, the UK 2017 agriculture sector's 45.6 MT CO_2e works out at about 0.68 t CO_2e per capita compared to US 2017 agriculture sector's 1.66 t CO_2e per capita (542.1 MT CO_2e).[6]

Another example close to the UK is the Republic of Ireland. In Ireland 34 per cent of the country's total emissions come from agriculture, with emissions projected to increase by 4 per cent up to 2030 to a total of 21 MT CO_2e.[7] Nearly 60 per cent of agriculture's total comes from enteric fermentation and a further 10.1 per cent from manure management. According to a report in Independent.ie (2019),[8] Ireland's agricultural-related greenhouse gas emissions rose 1.9 per cent in 2018, 'driven by an increase in dairy cows, whose number went up 27 per cent in the past five years'. The response to the rise from Ireland's then Prime Minister, Leo Varadkar, was that the country's agriculture emissions can be reduced without lowering the herd number, an option he saw only as a last resort. In doing so, he looked towards familiar alternative options cited by the livestock industry, such as forestry, biogas, slurry and fertiliser controls.

In making Ireland's case, Mr. Varadkar pointed out that the country exports 90 per cent of the food it produces and that 'the way emissions are categorised will have to be changed, as Ireland mass-produces food for other countries', and 'that [all that] food production gets accounted for in Ireland as a contribution to global warming, but that food production is going to have to happen, it's going to have to happen somewhere'.

Ireland's economy has become particularly dependent on livestock agriculture ever since the country turned to livestock intensification to get them out of a financial crisis back in 2010, and it's true that other countries benefit somewhat through offshoring a part of their own greenhouse gas emissions by importing Irish animal-based products. But animal food production doesn't have to happen elsewhere if Ireland reduces its herd size. We need to move on from animal agriculture, not create more. Furthermore, Ireland is itself an importer of livestock products from other countries (e.g. pork, chicken, eggs, beef and dairy), including the Americas, which have a whole host of associated environmental footprints, including greenhouse gases.

Like other countries, Ireland's options to tackle the climate emergency and other pollutants without reducing livestock are somewhat speculative. After spending millions of euros of public money since 2010 building an economy based around livestock intensification to bankroll its economy, the option of spending millions more to reverse that policy is politically unfavourable. Alternative mitigations, like tree planting, biogas plants, slurry and fertiliser controls and so, on can help reduce emissions given enough time, of which we have little, but throwing more livestock emissions (27 per cent in the last few years) on to the bonfire, which will also need to be mitigated, is nonsensical. More so as the environmental issues of livestock intensification in Ireland, mainly from the expansion of dairy farming, go well beyond greenhouse gas emissions, with the country also suffering a well-documented phosphorus water pollution crisis – and ammonia too!

Northern Ireland has similar issues to Ireland beyond greenhouse gas emissions, where livestock production, in particular intensively reared animals, has resulted in a build-up of phosphorus in the environment which exceeds the needs of the soil and crop on the land, which in turn has increased the risk of phosphorus loss to surface waters.

According to Doody et al. (2020),[9] sheep have the lowest phosphorus efficiency conversion to food (5 per cent), followed by cattle (16 per cent). Pig and poultry have the highest efficiency at 42 and 35 per cent respectively, with the greatest efficiency associated with crop production (grass and arable). Cattle are a particularly potent phosphorus source, given their size and dominance in the livestock sector. This phosphorus issue is acute in Northern Ireland (NI), where 93 per cent out of 75 per cent of NI's land is dedicated to grassland farming. Doody et al. (2020)[10] also report an upward trend in the phosphorus surplus since 2008, mainly due to the use of imported feed, from 13.4 to 16.8 kg of phosphorus per hectare (2008 to 2017). This increase is in spite of past efficiencies in phosphorus use, which shows how despite improvements on one hand the continued growth of the livestock sector often manages to give negatively on the other. The authors also reported that livestock manure phosphorus inputs in NI are 20 per cent higher than the total phosphorus demand for NI.

Britain and Ireland's agricultural system, dominated by grassland and grazing ruminants, is taking us on a perilous journey towards increased phosphorus pollution. Dairy and beef farms are becoming more intensive and concentrated in the wetter west and south west, with pig farms growing in the opposite direction. Dry parts of the country are not less polluting, as there is less water to dilute leakages from livestock farming. Poultry farms are also concentrating in catchments (e.g. the River Wye/River Lugg), in close proximity to their feed manufacturers/importers of rainforest-ready soya beans. We can expect many catchments where livestock farming concentrates to become phosphorus pollution hotspots. Conversely, arable farming is much more phosphate efficient and more likely to suffer a deficit. Long-term agricultural policy has relentlessly driven farmers away from small/extensive mixed farms towards concentrated/intensive livestock production. As such, we have far too much organic manure, which is handled or directly deposited where nutrients like phosphorus and nitrogen are not used efficiently and progressively enter our groundwaters, inland freshwaters or coastal waters, or into springs, wells or boreholes.

The phosphorus rises continued following NI's vaunted 2013 *Going for Growth Strategy*,[11] which, given NI's dominant indigenous livestock industry (14 million chickens, 1.8 million sheep, 1.6 million cattle and 425,000 pigs), has not done well for the environment in that respect. NI sought to grow, in part, by becoming one of the world's most carbon-efficient places by avoiding offshoring agricultural production to less environmentally efficient regions of the world, and by working livestock even harder. A similar strategy to Ireland.

When less carbon-efficient regions are mentioned, please remain mindful that a large reason for this is that they are carrying out land-use changes (e.g. deforestation) to accommodate their livestock, which the UK did a long time before 1990 and therefore is not included in the UK greenhouse gas inventory. Furthermore, the carbon footprint of soya bean meal imported from South America to feed livestock is attributed to the place of origin and not to the receiving country.

One way of becoming more carbon-efficient in the livestock sector – but which is not specifically expressed in the strategy, I must make clear – is increasing the milk yield by several thousand litres per dairy cow per year. Such a move would result in fewer dairy cows, but more imported feed for the remaining extra hungry cows, growing more erosion-ready maize, more cows housed all year and more slurry per cow, which would all act to counterbalance any savings. Housed cattle also increase the amount of ammonia emissions, unless abated, and the need for increased veterinary intervention to avoid painful conditions such as mastitis, amongst other issues.

There could also be a move somewhat in the other direction with smaller, more extensively farmed dairy herds, but to maintain and then increase NI's total milk yield the overall yield would need to increase – *to feed the world*. A move to extensively grazed livestock can lead to an all or mostly autumn-winter grazing regime, increasing the risk (i.e. from directly deposited excreta and keeping livestock on wet soil) of: nitrate-nitrogen leaching; preferential drainage of ammonium-nitrogen, phosphorus and microbial pathogens; and sediment and associated nutrient loss from an increased risk of soil compaction and run-off. Such a scenario would be perpetuated in NI, as 57 per cent of NI soils are classed as having a high run-off risk.[12]

Offsetting livestock greenhouse gas emissions, by including natural carbon sequestration from semi-natural grassland, dwarf shrub heath, bracken, mountain habitats, fens, marsh, swamp and bog, which all fall within the definition of 'Grassland' in the UK's Land Use, Land Use Change and Forestry inventory, would also be needed to cook the livestock sector's carbon ambitions book.

This problem is not just limited to the Ireland of Ireland, but to England, Wales and Scotland too, where livestock numbers are also rising beyond the capacity of the land. One example is the aforementioned River Axe Special Area of Conservation (a catchment covering and area of 308 km² across Devon, Somerset and Dorset), which is in an unfavourable and declining condition, owing to nutrient enrichment (i.e. high phosphorous concentration), resulting in good part from the intensification of dairy farming (and associated erosion-ready maize production). A livestock growth strategy of the kind developed in NI is not going to save the River Axe SAC any time soon, or anywhere else for that matter.

The most effective way to make a real difference is to drastically cut livestock numbers without increasing yield, but that is not on the industry's agenda – only *sustainable* growth and offsetting based on *ambitions*.

One of the UK's biggest sources of greenhouse gas emission offshoring occurs within the livestock sector, connected to the importation of animal feed from South America. In 2011/12 the UK imported around 1.83 million tonnes of soya bean meal from South America (e.g. Brazil, Argentina and Paraguay), equating to an overseas land use of about 900,000 hectares, all for livestock production.[13] At its height in 2015–2016, 2.25 million tonnes was imported, dropping back to 1.87 million in 2017–2018.[14]

According to the UK Roundtable on Sustainable Soya, the amount of soya bean meal imported was slightly higher at just over 2 million tonnes (2018).[15] For comparison, the embedded soya bean consumption in the EU-28 has been calculated at 30.7 million tonnes (26.6 million tonnes of soya bean equivalent, requiring about 10 million hectares of farmland).[16]

Added to soya bean meal, reported in the 2021 John Nix 51st Edition *Pocketbook for Farm Management*, is a further three-quarters of a million tonnes of imported whole soya beans which is also processed into animal feed.

Approximately 75 per cent of all soya beans grown globally is used as a protein feedstock for livestock production,[17] with the lion's share used to support factory-farmed animals. In Great Britain there are some claims that actual soya bean meal usage in livestock feed is only around 1.1 million tonnes (2019) based on Defra statistics.[18] This amount would nearly halve the 900,000 hectares I mention above. The conclusion jumped to is that the rest is used in the production of human food and drink, pet foods and some non-food products. The disparity is a victim of vast under-reporting.

Most of the soya bean meal imported into the UK comes from GM crops, which are not used in food products. Most comes from South America, where around 90 per cent of the soybean crop is GM, and what is supplied to the UK, and to the rest of Europe, for livestock feed, contains GM or GM-derived material. Most of it is fed to chickens, pigs and dairy cows, and a smaller percentage to beef, sheep and fish. We may have more grass-fed dairy cows than other EU countries, and perhaps use less soya bean meal on average than some other countries per animal in that respect, but when it comes to soya bean meal most of our livestock is fed similar amounts to those on the continent. GM and non-GM soya crops from South America are not always segregated at source as it is not commercially viable without paying a premium. Most farmers (e.g. non-organic) do not pay a premium for non-GM soya bean meal, as cheap, environmentally destructive feed is required to be commercially viable/competitive.

Data on soya bean meal usage is also complicated as many UK farmers and producers often use a home mix, where farmers buy and blend their own feed, and where soya bean meal used in compound feed is not an accurate reflection of actual livestock soya consumption.[19]

Around half of the soya bean meal imported into the UK is used to feed poultry. Of the remaining half, a third each is fed to dairy and pigs.[20] Poultry's share may be closer to 60 per cent when feed processed from imported whole soya beans is included.

In March 2018 the UK Round Table on Sustainable Soya (RTSS) was convened to address growing concerns about the link between soya and tropical deforestation.[21] The RTSS has suggested, once its scheme has been fully implemented, that around 400,000 hectares (2.5 times the size of London)[22] will be cultivated in a way that protects against conversion of forests and valuable native vegetation.[23] This figure relates to all soya (about 3.2 million tonnes soya bean equivalent), not just that used for feeding livestock, from around 1.4 million hectares.[24] This still leaves around 1 million hectares, if or when implemented. An unfortunate consequence of 'sustainable' soya is that other parts of the industry will inevitably make up the shortfall elsewhere in the name of commerce – whether that comes from the further expansion of soya-eating livestock in the UK or elsewhere in the world. It would also probably be too expensive to source/process separately and come mixed with other soya bean meal. The only way to achieve sustainability is to remove the actual causes of deforestation – livestock.

In September 2019 Greenpeace questioned twenty-three food sector companies on the extent to which their meat and dairy supply chain was driving deforestation. The companies included major supermarkets, fast food chains, food manufacturers and a number of well-known high-street brands. Of the nine prepared to fully engage, all 'admitted lacking even the most basic oversight of their soya supply chains', and none were 'able to demonstrate that it was tracking the full amount of soya consumed as animal feed in its supply chain'.[25] According to Greenpeace's findings, no 'company could demonstrate it was taking any meaningful steps to actually ensure its supply of soya-based animal feed was not contributing to forest destruction.

Ten of the twenty-three companies contacted even abstained from answering the most basic of questions put to them. Things do not bode well, at least in relation to UK RTSS implementation projections for soya bean meal fed to UK livestock. It is early days for the RTSS, and any suggestions that soya bean meal fed to livestock is being cultivated in a way that protects against conversion of forests and valuable native vegetation needs to be backed up by hard evidence. Until such a time, any public marketing of livestock industry green credentials in such respects should be regarded with great suspicion and be stringently challenged.

The disconnection was further exposed by Watt et al. (2020)[26] in a joint cross-border investigation reported in *The Guardian*. The authors uncovered that Tesco, Lidl, Asda, McDonald's, Nando's and other high-street retailers are 'selling chicken fed on imported soya linked to thousands of forest fires and at least 300 square miles (800 km^2) of tree clearance in the Brazilian Cerrado'. The centre of the investigation concerned chickens fed soya supplied by Cargill, the US's second biggest private business. The investigation further exposes the 'opaque supply chain', which Greenpeace uncovered in 2019.

Cargill was implicated through the collation of biome-level export figures by the supply-chain watchdog Trase, which indicated 'that nearly half of Cargill's Brazilian exports to the UK [1.5 million tonnes of soya to the UK over the previous six years] are from the Cerrado'.

The investigative collaboration between the Bureau of Investigative Journalism, Greenpeace Unearthed, ITV news, and *The Guardian* highlighted a 66,000-tonne soya bean shipment delivered to Cargill's Seaforth soya crush plant in Liverpool before being moved on to mills in Hereford and Banbury, 'where it is mixed with wheat and other ingredients to produce livestock feed [including chicken farms connected to the aforementioned retailers]'.

Whilst the use of soya bean meal is very high for chickens, UK beef and lamb do not consider their more meagre use as a driving force for deforestation.[27] However, it is a significant contributor at about 150,000 tonnes or 75,000 hectares deforestation equivalent.[28] A further amount is used in commercial fish farms.

Table 28: UK/England livestock rainforest soya bean meal footprint

Estimated UK/England livestock share of rainforest land-use change associated with imported soya bean meal(900,000 ha)[1]			
	Dairy(ha)	Pig(ha)	Poultry(ha)
UK	150,000	150,000	450,000
England[2]	91,500	120,000	375,000

[1] Excludes the landbank needed to grow another three-quarters of a million tonnes of soya bean also used for animal feed, as reported in 2021 John Nix 51st Edition Pocketbook for Farm Management. This would add another 350,000 ha in total and a further 190,000 tonnes to the UK poultry total (UK 540,000 ha and England 405,000 ha).

[2] The UK 2016 total dairy herd was 1.897 million or 1.156 million for England (61%); total pigs for England (c. 80%); and total poultry was 172,607 million or 128,879 for England (c. 75%). Source: Defra, Livestock data country split, 2016. From this, each dairy cow is associated with around 790 m^2 of rainforest or 12 ha for an average herd size of 150 dairy cows.

The UK is not fully self-sufficient in animal-based products and the net balance between imports and exports of animal-based products, primarily with the rest of the EU, also adds to the UK's soya bean meal footprint.

Table 29: UK pig and poultry self-sufficiency[29]

UK Pig and Poultry Self-sufficiency (2017)	
Product	% Self-sufficiency
Poultry meat	72%
Poultry eggs	85%
Pig meat	56.2%

Source: AHDB, *Pig and Poultry Pocketbook.*

According to research by a Efeca, a UK-based consultancy, the UK indirectly imports an estimated 500,000 to 750,000 tonnes (250,000 to 375,000 hectares equivalent) of soya in the form of meat and dairy products, the majority of which is embedded within imported chicken (43 per cent), pork (22 per cent), cheese (15 per cent) and beef (12 per cent).[30]

What does all this look like in terms of greenhouse gas emissions, expressed as MT CO_2e?

The average biomass of natural forests in Latin America is about 118 tonnes of carbon per hectare, around about 50 per cent of the total rainforest biomass.[31] To avoid exaggeration, this is the most conservative estimate I can find from a reputable source. To compare with CO_2e the carbon biomass is multiplied by 3.67. Taking a UK soya bean meal footprint of 900,000 hectares (450,000 hectares for poultry), this converts to 390 MT CO_2e or 3.9 MT CO_2e a year split over 100 years. Whole soya beans used for animal feed would add another 169 MT CO_2e or 1.69 MT CO_2e a year.

Furthermore, rainforest vegetation accounts for only part of the living carbon. The rest is found in the soil as living carbon in roots, organic matter built up over time (perhaps as a former peatland), mycorrhiza, etc., some of which will also be lost to the atmosphere following deforestation.

As an example, the above-ground living carbon stock in UK forests is about 65 per cent of total biomass (674 MT CO_2e) with about 242 MT CO_2e below ground. Soil carbon has also been calculated at 2,761 MT CO_2e to a depth of 0–100 cm. Further carbon is found in dead wood and litter.[32]

Taking into account below-ground living carbon loss in the rainforest scenario an additional 25 MT CO_2e can be reasonably added, giving a total of 415 MT CO_2e. Soil carbon loss also needs to be included, which would occur mainly in the top 35 cm of the soil (ploughed depth). Conservatively, over time I suggest at least a quarter of that carbon is returned to the atmosphere through oxidation, about 85 MT CO_2e worth, giving a total of 500 MT CO_2e; or around 700 MT CO2e a year in total including whole soya beans.

Not all of the CO_2e is attributed to soya bean meal (and whole soya beans), only 80 per cent or 400 (or 560) MT CO_2e. The rest is associated with the oil fraction, some of which is used in animal feed, especially for chickens and pigs. However, for simplicity, I assume only 80 per cent is used to produce soya bean meal and oil for animal feed markets.

This is a lot of greenhouse gas emission not reported in the UK agriculture inventory (or our European colleagues etc). Furthermore, the loss of 900,000 hectares from imported soya bean meal has a devastating impact on the density and species-richness of rainforest animal and plant groups that once lived within them. To this we also need to add up to an extra 700,000 hectares of rainforest to cater for whole soya beans used in animal feed, and from the meat and dairy food we import.

Using a Herefordshire broiler farm producing 1.3 million chickens per year as an example, the farm's share of rainforest would be around 400 hectares (560 Wembley Stadium football pitches), or 175,000 tonnes of CO_2e over 100 years (1,750 tonnes CO_2e per year). This fact appears inconsequential to the industry, who are firmly welded to a business model based on rainforest-ready chickens and turkeys; and who like to claim to consumers how much more sustainable UK livestock is compared to other countries like Brazil and Argentina, and how the rest of the world should buy their meat, fish, dairy and eggs from the UK instead. If UK livestock farms took responsibility for all their carbon, the UK GHG picture would appear less rosy and a more deadly nightshade.

The Cerrado, where evidence suggests at least some of the soya used in Herefordshire has come from, is home to 251 species of mammals, 856 species of avifauna, 262 reptile species, 204 amphibians and many rare plant species.[33] In fact the Cerrado's ecosystem profile lists 1,593 terrestrial and freshwater species classified as globally threatened by the International Union of Conservation of Nature (IUNC). We should all be disturbed by these figures, along with the cruelty caused for a Sunday roast chicken and a Christmas turkey dinner. If we do not want factory-farmed livestock, and many say they don't, the solution is very simple – stop eating them, and be healthier for it.

The total amount of soya embedded in the animal-based foods most of us eat in the UK, and across Europe, associated with deforestation, adds up to a tennis court each of ecocide. Contrariwise, soya used by the UK and our neighbours in plant-based foods only marginally contributes to rainforest loss.

To reduce reliance on soya beans the livestock industry could look for further efficiencies. This might include developing chickens that mature in thirty-five days instead of forty, and which eat less soya bean meal for good measure. A few decades back chickens were brought to market after sixty days, with a feed conversion rate of 2.5 compared to 1.5 today. However good this may look (turning a blind eye for a moment to the increased numbers and animal abuse), saving five days per chicken crop adds up to thirty-five days and before you know it there is room for nearly an extra crop a year. Can you see where things are going? Tens of thousands more chickens eating more soya and rainforest, and emitting more CO_2e, than previously. Intensification in a nutshell, where efficiency can create less rather than more for the environment.

The UK's offshore land-use change impact is not just limited to South America. It has strong ties to Southeast Asia too, mainly within Indonesia and Malaysia. In 2017–2018 the UK imported 452,760 tonnes of palm kernel meal.[34] About 750,000 hectares of land is needed to produce that amount of meal, given a palm kernel meal crop yield of about 600 kg per hectare. The remainder of the yield is made up of palm oil (4,000 kg per hectare) and palm kernel oil (500 kg per hectare). The meal is used to feed livestock, farmed fish and pets, with the bulk fed as a compound feed to livestock. In addition, 150,000 tonnes of crude palm oil is used.

Land-use change for palm oil/kernel meal production in Malaysia and Indonesia targets a combination of forest, forest peatland and some grassland. It has been estimated that a forest cleared on a mineral soil releases about 650 t CO_2e per hectare; and in forested peatland 1,300 t CO_2e per hectare released in the first twenty-five-year crop cycle, and a further 800 t CO_2e per hectare for each additional crop cycle as the peat continues to decompose.[35]

Despite the high level of CO_2e associated with deforestation, palm kernel meal only makes up about 12 per cent of the crop, compared to 80 per cent for soya bean meal. Taking a standing stock of 1,300 t CO_2e per hectare,[36] I estimate the greenhouse gas emission footprint for palm kernel meal fed to UK livestock to be about 1.2 MT CO_2e a year, or 120 MT CO_2 over 100 years, including the small amount of oil used.

Adding to the soya bean meal footprint gives a total greenhouse gas emission footprint approaching 5.2 MT CO_2e per year (4.0 + 1.2 MT CO_2e), from around one million hectares of previously forested land (520 MT CO2e per 100 years). This rises to 1.25–1.35 million hectares when you include imported meat and dairy, totalling more than 6.5 MT CO_2e per year; or up to 1.7 million hectares and 8.1 MT CO_2e in total when you include imported whole soya beans processed into animal feed.

This also assumes the UK's rainforest footprint has been static since 1990, but in reality it has increased by around 2 per cent a year (my estimate) to add another million-plus hectares to the footprint.

The livestock industry is keen to deflect and make its UK tofu-eating citizens the actual villains of soya bean deforestation in South America, but the reality is starkly different. Various sources suggest the annual UK intake of soya bean equivalent a year embedded in meat, dairy, eggs and other livestock foods is around 53–60 kg per person. For instance, research commissioned by the WWF (2015)[37] concluded that the average EU citizen consumes an embedded soya bean equivalent of 53 kg per year, which required 197 m^2 of farmland (the size of a tennis court). Also, according to WWF (2015),[38] 93 per cent of the soya consumed per year by Europeans is in the form of animal feeds.

Most of this soya is connected to deforestation and a whole host of other harmful footprints. UK citizens who eat a plant-based diet could consume 9–18 kg of soya bean equivalent a year. In the UK and in many other places this is non-GM soya, which is mainly from organic soya beans sourced from Europe, US and more recently China, which as a legume used in a crop rotation also helps fertilise soil and if used in the right way can help improve biologically dysfunctional soils. Another benefit of directly eating tofu is that you do not need to process it through, and be unkind to, an animal first, or suffer the other consequences of livestock's longer shadow. Tofu also leaves behind soy pulp (okara) that can be turned into a variety of tasty foods, so, provided it is used, this should not be discounted when assessing its footprint. The livestock industry also like to claim tofu is processed, but it is no more processed than cheese made from dairy cows, or cows turned into burgers.

When it comes to looking at how much soya bean is eaten without the help of the livestock industry, it's like looking for Wally in a Martin Handford book.

A further complication of the soya bean debate is that not all the land would have been deforested directly for soya bean production. Deforestation may have occurred initially for another purpose, which soya bean production then followed. A case study on the sustainability of Brazil's soya bean production captured the complex dynamics of land-use change and opinion and suggests that although it cannot 'be concluded that soybean cultivation caused deforestation, [*deleted*] it does seem to be one of the main factors'.[39] Some livestock-related deforestation is also a result of fresh incursions caused by biofuel production, displacing beef and soya bean farming into new areas.

But just how significant is all this compared to the 1990 benchmark? Turning the clock back, in 1964–1965 the UK imported only around 247,000 tonnes of soya bean meal. This rose to nearly 1.17 million in 1986.[40] By 1993–1994 the amount imported stood at 1.54 million.[41] The increase in soya bean meal at that time was in part due to the 1994 ban on the use of meat and bonemeal in livestock feeds, following the BSE crisis (see the fact box on page XXX), with the livestock sector turning to vegetable-based proteins instead.[42] According to the World Wildlife Fund, the Europe-wide ban on the use of processed animal protein in feed rations removed the protein equivalent of 3.7 million tonnes of soya bean meal in the UK alone.[43]

Imports increased still further to 1.95 million tonnes by 1999–2000 before falling back dramatically to just over 1.17 million the following year,[44] this time coinciding with a 2001 the foot-and-mouth outbreak (see the fact box on page 245).

BSE. The government announced on 21 March 1996 a suspected link between bovine spongiform encephalopathy (BSE) and the human equivalent, Creutzfeldt-Jakob disease (variant CJD).

In response from 1996 an over thirty-month cattle cull started, with 975,296 slaughtered in that year. Six days later, on 27 March, the EU placed a worldwide ban on all British beef, leading to a period of conflict with the EU that lasted until late June 1996.

In June, under the EU Florence Framework, amongst other measures, the UK agreed to legislate for the removal of meat and bonemeal from feed mills and farms and to the implementation of the Over Thirty Months (OTM) slaughter scheme. The OTM policy was introduced earlier as policy in April 1996. During the eradication programme, which actually started in 1986, when BSE in cattle was first acknowledged, about 4.6 million cattle were slaughtered and removed from the food chain by 2000.

According to a Parliamentary Office of Science and Technology (POST) Technical Report issued in October 1997, 60 per cent of 170,000 cattle diagnosed with BSE were dairy cows (and 16 per cent of suckler herds). Unsurprisingly then the highest percentage of cattle slaughtered came from the dairy sector, a large number of which would have been from uneconomical milking cows that would previously have entered the food chain.

The use of meat and bonemeal at the time of BSE was far from a new phenomenon. In the 1960s, and before, a variety of other animal feedstuffs, e.g. white fish meal, oily fish meal or herring meal and even whale meat, were used as a livestock protein source, in addition to meat and bonemeal feed stripped from slaughtered cattle carcasses and rendered down in steam-jacket melters.[45]

Foot-and-mouth. On 19 February 2001 foot-and-mouth was discovered at an Essex abattoir and quickly spread across the UK. Two days later the government banned all exports of live animals, meat and dairy products, coinciding with an EU ban the same day. The movement of livestock within the UK from infected areas was banned on 23 February, followed by the mass slaughter of livestock, including cattle and pigs, in England on 24 February. So big was the slaughter that the government called in the army to organise the cull. The end of the outbreak was recorded on the 30 September 2001, with a final case in Little Asby, Appleby, Cumbria. The cull ended on 1 January 2002, with a flock of 2,000 sheep in Bellingham, Northumberland. The restriction on livestock movements continued into 2002.

> *The government recorded that about 6 million animals were culled (4.9 million sheep, 0.7 million cattle and 0.4 million pigs), 4 million as a direct result of the cull and a further 2 million for welfare reasons (e.g. lack of feed and space). A further 4 million young animals were also slaughtered 'at foot', meaning they were slaughtered with their mothers but not counted.*[46]

In respect to palm oil kernel, around 360,000 tonnes was imported in 1995–1996. The closest record I could research for pre-1990 was for 'other' oil cake, totalling 388,000 tonnes.[47]

Despite an apparent pre-1990 rainforest footprint of around 500,000 hectares, this can't all be discounted against UK emissions post-1990. It is likely much of the pre-1990 land-use change that once supported UK livestock is not used today; not least since the rapid expansion of soya bean production, which occurred after the introduction of transgenic (genetically modified) soya beans in South America during the 1990s.

In 2006 the Food and Agriculture Organisation of the United Nations published a landmark report titled: *Livestock's Long Shadow: environmental issues and options*. The report stated that animal agriculture was responsible for 18 per cent of global greenhouse gas emissions (7.1 billion tonnes of CO_2e), calculated using a global life-cycle assessment (LCA). This was subsequently revised down to 14.5 per cent, though the amount of emissions, 7.1 billion tonnes of CO_2e, remains the same today.[48] 18 per cent is still often quoted.

The role of livestock in carbon dioxide, methane and nitrous oxide emissions, reproduced from the report, is set out in Table 30.

Table 30: Role of livestock in carbon dioxide, methane and nitrous oxide emissions

Gas	Source	Mainly related to extensive systems (10^9 tonnes CO_2e)	Mainly related to intensive systems (10^9 tonnes CO_2e)	Percentage contribution to total animal food GHG emissions
CO_2	Total anthropogenic CO_2 emissions	24 (~31)		
	Total from livestock activities	~0.16 (~2.7)		
	N fertiliser production		0.04	0.6
	On-farm fossil fuel, feed		~0.06	0.8
	On-farm fossil fuel, livestock-related		~0.03	0.4
	Deforestation	(~1.7)	(~0.7)	34
	Cultivated soils, tillage		(~0.02)	0.3
	Cultivated soils, liming		(~0.01)	0.1
	Desertification of pasture	(~0.1)		1.4
	Processing		0.01 – 0.05	0.4
	Transport		~0.001	
CH_4	Total anthropogenic CO_2 emissions	5.9		
	Total from livestock activities	2.2		
	Enteric fermentation	1.6	0.20	25
	Manure management	0.17	0.20	5.2
N_2O	Total anthropogenic CO_2 emissions	3.4		
	Total from livestock activities	2.2		
	N fertiliser application		~0.1	1.4
	Indirect fertiliser emission		~0.1	1.4
	Leguminous feed cropping		~0.2	2.8
	Manure management	0.24	0.09	4.6
	Manure application/deposition	0.67	0.17	12
	Indirect manure emission	~0.48	~0.14	8.7
	Grand total of anthropogenic emissions	33 (~40)		
	Total emissions from livestock activities	~4.6 (~7.1)		
	Total extensive vs. intensive livestock system emissions	3.2 (~5.0)	1.4 (~2.1)	
	Percentage of total anthropogenic emissions	10 (~13%)	4.1 (~5%)	

Note: All values are expressed in billion tonnes of CO_2 equivalent. Values between brackets are or include emissions from land use, land-use change and forestry category.

The revised 14.5 per cent of global greenhouse gas emissions takes account of a variety of livestock farming activities dissimilar to the EU and UK. The calculated percentage is also different to the sector specific way individual countries calculate their national GHG emissions, including agriculture.

In recognition of this difference, a project was undertaken by the European Union in 2010 to provide a realistic comparison for the EU-27 livestock sector, including a combined EU-27 'national' inventory.[49]

The project reported a combined emission of 661 MT CO_2e (land use and land-use change included). Of this, 29 per cent each came from beef and dairy milk production and 25 per cent from pig production. A further 17 per cent came from other livestock.

Total emissions for EU-27 livestock as a percentage of national GHG inventories (life cycle, land use and land-use change excluded) amounted to 9.1 per cent, and when land use and land-use change emissions are included 12.8 per cent. Please keep in mind that these percentages are relative to other inventory emissions.

Here land use referred to the balance of carbon sequestration and CO_2 emissions from the cultivation of organic soils. Reported by the project, land use adds little to UK emissions, which is 'mainly due to the carbon removal credited to the grassland used [...] which offsets most of the foregone carbon sequestration for the cultivation of feed crops'. For Ireland the calculation is positive, with more carbon sequestration in grassland than carbon lost in cropland. The UK's land-use change emission percentage is bettered by eleven of the EU-27.

Specific to livestock meat (bone-in carcass weight) and other products the project calculated, according to cradle-to-grave life-cycle assessment to the farm-gate, using the Common Agricultural Policy Regional Impact (CAPRI) modelling system, the following GHG emissions (on EU average) are:

Beef meat	22.2 kg CO_2e/kg
Sheep and goat meat	20.3 kg CO_2e/kg
Pig meat	7.5 kg CO_2e/kg
Poultry meat	4.9 kg CO_2e/kg
Eggs	3 kg CO_2e/kg
Cow milk	1.4 kg CO_2e/kg

Overall, using CAPRI, the UK livestock industry came out as one of the higher performers across the EU-27, with the UK having a lower GHG intensity than the EU-27 average for all but sheep and goat meat.

On a more personal level, according to a 2009 farm-gate life-cycle assessment led by EBLEX (English Beef and Lamb Executive, now AHDB Beef and Lamb), the English average for beef cattle is 13.89 kg CO_2e per kg. The lowest footprint is dairy beef at 10.97 kg CO_2e per kg, representing 51 per cent of England's prime carcass beef (dairy beef is lower, as a proportion of GHG emissions is accounted for within dairy production); 17.12 kg CO_2e per kg for lowland suckler beef (19 per cent); and 16.98 kg CO_2e per kg for hill and upland suckler herds (30 per cent).

English sheep production, excluding goats, also came out lower than the EU-27 figure, averaging at 14.64 kg CO_2e per kg. The highest emissions arose from hill flocks at 18.44 kg CO_2e per kg, representing 39 per cent of prime carcass lamb; 13.82 kg CO_2e per kg for upland flocks (30 per cent) and 12.62 kg CO_2e per kg for lowland flocks (31 per cent).[50]

Taking a step back, though, to avoid argument I choose to use the rounded figures that appear to be accepted by the UK industry, with a kilo of British beef having an estimated 17.12 kg CO_2e per kg carbon footprint and a kilo of lamb produced in England and Wales of around 14.6 kg CO_2e. This compares to a global average of 46 kg CO_2e and 24 kg CO_2e per kilo of beef and lamb respectively.[51]

For dairy the weighted average footprint for British milk has been calculated at 1.232 kg CO_2e per litre, a little under the EU-27 average.[52]

Calculated on a comparable basis, other EU GHG intensities would also be expected to be lower.

However, for livestock farming as a whole the EU report and UK specific GHG life-cycle assessment support claims made by UK livestock representatives that the UK is a higher performer within the EU-27 (UK 7.2 per cent vs. EU-27 9.1 per cent), with only two countries bettering the UK on land use and land-use change emissions combined.

Another way to compare GHG performance is to look at the amount generated per kg of edible solid product across the full supply chain, rather than the bone-in carcass weight. A 2020 Committee on Climate Change report, *Land use: Policies for Net Zero UK*, includes an estimate of around 48 kg CO_2e per kg of meat from dedicated beef herds (c. 26 kg CO_2e for dairy-beef herds). This compares to a global average for dedicated beef herds of around 99 kg CO_2e per kg and up to four times less than Brazil.

Brazil and a few other countries' GHG intensity is much higher due to land-use change emissions associated with the clearance of carbon-rich land for grazing (GHG accounting for UK and European production does not include historical deforestation, which mostly happened centuries ago). The report concludes that other EU countries have similar figures for beef, but on average it is around 14 per cent higher than the UK and that the GHG-intensity for lamb is also likely to be favourable against other European producers. Nevertheless, UK lamb is likely to be more GHG-intensive than New Zealand lamb.[53]

Other examples of UK edible solid product emissions are 37.4 kg CO_2e per kg lamb meat, 11.9 kg CO_2e per kg pig meat, 9.8 kg CO_2e per kg chicken meat and 12.6 kg CO_2e per kg turkey meat.[54]

Whilst UK and global figures may vary depending on how GHG intensity is calculated, the UK livestock industry's assertion of being less GHG-intense than most countries is supported by the available evidence (within the UK's own context). The UK also benefits from not having to include pre-1990 land-use change emissions, whilst developing countries do as their economies mature.

The livestock industry, particularly the red meat sector, is agitated if balance is not given to the 'health benefits' of meat or to the environmental footprint of fruit, vegetables and other plant-based foods, including their sea and air miles. I have already addressed livestock sector health claims in my first two chapters. Essentially, if you do not have access to a balanced range of plant foods, and meat, dairy, fish and eggs are largely all you have to eat, then meat is indeed better for you than nothing at all. However, the more animal-based foods you replace with whole plant-based foods, the healthier you will be.

To help illustrate the carbon and land-use footprint of different foods, Hannah Ritchie and Max Roser (2020), of Our World in Data, drew together the most relevant data on the subject in the presentation 'Environmental impacts of food production',[55] most of which is based on the largest meta-analysis of global food systems published to date, by Joseph Poore and Thomas Nemecek (2018) in *Science*.[56] The data is freely available and I would urge you to take time to look through their presentation. Some examples are given in Tables 31 and 32, but do bear in mind that the data is not UK-specific for neither animal nor plant foods.

Table 31: Environmental impacts of food production (1)

Food	GHG emissions across the supply chain kgCO$_2$e per kg product	GHG emissions per 100g protein kgCO$_2$e per 100g protein	GHG emissions per 1,000 kcal kgCO$_2$e per 1000kcal
Beef (beef herd)	60.00 kg	49.89 kg	36.44 kg
Beef (dairy herd)	21.00 kg	16.87 kg	12.20 kg
Milk	3.00 kg	9.50 kg	5.25 kg
Cheese	21.00 kg	10.82 kg	6.17 kg
Pig meat	7.00 kg	7.61 kg	5.15 kg
Poultry meat	6.00 kg	5.70 kg	5.34 kg
Eggs	4.40 kg	4.21 kg	3.24 kg
Lamb & mutton	24.00 kg	19.85 g	12.53 kg
Prawns (farmed)	12.00 kg	18.19 kg	26.09 kg
Fish (farmed)	5.00 kg	5.98 kg	7.61 kg
Apples	0.40 kg	14.33 kg	0.90 kg
Bananas	0.70 kg	9.56 kg	1.43 kg
Berries & grapes	1.10 kg	15.30 kg	2.68 kg
Brassicas	0.40 kg	4.64 kg	3.00 kg
Citrus fruit	0.30 kg	6.50 kg	1.22 kg
Groundnuts	2.50 kg	1.23 kg	0.56 kg
Nuts	0.30 kg	0.26 kg	0.07 kg
Oatmeal	1.60 kg	1.91 kg	0.95 kg
Onions & leeks	0.30 kg	3.85 kg	1.35 kg
Other pulses	1.50 kg	0.84 kg	0.52 kg
Peas	0.90 kg	0.44 kg	0.28 kg
Rice	4.00 kg	6.27 kg	1.21 kg
Root vegetables	0.40 kg	4.30 kg	1.16 kg
Tofu (soya beans)	3.00 kg	1.98 kg	1.17 kg
Tomatoes	1.40 kg	19.00 kg	11.00 kg
Wheat & rye	1.40 kg	1.29 kg	0.59 kg

Source: Poore and Nemecek (2018), from OurWorldinData.org.

Table 32: Environmental impacts of food production (2)

Food	Land use per kg of food product m² per kg	Land use per 100g protein m² per kg	Land use of foods per 1,000 kcal m² per kg
Beef (beef herd)	326.21 m²	163.6 m²	119.49 m²
Beef (dairy herd)	43.24 m²	21.90 m²	15.84 m²
Milk	8.95 m²	27.10 m²	14.92 m²
Cheese	87.79 m²	39.80 m²	22.68 m²
Pig meat	17.36 m²	10.70 m²	7.26 m²
Poultry meat	12.22 m²	7.10 m²	6.61 m²
Eggs	6.27 m²	5.70 m²	4.35 m²
Lamb & mutton	369.81 m²	184.8 m²	116.66 m²
Prawns (farmed)	2.79 m²	2.00 m²	2.88 m²
Fish (farmed)	8.41 m²	3.70 m²	4.70 m²
Apples	0.63 m²	21.00 m²	1.31 m²
Bananas	1.93 m²	21.40 m²	3.22 m²
Berries & grapes	2.41 m²	24.10 m²	4.23 m²
Brassicas	0.55 m²	5.0 m²	3.24 m²
Citrus fruit	0.86 m²	14.30 m²	2.69 m²
Groundnuts	9.11 m²	3.50 m²	1.57 m²
Nuts	12.96 m²	7.90 m²	2.11 m²
Oatmeal	7.60 m²	5.80 m²	2.90 m²
Onions & leeks	0.39 m²	3.00 m²	1.05 m²
Other pulses	15.57 m²	7.30 m²	4.57 m²
Peas	7.46 m²	3.40 m²	2.16 m²
Rice	2.80 m²	3.90 m²	0.76 m²
Root vegetables	0.33 m²	3.30 m²	0.89 m²
Tofu (soya beans)	3.52 m²	2.2 m²	1.30 m²
Tomatoes	0.80 m²	7.3 m²	4.21 m²
Wheat & rye	3.85 m²	3.2 m²	1.44 m²

Source: Poore and Nemecek (2018), from OurWorldinData.org.

In general, transport makes up about 6 per cent of emissions, which would also apply to the transport of UK livestock products (alive, dead or processed) around the world.

When comparing livestock GHG emissions as a percentage of the UK's total (or anywhere else in the world), the picture is complicated. Not all the CO_2 lost to the atmosphere from fossil fuels, and other direct CO_2 emissions, stays in the atmosphere in any given year.

Past studies have shown that losses to the oceans (i.e. contributing to ocean acidification) and terrestrial ecosystem net gain (i.e. through photosynthesis) leave only about 40 per cent of anthropogenic CO_2 emissions in the atmosphere each year.[57] A recent assessment puts that figure at 45 per cent.[58]

In our oceans, CO_2 is captured by floating phytoplankton, which use it for photosynthesis. These organisms eventually die or are eaten by zooplankton, and so on, up the food chain. This is why you sometimes hear whales being cited as carbon stores, which in turn also fertilise the ocean with their excrement, along with other sea animals, which feeds the phytoplankton and so forth. Much is also drawn down to the ocean floor, where it becomes buried in sediment.

Recent research suggests that the amount of carbon captured in the ocean in this way may be greater than previously thought, which has implications for climate assessments.[59] The research focussed on identifying the true edges of the active euphotic zone (penetrating depth of sunlight), using chlorophyll sensors rather than the usual present-day method which relies on measuring data from fixed depths. The depth of the euphotic boundary was found to vary around the world, was much deeper than previously thought as a whole and as such captured twice as much carbon a year.

According to IPCC data (2019) total CO_2 emissions for the world was 39.1 Gt CO_2. Correcting this to reflect how much CO_2 remains in the atmosphere each year reduces the total. Without making an adjustment for CO_2 left in the atmosphere, methane from livestock (enteric fermentation and manure management) contributes 5.6 per cent of global CO_2 emissions (2.2 Gt CO_2); but when adjusted as a proportion of CO_2 emissions left in the atmosphere, it raises methane's share from livestock.

The IPCC also measures methane produced over a 100-year timeframe, which some suggest underestimates the more immediate impact of this gas.[60] For example, over a twenty-year timeframe methane has a global warming potential (GWP) of 86, compared to 32 for 100 years.

Using a twenty-year accounting period for methane rather than a 100-year timeframe creates a situation where the percentage of methane is more than several orders of magnitude higher. If the ocean captures more carbon than previously thought then the percentage would be greater.[61] Should we then worry more about methane from livestock than CO_2 from other sectors? No, as we need to view and deal with each GHG emission in its own context.

Methane also only stays in the atmosphere for around ten to twelve years, leading to calls by some to modify the way methane is calculated in GHG inventories. The argument presented is that provided livestock numbers are stable, or better still reducing, they do not put additional methane into the atmosphere and as such should be discounted. However, although short-lived, methane is a powerful GHG when it is in the atmosphere. All GHG emissions need to be dealt with urgently by genuinely reducing/removing the problem, not by changing the maths.

Furthermore, methane also leads to the formation of near-surface-level or tropospheric ozone, another GHG and a potent local air pollutant which is harmful to people and ecosystems, and perhaps double methane's warming effect.[62] It is not enough to offset methane emissions from livestock through measures like biogas production, putting biochar into soil, feeding cows seaweed, genomics, etc., which themselves could lead to other consequences (e.g. pollution swapping, harm to soil health, further animal abuse, etc). More drastic mitigations are required, including at least halving cattle and sheep numbers and better still cutting them by 75 per cent or more. Animals left should be used to help manage a biodiverse and varied landscape, without recourse to eat them. Instead, the livestock industry appears set on a pathway to actually grow livestock production (either by number, production per head of livestock or both), which in turn would cut into any savings that could be made through the methods they propose, and in turn would be heavily reliant on taxpayer support to come to fruition. The latter is also a major shortfall of industry Net Zero ambitions, which already relies on *ifs*, *buts*, *speculations* and substantial *consequences* for land managers which have not been explained. Any increase in livestock production may also involve a further land grab to feed them.

Ultimately, we must stop trying to shoehorn livestock production, existing and future, into climate change mitigation, and other forms of mitigation, and instead make the transformative changes we need, however unpalatable that may be to many. We have to stop farming to *wants* and begin farming to our *needs*. Livestock farming in the West does not feed a hungry world, and our obsession and unintelligent defence and financial support of it instead makes the world a hungrier and more polluted place.

The impact of livestock farming is more than methane. Table 30 showed how livestock farming contributes to direct CO_2 emissions and nitrous oxide emissions too, together making up the 14.5 per cent, for the world as a whole. However, not all of these emissions would disappear in a WFPB world. We will still need a number of ruminant herbivores, which would be crucial to biodiversity conservation rather than for meat production (e.g. stock-free farming). This does not mean the reduced grazing practices which are happening in some places today and which provide a limited biodiversity improvement, but rewilding on a grand scale.

Evolution to stock-free farming would be a long, drawn-out process. An initial step would be to move away from factory farms and towards giving livestock a more worthwhile life as we transition our food and related economy. Mixed farms spread back out across the country and largely producing farmyard manure (ideally composted), rather than slurry, would be helpful, slowly but surely improving soil health and wealth to a state which bridges the gap with chemically grown crops.

A move back towards a more mixed farming system is beginning to be considered as an option within parts of the livestock industry, including populating the east of the country with dairy cows (unfortunately intensively). Parts of the livestock industry also see the necessity of reducing ruminant numbers to reduce emissions, which sounds positive until you realise it would come with further increases in *per-head* meat, egg and dairy yield production. Translated, this means further intensification, working livestock even harder and being even unkinder to them.

Our soil health is in such a state that it will take decades to regenerate. Repairing soil is a lot more complicated to achieve in *modern* agriculture than simply turning to zero/minimum tillage and planting cover crops within mixed rotations. Decades of chemical drenches (pesticides and inorganic fertilisers), soil compaction, and soil humus, fauna and flora loss, need to be corrected. Many farmers are no longer farming soil but instead farm dirt. We need to nurture this dirt and turn it back into soil by creating the right balance and conditions for soil biology to return and prosper, which will then help feed plants naturally, protect them from disease and rebuild soil structure. We need to put back the soil's natural scaffolding we have so erroneously dismantled through years of intensive agriculture and neglect, the sooner the better. Correcting dirt is a lot more complicated than just spreading it with slurry and other organic manures.

This will involve removing harmful chemicals and anaerobic soil conditions, which kill and/or inhibit the beneficial bacteria and fungi; bacteria and mycorrhizal fungi which plants feed and propagate with the root exudates (e.g. carbohydrates and proteins) they release into the surrounding soil. Root exudates target the good, and not the bad, microorganisms which in turn help build microaggregates around the crop's roots (soil microaggregates initially form through the flocculation of clay minerals, which then bond with organic matter). These microaggregates are then combined into macroaggregates, supported by the secretions of mycorrhizal fungi (glomalin), giving the soil structure, aeration and drainage. Worms also form soil aggregates (worm castings). The process of soil aggregation is more complex than that, but this is not a soil textbook, so I will leave it at that. The microorganisms in soil are then targeted by the next trophic level, a combination of shredder arthropods and predators and grazers (e.g. nematodes and protozoa), whose excretions provide all the micronutrients and macronutrients a plant needs in natural situations to grow in the right balance, at the right time and in the right place within the root zone.

When soil biology works as nature intended, plants are protected from diseases and pests. This trophic level is then fed upon by the next, and so on, up the food chain. The relationships also greatly increase crop nutrient density. Life starts in the soil, and when we kill soil and turn it into dirt, we kill life. Nature does not leak nutrients and pollute soil, air and water either, but rather we do, with livestock farming and chemical warfare, which breaks down soil and plant functions and relationships. Farmers need to start counting and farming soil biology instead of favouring terrestrial livestock.

For a UK-wide WFPBD it may be necessary to convert some permanent pasture to arable land, but that does not mean ploughing up the land and releasing large quantities of CO_2 and nitrate. Land-use change can be managed sensitively, but there would inevitably be a temporary mix of gains and losses as we move towards a new equilibrium. New crops, and increased hectares of more specialised crops such as lentils, soya bean, hemp and quinoa, already grown in the UK, along with other protein crops like amaranth, will be needed to lower food imports.

GHG emissions are only one part of livestock's longer shadow. The necessity to farm less livestock, and to end livestock farming over the longer term, is overwhelming when all things are added together. livestock's longer shadow must be addressed as a whole, not in silos.

Citations:

1. United Nations Framework Convention on Climate Change (1997), Art. 2
2. Defra (2019), *2017 UK Greenhouse Gas Emissions, Final Figures*. Statistical Release, National Statistics
3. Defra. Based on Tables 3 to 7 of *Final UK greenhouse gas emissions national statistics 1990-2017*
4. Based on National Atmospheric Emissions Inventory. http://naei.beis.gov.uk/reports/reports?report_id=991
5. United States Environment Protection Agency, *Greenhouse Gas Emissions*. Sources of Greenhouse Gas Emissions. https://www.epa.gov/ghgemissions/sources-greenhouse-gas-emissions
6. United States Environmental Protection Agency (2019), *Inventory of US Greenhouse Gas Emissions and Sinks, 1990-2017 – 5. Agriculture*. https://www.epa.gov/sites/production/files/2019-04/documents/us-ghg-in ventory-2019-chapter-5-agriculture.pdf
7. [Irish] Environment Protection Agency, *Greenhouse Gas Emissions from Agriculture*. http://www.epa.ie/ghg/agriculture/
8. Moore, A. (2019), 'Reducing cow herd 'last resort' in tackling carbon emissions, say Varadkar.' *Farming Independent*, 19 November 2019 https://www.independent.ie/business/farming/forestry-enviro/environme nt/reducing-cow-herd-last-resort-in-tackling-carbon-emissions-says-varadkar-38649340.html.
9. Doody, D. G. et al. (2020), *Phosphorus Stock and Flows in the Northern Ireland Food System* [RePhoKUs Project Report]
10. Ibid.
11. Ag i-Food Strategy Board (2013), *Going for Growth: A Strategic Action Plan in Support of the Northern Ireland Agri-Food Industry*
12. Doody, D. G. et al. (2020), *Phosphorus Stock and Flows in the Northern Ireland Food System* [RePhoKUs Project Report]
13. Jones, P. J. et al. (2014), *Replacing soya in livestock feeds with UK-grown protein crops: Prospects and implications*. CAS Report 19
14. AHDB, *Cereals and Oilseeds market*. https://cereals-data.ahdb.org.uk/archive/import.asp.
15. EFECA (2019), *UK Roundtable on Sustainable Soya. Annual progress report*.
16. Kroes, H. and Kuepper, B. (2015), *Mapping the soy supply chain in Europe: A research paper prepared for WNF*
17. WWF (2016), *The Story of Soy*. https://www.worldwildlife.org/stories/the-story-of-soy

18. AHDB (2019), *GB animal feed production.*
   https://ahdb.org.uk/cereals-oilseeds/cereal-use-in-gb-animal-feed-production

19. EFECA (2018), *UK Roundtable on Sustainable Soya. Baseline study*

20. WWF (2011), *Soya and the Cerrado: Brazil's forgotten jewel*

21. EFECA. *The UK Roundtable on Sustainable Soya.*
   https://www.efeca.com/the-uk-roundtable-on-sustainable-soya/

22. EFECA (2019), *UK Roundtable on Sustainable Soya: Annual progress report*

23. EFECA (2018), *The UK Roundtable on Sustainable Soya.*
   https://www.efeca.com/the-uk-roundtable-on-sustainable-soya/

24. Greenpeace (2020), *Winging it: How the UK's Chicken is Fuelling the Climate and Nature Emergency*

25. Ibid.

26. Watt et al. (2020), 'Revealed: UK supermarket and fast food chicken linked to deforestation in Brazil.' The Guardian, 25 November 2020
   https://www.theguardian.com/environment/2020/nov/25/revealed-uk-supermarket-and-fast-food-chicken-linked-to-brazil-deforestation-soy-soya

27. AHDB. *Livestock and Climate Change: The Facts*

28. Fraanje, W. (2020) '*Soy in the UK: what are its uses?* Food Climate Research Network, UK.
   https://www.tabledebates.org/blog/soy-uk-what-are-its-uses

29. AHDB, *Pork* and *Poultry.*
   http://pork.ahdb.org.uk/media/273703/pig-pocketbook-2017.pdf and
   http://pork.ahdb.org.uk/media/273704/poultry-pocketbook-2017.pdf

30. EFECA (2018), *UK Roundtable on Sustainable Soya: Baseline study*

31. Houghton, R.A. (2005), 'Aboveground Forest Biomass and the Global Carbon Balance.' *Global Change Biology* 11(6), pp. 945-58

32. Forestry Commission (2019), *Forestry Statistics: A compendium of statistic about woodland, forestry and primary wood processing in the United Kingdom.* National Statistics at Gov.uk

33. Critical Ecosystem Partnership Fund (2017), *Ecosystem Profile: Cerrado Biodiversity Hotspot*

34. AHDB, *Cereals and Oilseeds markets.*
   https://cereals-data.ahdb.org.uk/archive/import.asp.

35. Germer, J. and Sauerborn, J. (2008), 'Estimation of the impact of oil palm plantation establishment on greenhouse gas balance.' *Environment, Development and Sustainability* 10, pp. 697-716

36. Calculated from Morel (2009), cited in Dr. F.W. Klaarenbeeksingel (2009), *Greenhouse Gas Emissions from Palm Oil Production: Literature review and proposals from the RSPO Working Group on Greenhouse Gases.* Final report.

37. Kroes, H. and Kuepper, B. (2015), *Mapping the soy supply chain in Europe: A research paper prepared for WNF*

38. Ibid.
39. van Berkum, S. and Bindraban, P. S. (2008), *Towards sustainable soy: An assessment of opportunities and risks for soybean production based on a case study of Brazil*
40. Marks, H. F. (1989), *A Hundred Years of British Food & Farming: A Statistical Survey*, edited by D. K. Britton. Taylor & Francis, London, UK
41. AHDB, *Cereals and Oilseeds markets.* https://cereals-data.ahdb.org.uk/archive/import.asp.
42. Jones, P. J. et al. (2014), *Replacing soya in livestock feeds with UK-grown protein crops: Prospects and implications.* CAS Report 19
43. WWF (2011), *Soya and the Cerrado: Brazil's forgotten jewel*
44. Marks, H. F. (1989), *A Hundred Years of British Food & Farming: A Statistical Survey*, edited by D. K. Britton. Taylor & Francis, London, UK
45. MAFF (1960), *Bulletin No. 48: Rations for Livestock.* HMSO, London
46. Defra (2004), *Animal Health and Welfare:* 'FMD Data Archive. Foot and Mouth Disease.' http://footandmouth.fera.defra.gov.uk/
47. Marks, H. F. (1989), *A Hundred Years of British Food & Farming: A Statistical Survey*, edited by D. K. Britton. Taylor & Francis, London, UK
48. Mottet, A. and Steinfield, H. Thomson, *'Cars or Livestock: which contribute more to climate change?' Reuters Foundation News*, 18 September 2018. http://news.trust.org/item/20180918083629-d2wf0
49. European Commission Joint Research Centre, Weiss, F, et al. (2010b), *Quantification of the GHG emissions of EU livestock production in form of a life cycle assessment (LCA)*, in: Leip, A. et al. (2010), *Evaluation of the Livestock Sector's Contribution to the EU Greenhouse Gas Emissions (GGELS) – Final Report*
50. AHDB (2010), *Change in the Air: The English Beef and Sheep Production Roadmap*
51. NFU (2020), *The facts about British red meat and milk.* https://www.nfuonline.com/nfu-online/sectors/dairy/mythbuster-final/
52. AHDB (2014), DairyCo. *Greenhouse gas emissions on British dairy farms*
53. Poore, J. and Nemecek, T. (2018), 'Reducing food's environmental impacts through producers and consumers.' *Science*, 360(6392), pp. 987-92
54. Centre for Innovation Excellence in Livestock (2020), *Net Zero Carbon & UK Livestock.* https://www.cielivestock.co.uk/wp-content/uploads/2020/09/CIEL-Net-Zero-Carbon-UK-Livestock_2020_Interactive.pdf. Figures from Poore J and Nemecek T (2018)
55. Ritchie, H. and Roser, M. (2020), 'Environmental impacts of food production'. *OurWorldInData.org.* https://ourworldindata.org/environmental-impacts-of-food?country=#carbon-footprint-of-food-products
56. Poore, J. and Nemecek, T. (2018), 'Reducing food's environmental impacts through producers and consumers.' *Science*, 360(6392), pp. 987-92

57. Knorr, W. (2009), 'Is the airborne fraction of anthropogenic CO_2 emissions increasing?' *Geophysical Research Letters*, Volume 36, L21710

58. Rao, S. (2019), *Animal Agriculture is the Leading Cause of Climate Change – A Climate Healers Position Paper*

59. Buesseler, K. O. et al. (2020), 'Metrics that matter for assessing the ocean biological carbon pump. *PNAS* 117(18), pp. 9679-87

60. Rao, S. (2019), *Animal Agriculture is the Leading Cause of Climate Change – A Climate Healers Position Paper*

61. Buesseler, K. O. et al. (2020), 'Metrics that matter for assessing the ocean biological carbon pump. *PNAS* 117(18), pp. 9679-87

62. NASA. Krishna Ramanujan.

32. Carbon Sequestration

Today the UK has only around 3.19 million hectares of woodland, representing 13 per cent of the total land area.[1] This is about half of other European countries and about a third of the European average.

According to the Forestry Commission, an area covering almost 8,300 hectares is needed to sequester 3.4 MT CO_2 (c. 410 t CO_2e per hectare). The amount equates approximately to the sum of above- and below-ground biomass, carbon in dead wood and carbon in litter. It also assumes that all that carbon will remain locked up. For a native mixed woodland, this means planting around 1,600 trees per hectare (average).[2] It is also calculated over a 100-year lifetime, and we must be mindful that we are already in a climate emergency, which cannot wait 100 years.

If trees are used to help offset UK livestock emissions, planting needs to be repeated each year to compensate for future releases. This would still leave the UK with over a one-million-hectare offshored rainforest footprint to address. A rainforest holds at least twice as much carbon as a UK woodland so would require planting at least two million hectares of land or 3.2 billion trees. That's over one million hectares worth of trees the chicken meat and egg industry needs to plant; an eye-watering figure for an industry that likes to claim how little they impact the climate emergency and biodiversity compared to other livestock sectors. Trees planted must also be the right ones in the right places, to achieve sequestration, protect the environment and enhance biodiversity.

Pledges to plant trees often fall behind. In the 2015 Conservative Party manifesto they pledged to plant a net figure of 11 million trees over the following five years, as part of government schemes in

England.[3] In the first two years only 1.44 million had been planted, leaving a lot to do. Furthermore, according to a warning made by the Woodland Trust, reported by the *The Independent* (2018),[4] England is likely to actually be experiencing net deforestation. Things do not bode well, not least as these schemes need to be taken up and financed over the lifetime of the trees. For instance, you can provide a subsidy to landowners to plant trees, but if that subsidy is removed and another land use is more economical, then trees may be felled. Furthermore, any deforestation, including ancient woodlands, inevitably releases CO_2, the amount depending on what happens to the wood (and soil). And, as already mentioned, we can't wait 100 years. We need to plant enough trees to start sequestering the CO_2 we need to remove now, and over a much shorter timescale.

In the 2019 election campaign Prime Minister Boris Johnson announced a £640 million Nature for Climate Fund, with an 'ambitious' aim of, working with the devolved administrations, planting 30,000 hectares every year. The 30,000-hectare annual planting figure is derived from a recommendation made to government by The Committee on Climate Change (CCC) advisory group. Politicising, the Liberal Democrats countered that with a pledge to plant 40,000 hectares a year by 2025. At these rates, it would take between 50 to 67 years just to replace the UK's rainforest footprint. This also assumes all the trees planted survive and other trees are not cut down/harvested, which is very unlikely. The only political party that gets the tree issue is the Green Party.

To give some sense of current planting across the UK, excluding deforestation, about 13,400 hectares were planted in the year to March 2019.[5] Broken down across the UK, the figures for new plantings are as follows: England 1,420 hectares, Wales 520 hectares, Scotland 11,210 hectares; and Northern Ireland 240 hectares. Should Scotland ever be given full independence, achieving Net Zero in England and Wales is going to be extremely challenging, to say the least.

I would suggest, for England, a good ambition to start with would be to rewild at least two million hectares of the least productive grassland, with one million hectares worth of new tree plantings (1.6 billion trees). More CO_2 would be sequestered in shrubs and so on, with a proliferation of biodiversity to accompany it.

We can certainly afford that if we adopted a whole-food plant-based diet, although it will take time to untangle and rebuild our food economy. Trees, like walnut, hazel and chestnut, fruit trees, and even oak trees, can also provide food. However, farmers would need to be properly compensated over the long term. Another benefit would be a reduction in the number of ruminant livestock, and therefore a respective decrease in methane and other emissions. Unfortunately, reducing ruminant livestock is not a feature of the UK livestock sector's Net Zero ambitions, Ireland's or of any other countries solution. Instead, offsetting, efficiency saving and increasing output remains the order of the day.

This poses another problem with industry Net Zero ambitions, the continued expansion of production and consumption of the meat, fish, egg and dairy sectors, which 'is incompatible with the goals of the Paris Climate agreement', as stated by Prof Pete Smith, chair in plant and soil science at the University of Aberdeen, and a lead author of a UN report on the impact of land use and agriculture on climate change. If this happens, and current projections suggest it will, it would undermine what greenhouse gas reductions can be made/offset for each unit of meat, fish, eggs or dairy produced.[6] According to a Greenpeace report, if left unchecked agriculture is projected to produce 52 per cent of global greenhouse gas emission in the coming decades, 70 per cent of which will come from the meat and dairy sector.[7]

Things do not bode well on our current trajectory, with total UK meat production increasing by 0.7 per cent to 4.06 million tonnes in 2019.

Table 33: UK meat production

Home-fed production (tonnes, in thousands)	2015	2016	2017	2018	2019
Cattle	880	916	904	901	917
Pigs	861	887	867	891	922
Sheep	309	300	309	299	318
Poultry	1,733	1,805	1,840	1,939	1,901
Total production	3,784	3,908	3,920	4,029	4,058

Source: Defra, 2020. Agriculture in the United Kingdom 2019.

In 2020 Defra also reported a milk production increase of 1.6 per cent to 15.2 billion litres, and this despite a fall of around 0.5 per cent in the dairy herd, the shortfall indicating an increase in the average milk yield per cow of 162 litres to 8,122 per year.[8] We may be drinking less milk per capita, but the shortfall is being met and exceeded by 'value'-added dairy products (e.g. butter, cream, yoghurt and milk powders), population growth and exports. In terms of egg production, the number of laying hens, and with it the number of rainforest-ready eggs laid for UK consumption, is also on the march, including increased exports to the EU and to the rest of the world. In 2019 there were over 2,000 million more layers compared to 2017.[9]

More hungry livestock mouths to feed risks turning even more of our finite land resource over to livestock farming, increasing the intensity at which land is farmed and increasing the number of high-risk crops grown (e.g. erosion-ready maize) to feed them. Our share of deforested land may also increase (e.g. for soya bean meal feed).

UK agriculture boasts about having one the most greenhouse-gas-efficient livestock production systems in the world, and so buying livestock products from UK livestock will help reduce global emissions. Not exactly good news for the UK greenhouse gas inventory, and for the extra land and food needed to feed all these extra animals.

Even the most defensive farmer and consumer, for the sake of their health and for their children's, and if not for the planet, should recognise the need to produce and eat much less meat, fish, eggs and dairy products. We cannot deliver the goals of the Paris Agreement unless we do, but farmers are not going to produce less without a corresponding shift in our food production and subsidy system. With governments unwilling to face up to the situation, change is left to consumers.

However, even if that happened, should domestic consumption of meat, fish, egg and dairy products fall, the industry will simply look to sell their products abroad; and expect the government, in the form of taxpayer handouts, to help them build the export market.

Grazing land is often touted as a potential storage medium for carbon and has even been hyped as being as good as tree-planting by the National Sheep Association (NSA). The NSA also promotes the benefit of grasslands to landscape, soil life and public well-being.

Quoted in FarmingUK,[10] NSA Chief Executive Phil Stocker said: 'For too long the importance of grassland has been under-acknowledged and it has regularly been treated as a poor cousin, both agriculturally and environmentally. [...] If you consider the attention that trees have received for their ability to capture carbon and improve conditions for nature it is frustrating that few policy makers understand that well-managed grassland can be just as effective at storing carbon while also providing a multitude of other public goods.'

The organic matter we can readily see in soil is particulate organic matter in various states of decomposition, which should not be confused with *genuinely* sequestered soil carbon captured within a stable system (sum of interactions, involving processes like photosynthesis, decomposition and soil respiration, which are in turn affected by land use and land-use change). Particulate organic matter consists of non to partially rotted organic matter, arising from directly deposited or handled manures, roots, leaves and other fragments. Particulate organic matter is not a stable form of soil carbon. It is exposed to decomposition and transformative processes, which in turn are influenced by geography and the soil's physical, chemical and biological properties.

Whilst you can raise the amount of soil organic carbon in grasslands through better land, grazing and manure management, keeping it in the soil bank is an entirely different matter. A lot of what's banked can eventually get spent, and even if you do manage to save a sizeable quantity, the amount of organic carbon you can accumulate over the years will depreciate until an equilibrium is reached. At that point, the soil bank shuts and you are back at square one (i.e. lots of ruminants emitting methane without any further soil carbon sequestration). It takes annual additions of large amounts of organic matter, over decades, to build an appreciable humus soil bank.

Consider comparing particulate organic matter to your monthly pay packet, money you put into your account, which over the month is spent on day-to-day living. If you are not very good with money it can be spent unwisely and you stand still or, even worse, go overdrawn. If you keep your finances sound you can save some money, but some will always be spent on the cost of living and luxuries, and if you do not invest it wisely its value will also depreciate over time. When you stop earning money, you begin to draw on your savings, and if the situation continues long enough you will have nothing left.

Particulate carbon in soil is not *saved* or *sequestered* carbon (where more carbon is absorbed than is released, thereby lowering the amount of CO_2 in the atmosphere), at least in my book, as only a small amount of what you put in annually ends up as stored carbon at the end over time.

Humus, or mineral-associated organic matter, is less available and can stay in a well-managed soil for millennia. However, if you do get access to it and spend it quickly it can be gone within a decade or so too. Humus takes a very long time to build up, and on some light highly oxygenated soils, such as sandy soil, most will always be spent without a major change in land use. Soil biology (microbes and mycorrhizal fungi) is also extremely important in forming and holding onto carbon within soil aggregates, along with increased rootage.

Differences between particulate organic matter and humus:[11]

Particulate organic matter	Humus
Relatively less complex	Relatively more complex
Fresh and partially decomposed	Highly decomposed
No new products are formed	New products are formed
It is unstable	It is fairly stable
C:N ratio is high	C:N ratio is low
Carbohydrate content is high	Carbohydrate content is low
Protein and lignin content is less	Protein and lignin content is more
Undergoes further decomposition	Does not further decompose
Original tissues can be recognised	Original tissues cannot be recognised
It is less darkened in colour	It is more darkened in colour
Not colloidal in nature	Colloidal in nature

From the Royal Commission on Environmental Pollution,[12] it takes about ten times longer to build up soil carbon following conversion to pasture than it takes to deplete carbon stocks after pasture has been ploughed. For example, when an old pasture is ploughed, a little under one half of the carbon can be lost in as little as twenty-five years, with the majority released in the first few years. In the reverse it can take well over 250 years to replace that loss, with the greatest gain, around 50 per cent of what was lost, made up in the first few decades,[13] after which the process greatly slows. In isolation this promotes the value of grassland livestock farming, but only where the soil has been managed well over time, which sadly is less often the case, conversion of hay meadows to monocultured pastures being a case in point. Succession to woodland, e.g. through a rewilding project, will increase soil organic matter over conventionally grazed pastures but would take time.

However, although ultimately rewarding, planting woodland into previously grazed pasture can also take a long time before it registers net soil carbon sequestration. Particulate organic matter in the soil can continue to decompose and release its CO_2 into the atmosphere, at a rate greater than the young trees can absorb. Eventually, this situation would reverse, and more carbon would begin to be sequestrated than lost. The situation would be quicker for fast establishing/growing coniferous woodland. Trees are great, an awful lot need to be planted, but are not a quick fix. The quickest fix in the time we still have available is to drastically reduce the number of ruminants we farm.

As a consequence of climate change, warming would increase the rate organic matter is decomposed in soil and in turn release more carbon dioxide into the atmosphere (and dissolved organic matter and inorganic carbon to water). Conversely, a higher amount of CO_2 in the atmosphere could somewhat offset that loss.[14]

In grassland systems there is only so much soil carbon you can save before you reach a state of equilibrium and the account is closed. If it didn't, all soil would end up looking like peatland, which it doesn't, though in the right circumstances you can create an exposed 'peaty-like' layer at the surface. More carbon could be stored if the land eventually transforms into woodland. In natural woodland and forest systems a considerable amount of carbon is locked within the above- and below-ground matter.

Successful farming requires all forms of organic matter, including carbon from plant exudates which plants trade with mycorrhizal fungi for nutrients (incidentally, fungi is vital to make best use of existing soil phosphorus reserves and reduce the need for finite resource inputs too). Particulate organic matter contains nitrogen in a form that more easily converts to crop-available nitrogen. Humus contains more nitrogen, but as it is much more stable and harder to break down, only a very small amount becomes crop-available.

Continuous, poor grazing of fresh grass growth slowly depletes and exhausts the soil – killing the grass and soil life with it. Unless corrected, the soil will not recover and may deteriorate further. Maintaining soil health and sufficient soil organic carbon is essential to avoid this.

In a *livestock* farming world, grazing must be properly managed to maintain and improve soil. This can include measures like avoiding overgrazing (e.g. reduced stock level) and moving livestock as one mixed-age herd between grazing paddocks to give the previously grazed land time to recover (e.g. mob grazing). Mob grazing is an intensive grazing method rather than a necessarily intensive production system, where livestock grazes smaller areas of land in quick succession with vacated areas left for an appreciable length of time to recover. Rather than the supplementary use of chemicals, mob grazing relies on the manure from the animals and nitrogen from legumes to help feed the land, along with the avoidance of plant protection chemicals. It is designed to mimic, as far as is possible, migratory herd behaviour, which has been shown in some instances to build organic carbon in soil.

Weeds are controlled through being eaten and then suppressed through vigorous grass grow-back. Without the use of pesticides and herbicides, soil microbiology can improve too and with the right biological seeding inputs enable arbuscular mycorrhizal fungi to regenerate an extensive network of hyphal mycelia. Arbuscular mycorrhizal fungi is important as the fungi has a symbiotic relationship with plants that support plant health (e.g. act as root extensions, which give plants access to nutrients well beyond the range of their roots; pest resistance; a stable soil environment; and help plants bear environmental stresses, such as drought) and produce a sticky glycoprotein called glomalin, which cements/clumps soil into stable aggregates and thereby improves soil structure. The fungi also help moderate other physical, chemical and biological processes within the soil[15] and boost soil carbon. In return plants provide the fungi, and bacteria, with carbon-rich exudates from their roots and plant litter. Around 80 to 90 per cent of plant matter nutrient acquisition is facilitated microbially, including unlocking phosphorus and calcium for plants, at some point.[16]

Mob grazing in a nutshell:[17,18]

The practice of mob-grazing involves high-density (e.g. up to 250 dairy cows per hectare), short duration (per day) grazing on highly varied and herbal rich leys. The leys include a range of shallow to deep rooting species, which are allowed to grow to maturity/flowering. The livestock, which can also be beef and sheep, single out the most nutritious bits, leaving behind a trodden mass of lignified stems, which in turn feed the soil biology. The breakdown of the stems can be slow in the first season but after a few years becomes much quicker as the soil biology begins to thrive. The daily stocking density is not uniform and will vary throughout the growing season in relation to the amount of forage available.

The animals are regularly moved, perhaps twice a day for dairy cows, and it can be 40–50 days before they return to the original paddock. Organic matter builds up more quickly given the diverse range of plant species and rooting depths which the duration of grazing (daily, and the period before returning) allows to build up, along with the organic matter generated from the breakdown of the stems and the addition of livestock excreta.

Maxing out with 250 dairy cows would return about 15 m³ of manure, containing around 70 kg of nitrogen, to each hectare of land grazed. With legumes included in the ley, there is no need for the addition of manufactured nitrogen fertiliser. Biodiversity above and below ground is higher and allows the soil structure to begin to recover, provided it is not then destroyed again by ploughing at a later date (within a rotation). The livestock are normally much healthier, as, like people enjoying a varied whole-food plant-based diet rather than meat, the nutritional quality is much higher and medicinal.

Improved carbon capture in grassland, and particularly at depth, is also associated with extensively managed practices on species-rich swards. Where soil organic carbon is generated in mob-grazed and extensive systems, including at depth, it cannot all be counted as sequestrated carbon in any one year as a percentage will be lost over time. Importantly, the percentage of organic matter measured does not equal sequestrated carbon too.

On the opposite end of the scale, intensive rye-grass livestock-based systems which overgraze or over-cut (e.g. two or more cuts of silage per year) can deplete soil carbon by over-stimulating plant decomposition.[19]

Whilst mob-grazing executed 'properly' can improve soil health and sequester carbon up to a point, livestock farming is not necessary to improve soil health and store carbon. Soil carbon can also be effectively generated through rewilding in a closed system, with the added benefit of allowing nature to do what it does best when untamed and allowed to function – build biodiversity. And when you eventually reach soil carbon equilibrium, you will not be left with lots of ruminants still emitting methane. Such rewilding is not about abandoning land or excluding grazers and rooters, which contribute to, rather than harm, biodiversity in a natural setting.

Most livestock systems currently have dominion over nature, where we are in control of what biodiversity is allowed to exist within the margins of livestock farming. Nature does not create dominions of biodiversity, and we need to take notice of that.

The popularity of modern-day mob grazing rests with the ecologist Allan Savory, following a TED Talk he presented in 2013. The notoriety Savory has enjoyed since is born out of an element of truth, in that in certain situations mob grazing can be an improvement on the destructive nature of conventional grazing systems.

However, despite Savory's enthusiasm for mob grazing (aka 'holistic management'), including the method's ability to restore land and to feed the world, much of the world's land is simply too fragile and drought-ridden to support such a system. Examples he shows as evidence of land regeneration can be achieved simply by removing the damaging grazing practices which previously plagued the land and allowing nature to rejuvenate and rewild the land. Savory is also not shy in making speculative, unsupported and outlandish assumptions on how much carbon can be sequestered by the method, including a claim of storing 2.5 tonnes of carbon per hectare per year (for almost forty years) on a third of the world's land (c. 5 billion hectares) in order to reverse climate change.[20]

The lack of scientific evidence Savory has to support this and other claims has led to a lot of open criticism. One substantial critical review of 'holistic management' (HM) was conducted by Maria Nordborg (2016).[21] Nordborg's review concluded that the total carbon storage potential in pastures for HM would not exceed an average of 0.8 tonnes of carbon per hectare per year globally, a calculation itself based on very optimistic assumptions. A more conservative estimate of 0.35 tonnes of carbon per hectare per year was suggested.

Presented as a global average, the potential to sequester carbon in some UK soils could be larger, but the evidence base is currently threadbare and anecdotal in respect of time. To help put this in context, based on a life-cycle assessment (LCA), a dairy cow with an average yield of 8,000 litres of milk per year would emit nearly 10 tonnes of CO_2e per year, or twice that for the average stocking density on a lowland dairy farm of two 550 kg dairy cows per hectare. This could be reduced by about 30 per cent to 14 tonnes CO_2e, excluding any offshored land-use emissions (e.g. not feeding dairy cows soya bean meal from South America).

Another critical review of Allen Savory's HM, by John Carter et al. (2014), also found little to be optimistic about:

This review could find no peer-reviewed studies that show that this management approach is superior to conventional grazing systems in outcomes. Any claims of success due to HM are likely

*due to the management aspects of goal setting, monitoring,
and adapting to meet goals, not the ecological principles
embodied in HM. Ecologically, the application of HM principles
of trampling and intensive foraging are as detrimental to
plants, soils, water storage, and plant productivity as are
conventional grazing systems. Contrary to claims made that
HM will reverse climate change, the scientific evidence is that
global greenhouse gas emissions are vastly larger than the
capacity of worldwide grasslands and deserts to store the
carbon emitted each year.*

These particular findings have to be placed in context with the
contrasting UK temperate maritime climate, which is vastly different to
the western North American arid and semiarid ecosystems the study
focussed on. Nevertheless, the UK has its own set of problems. For
example, where grazing is extended into the autumn and winter months
the risk of nitrate-nitrogen, ammonium-nitrogen, phosphorus and
microbial pathogens, present in directly deposited excreta, finding their
way into groundwater, inland freshwaters or coastal waters, or into
springs, wells or boreholes, increases. As soils wet into winter, poaching
and associated run-off acts as an additional pathway for pollutants,
including sediment. Then we have erosion-ready maize.

Examples of reasonable precautions farmers can consider are moving
livestock regularly, reducing stocking density and wintering livestock on
well-drained level fields. However, free-draining soils are leaky and can
swap one pollution route for another. Fields with a medium to heavy,
poorly drained soil, and fields that are land drained, should be avoided.
Field drains and drainage ditches are considered inland freshwaters, and
as such are factors farmers must take into consideration to avoid
causing direct or diffuse pollution.

Another issue with HM is that the method would not be able to
support the current level of demand for meat and dairy, or at a price
most people are willing or able to pay. Livestock systems are
land-intensive and unsustainable compared to plant-based cropping and
nutrition, whether the animals are kept in or outdoors, or a bit of both.

According to the Forestry Commission calculation mentioned at the
beginning, one hectare of woodland can potentially sequester around
410 t CO_2e per hectare, and according to the NSA a well-managed
grassland can be just as effective. I cannot find any evidence to support
the NSA claim, and it does not appear to offer anything scientific in
support its statement. According to industry LCA studies, grassland

livestock farming provides a net carbon loss. Furthermore, the carbon lost from cropland used to feed livestock counteracts what carbon may be sequestered in grassland.

In a recently published study (Poulton et al., 2018),[22] the authors concluded that it was:

> [...] more realistic to promote practices for increasing [soil organic matter] based on improving soil quality and functioning [over soil carbon sequestration] as small increases can have disproportionately large beneficial impacts, though not necessarily translating into increased crop yield.

The beneficial impacts include soil biological activity, water infiltration, aggregate stability and ease of tillage.

An important observation they make early on in the study is that an annual application of 35 tonnes per hectare of farmyard manure (FYM) only supplies around 3.2 tonnes organic carbon. Not as much as you may think. Following on, when you come to measuring the amount of organic carbon in the soil over time, it is 'important to sample the same weight of mineral soil each time'. Essentially, as you add organic carbon to soil you decrease the bulk density of the soil, similarly when you convert land to pasture for a period, and as a result it is important to sample the soil to a 'slightly greater depth so that the [soil organic carbon] is determined in the same weight of soil'. You can also cause an underestimation, depending on how the soil and carbon has been measured. It's important to get the measurement right either way when evidencing soil organic carbon (SOC) to ensure you are measuring like for like, and if that's not possible to make an educated and transparent adjustment for bulk density changes.

The primary reason for preferencing soil functioning and quality over SOC is that most of the carbon added in bulky organic manures, like FYM, is lost over time and not sequestered. In two long-term treatments the authors found only 5 per cent of the organic matter added to a sandy loam and 11 per cent added to a silty clay loam soil had been retained, and 35 per cent on a shorter-term treatment, on a soil yet to reach an SOC equilibrium. The greatest increases occurred on land starting out with a very low SOC level.

Whenever assessing soil carbon sequestration, it is important to measure stored/stable carbon within a complex system only and not simply what has been added. The authors also state that 'it is essential that the additional [carbon] sequestrated in soil would otherwise have been in atmospheric CO_2 and is not simply being transferred from one terrestrial location to another'. Furthermore, that as the rate of SOC increases, the amount of carbon sequestrated will slow over time, as it reaches a point of equilibrium, with less carbon sequestrated thereafter; and that any improvements can easily be reversed with a change in land management. In general, increasing SOC is very good for soil health but will do little towards climate mitigation unless it is directed towards low-SOC soils where it can do a little climate good.

The experimental plots the authors drew their conclusion from are also not representative of actual farm practices. As farming has intensified there is far less FYM being produced than in the past, and what FYM is produced is less widespread across the country. For instance, most of the organic manure produced on dairy farms is slurry, which has a much lower, and stable, carbon content, and what carbon is available will be more readily 'burnt' off in the soil. With fewer but more intense farms, the amount of nitrogen and phosphorus in the organic manure produced can also exceed soil and crop need on the available land and result in a significant risk of diffuse pollution. Wet manures are also expensive to transport and are mostly used as close to the source as possible, concentrating pollution.

To make the most of organic manures on land you ideally want farms to use well-composted FYM, have a mixture of arable and grassland or have ready access to arable land to exchange muck for straw. Since World War II, arable and grass-based livestock farms have become increasingly divorced from each other and this is a trend which continues to evolve. This will make it increasingly more difficult to improve soil health across the country and to maximise carbon storage in soils without creating a polluting soil phosphorus legacy. Moving of organic manures is also expensive and contributes to greenhouse gas emissions and traffic pollution.

Concentrating manures from cattle, pigs and poultry increases the risk of pollution. For instance, many soils in the UK have already become too rich in phosphorus, with further applications exceeding soil and crop need. If soil phosphorus leaches into groundwater, runs through into field drains or is washed off fields into fresh and coastal waters it can lead to the eutrophication of aquatic ecosystems. Terrestrial ecosystems directly depending on aquatic ecosystems may also be harmed. One of the reasons for concentrating manure is the importation of feed to enable more livestock to be farmed at a particular place (e.g. dairy and pigs), with the output, the extra manure generated, not been exported and ending up being used on too small an area of land, resulting in nutrient inflation.

Phosphorus leaching is actually low on most soils but can be a higher risk on coarsely textured, sandy, phosphorus-rich, soils with a low clay content. In clay-rich soils phosphorus readily binds to the clay surfaces, or to iron, aluminium oxides and hydroxides present in the soil. However, high concentrations of phosphorus in solution can bypass soil particles and end up in groundwater if the soil is deeply cracked. In most circumstances in the UK phosphorus finds its way into surface water in run-off, in which it dissolves more readily.

Even if FYM is transported to low-SOC soils, where it can do some good towards helping the climate emergency, that in itself creates greenhouse gas emissions, which would undermine any benefit.

The UK could introduce policies to reverse this trend, but that would not solve many of the other environmental problems that go hand in hand with livestock farming, including its part in our climate emergency. The world is also starting to move on, with more whole plant-based foods, meat substitutes, lab-grown meat and many other alternative sources of calories and proteins in development. We need to invest in a better future and not in our failed past. Livestock farming is also the UK's least profitable sector, propped up with subsidies to give us meat, eggs, fish and dairy foods, which we do not need, cause pollution and are associated with many illnesses we succumb to.

Before leaving this topic, it is worth returning to claims that grazing land is a potential storage medium for carbon sequestration. The NFU publication *The facts about British red meat and milk* (2020), 'Myth 1' states that:

- Cattle (beef and dairy) and sheep accounted for 26.2 MT CO_2e. That's 5.7 per cent of the UK's total emissions (enteric fermentation and wastes only).
- Carbon sequestered in UK grasslands accounted for a net reduction of 9.2 MT CO_{2e}.
- Taking into account grassland sequestration, cattle and sheep account for 3.7 per cent of emissions. Excluding grassland sequestration, cattle and sheep account for 5.7 per cent.

With the stroke of a pen, including grassland sequestration removes 2.0 per cent of emissions, or 9.2 MT CO_2e, but is that the case? If nature produced dysfunctional soils and ecosystems in which nothing grew on land except the monoculture grasslands created by livestock farming, the answer may be yes. The reality, of course, is no, as nature would otherwise occupy the land and carbon sequestration, and biodiversity would be even greater for it. The 'myth' buster also fails to include carbon lost in cropland used to feed livestock.

Furthermore, the UK uses a wide definition of Grassland in its Land Use, Land Use Change and Forestry (LULUCF) inventory, which includes 'semi-natural grassland and other habitats which may not have grassy vegetation including dwarf shrub heath, bracken, montane habitats [near-natural vegetation which lie above the natural treeline], fens, marsh, swamp and bog',[23] which would make up a large chunk of the 9.2 MT CO_2e. The Grassland left is largely 'improved grassland', which has lost rather than gained carbon in comparison to its original status, though it does have potential through land management change to reverse some of that loss. Also, according to the Centre for Ecology and Hydrology: 'A lack of field data on the response of Grassland soils to management activities means that it is not currently possible to include estimates of the effects of Grassland Management on soils in the LULUCF [land use, land-use change and forestry] inventory.'[24] This includes a lack of data on the impact of soil erosion, grass–crop rotation cycles and intensification.

A similar claim has been made in an Auckland University of Technology project report, commissioned by Beef and Lamb New Zealand, which suggests the standard method for calculating carbon sequestration should be changed to include below- and above-ground carbon stocks on farms (grassland, hedgerows, trees, etc).[25]

Merit can only be found in relation to *new* permanent plantings and *changes* in land management practices which would genuinely generate more *real* carbon storage than presently happens. Livestock farming should not be carbon-credited for hedges, trees etc. still left standing after years of damaging the rest of the land for grazing livestock. Any *real* carbon stored would also need to be offset against the offshored land-use change emissions associated with livestock production and deforestation, for example the use of soya bean meal and beans from South America.

Also on this point, the UK agriculture emissions, 10 per cent of total greenhouse gas emissions, do not capture soil and peatland emissions.[26] Counting emissions from soil and peatland, and other exclusions, would raise agriculture's share to around 15 per cent,[27] which would need to be included when making Net Zero subtractions. Ultimately, for transparency and fairness, Net Zero calculations need to be based on life-cycle assessment, which in turn would include offshored emissions resulting from deforestation and the annual carbon sequestration lost as a consequence of that deforestation.

The NFU publication, *The facts about British red meat and milk*, Myth 1, additionally states that:

> *Emissions from UK livestock are estimated at around 5 per cent of the country's total GHG emissions, significantly lower than the estimated EU-wide figure for livestock of around 9.1 per cent of all emissions. This is part due to the UK's efficient production system.*

By using the last sentence this is true, but without further context it leaves a misleading impression that the difference is much greater than it really is. As I've already covered, care must be taken when quoting greenhouse gas emission figures as the percentages are proportionate to the level of each sector activity that occur in England, the devolved administrations, the EU and elsewhere. For instance, England's, Wales's, Scotland's and Northern Ireland's livestock sectors produce, respectively, 5.5, 10.33, 14 and 21.9 per cent each out of the individual countries' *total* GHG emissions.

England is much more industrialised than the devolved administrations and therefore it cannot be read that England is much more GHG-efficient per unit of animal-based production. If we calculated emissions in the way the publication suggests then America's livestock GHG footprint would be even better than the UK's, which it isn't. The more transparent and fair way, as I've already alluded, is to calculate GHG emissions between each country from life-cycle assessments. The UK agriculture industry, like other developed nations, already profits from not having to make up for pre-1990 emissions, created by the heavy industrialisation of farmed land, and as such there's only so much generosity which should be afforded.

Citations:

1. Forestry Commission (2019), *Forestry Statistics: A compendium of statistic about woodland, forestry and primary wood processing in the United Kingdom*. National Statistics at Gov.uk
2. Coed Cymru (2017), *A Basic Guide to Tree Planting*
3. BBC News (26 October 2017), *Reality Check: Are millions of trees being planted?* https://www.bbc.co.uk/news/science-environment-41551296

4. Beament, E. 'England 'highly likely' to be suffering from deforestation, campaigners warn.' *Independent*, 15 June 2018 https://www.independent.co.uk/environment/deforestation-england-forestry-commission-woodland-trust-environment-climate-change-a8399726.html

5. Forestry Research (2020), *New Planting and Restocking* [2020 data] via https://www.forestresearch.gov.uk/tools-and-resources/statistics/statistics-by-topic/woodland-statistics/

6. Wasley, A. and Heal, A., 'Revealed: development banks funding industrial livestock farms around the world.' *The Guardian*, 2 July 2020. https://www.theguardian.com/environment/2020/jul/02/revealed-development-banks-funding-industrial-livestock-farms-around-the-world

7. Greenpeace (2018), *Halve meat and dairy Production to protect climate, nature and health.* https://www.greenpeace.org/eu-unit/issues/nature-food/1100/halve-meat-and-dairy-production-to-protect-climate-nature-and-health/

8. Defra (2020), *Agriculture in the United Kingdom 2019.* National Statistics at Gov.uk

9. Ibid.

10. FarmingUK (15 March 2020). *Grasslands as vital as trees for environment, sheep farmers say.* https://www.farminguk.com/news/grasslands-as-vital-as-trees-for-environment-sheep-farmers-say_55207.html

11. Accessed at https://agriculturistmusa.com/difference-between-organic-matter-and-humus/

12. Royal Commission on Environmental Pollution (1996), Royal Commission on Environmental Pollution: nineteenth report: sustainable use of soil. HMSO, London, UK

13. Cannell, M. G. R. et al. (1994), *Carbon pools and sinks in British vegetation and soils.* NERC Annual Report 1993/94

14. Royal Commission on Environmental Pollution (1996), Royal Commission on Environmental Pollution: nineteenth report: sustainable use of soil. HMSO, London, UK

15. Borie, F. et al. (2008), 'Arbuscular Mycorrhizal Fungi and soil Aggregation.' *J. Soil Sc. Plant Nutr.* 8(2), pp. 9-18

16. Phillips, M. (2017), *Mycorrhizal Planet: How Symbiotic Fungi Work with Roots to Support Plant Health and Build Soil.* Chelsea Green, White River Junction, VT

17. Wilkinson, I., Lane, S. and Mountain, F. (N. D.), *Mob Grazing: A Farmer's Guide.* Cotswold Grass Seeds Direct

18. Zaralis, K. (2015), *SOLID participatory research from UK: Mob Grazing for Dairy Farm Productivity.* The Organic Research Centre

19. Ward, S. E. et al. (2016), Legacy effects of grassland management on soil carbon to depth. *Global Change Biology* 22(8), pp. 2929-38

20. Nordborg, M. (2016), *Holistic Management: A Critical Review of Allan Savory's Grazing Method.* SLU/EPOK, Uppsala, Sweden

21. Ibid.

22. Poulton, P. et al. (2018), 'Major limitations to achieving "4 per 1,000" increases in soil organic carbon stock in temperate regions. Evidence from long-term experiments at Rothamsted Research, United Kingdom.' *Global Change Biology* 24(6), pp. 2563-84

23. Moxley, J. et al., Centre for Ecology & Hydrology. Reporting the effects of Grassland Management on carbon storage in soils and biomass (pdf)

24. Ibid.

25. Priestley, M. 'New Zealand sheep and beef farms "close to carbon neutral".' *Farmers Weekly*, 9 October 2020. https://www.fwi.co.uk/livestock/grassland-management/new-zealand-sheep-and-beef-farms-close-to-carbon-neutral

26. UCL (2021), *Towards Net Zero in UK Agriculture: Key information, perspectives and practical guides.* https://www.sustainablefinance.hsbc.com/carbon-transition/towards-net-zero-in-uk-agriculture

27. Dieter Helm CBE (2021), *The future of Farming.* John Innes Centre talk. http://www.dieterhelm.co.uk/natural-capital/environment/the-future-of-farming-john-innes-centre-talk/

CHAPTER V

Rainforests

*Politicians have a clean conscience ... because they
never use it!*

Formosan subsistent farmers, Argentina

33. Deforestation & Exploitation

As discussed previously, UK livestock production today has a South
America and Southeast Asia land-use change footprint of at least one
million hectares, excluding any imported animal-based meat foods or
directly imported soya beans. The majority, over 900,000 hectares, is
associated with soya bean meal imports.

Annihilation of native rainforests though is far from a new
phenomenon. During the nineteenth and early twentieth centuries
British businesses employed local rural workers to clear large tracks of
Argentina's rainforest to make way for sugar production and other
commodities.[1] The soya bean incursions began in the main after World
War II to support UK livestock intensification, with MAFF records
showing soya bean meal use averaging nearly 250,000 tonnes between
1964–65 and 1975, before rapidly expanding into the 1980s.[2]

The problems caused by forest clearance though are not limited to
greenhouse gas emissions. Soya bean production in South America
comes with a whole set of other troubles too:[3]

- Large-scale economic concentration, strongly integrated with, and
  dependent upon, global commodity markets, economic debt and
  a dependency on patented seed and potentially harmful
  pesticides.
- Serious environmental damage.
- Severe consequences for traditional rural communities'
  livelihoods and health and well-being.

In Argentina production of soya bean meal accelerated from the
mid-1990s, following the introduction of transgenic/genetically

modified soya beans. As it did, the economic benefits of soya bean production increasingly transferred into the hands of a smaller number of farmers, employment in agriculture fell and a soya bean monoculture progressively displaced the diverse array of traditional crops. The growth of soya beans also resulted in a reliance on patented GM seeds and the associated herbicide glyphosate, which has markedly contributed to national debt and economic dependence.

In Argentina the use of transgenic/GM herbicide resistant soya bean crops was approved in 1996.[4] Back then about around 18.7 million tonnes worth of soya bean were grown, rising to 47.5 million tonnes by 2006–2007; of which 90 per cent was estimated to come from GM crops.[5]

Serious environmental damage comes in many forms:

Deforestation with disastrous ecological impacts. The rainforests of South America are exceptionally species-rich. Conversion of forests to a soya bean monoculture has caused major biodiversity loss, with little to no provision for the conservation of the many rare and abundant species that inhabited the land previously. In their place, monoculture soya bean fields can stretch as far as the eye can see.

An exponential increase in, and uncontrolled use of, herbicides.

Chemical water pollution.

Appearance of herbicide-resistant super-weeds.

Agrochemical drift damaging and killing traditional crops. GM herbicide ('glyphosate') tolerant soya beans occupy much of the crop. The growth of glyphosate, and the dependence on the chemical, has expanded exponentially over the years. Monsanto, the company who supply the 'RoundUp Ready' GM glyphosate tolerant soya bean seed, only make a small amount of their profit from them. The majority of their profits come from the sale of their herbicide, which is profusely used. There is little let-up in their determination to increase herbicide sales.

The repeated use of the herbicide, and other types, has inevitably led to herbicide-resistant superweeds. As a consequence, chemical warfare has intensified, with farmers using heavier applications of glyphosate and having to resort to mixing in 2,4-D and other highly toxic chemicals, such as dicamba, paraquat, atrazine and endosulfan,[6] in the fight against the super-weeds.

These excessively used and poorly managed chemicals leach or are otherwise washed into and contaminate streams, groundwater and water supplies. They also damage soil life. Furthermore, given the vast expanse of soya bean crops, they lend themselves to aerial application. Such applications are prone to drift, with well-documented devastating impacts on adjacent local communities, animals, natural vegetation, traditional crops and water supplies.

Depletion of soil nutrients.

Soil degradation and erosion. Soil wealth and health is vital towards ecological sustainability and food security. The growth and poor management of soya bean cultivation has resulted in severe soil nutrient deficiency, through the continuous mining of soil fertility. Chemicals used on the land kill soil biology and turn the soil into dirt. Removal of the forest canopy has exposed the soil, changing it from a once stable substrate into a highly erodible one, with soil loss levels in Brazil and Argentina estimated at 19–30 t per hectare per year.[7]

Severe human impacts come in the form of:

Enforced dispossession/eviction of small farmers, agricultural labourers and indigenous families from their customary land.

Violence, including the killing of indigenous activists, changes to cultural reference.

Poverty, and hunger from the transference of traditional and diverse foods to exported animal feeds for use in faraway countries.

Poor health, from agrochemical drift over populated areas including schools and contaminated water. Health issues range from headaches, soreness in throats and faces, muscular pain and skin rashes to cancer, organ failure and birth defects alleged from the indiscriminate use of agrochemicals.[8]

Enforced migration towards cities and shanty towns, often with little job opportunity and/or transferable skills to find work.

Human rights violations.

Today the majority of wealth generated from soya bean production rests in the hands of a few transnational corporations, large national companies, and concentrated sectors of the economy.[9]

One would hope local and national politicians would intervene, but as commonly communicated by indigenous subsistence farmers, they see politics as a 'dirty business, driven by personal, selfish interests, and tainted by corruption'.[10]

The UK has looked at alternatives to soya bean meal, including home-grown protein feeds. Back in 2009, DairyCo published the Lupins in Sustainable Agriculture (LISA) report, which demonstrated that lupins can provide a valuable source of dietary protein and energy for ruminants, offering a traceable and non-GM alternative to imported soya. Further research into the issue was also undertaken and reported on in 2014, by Reading University,[11] and by the Defra sponsored Green Pig Project, LK0682.

However, unless the economic outlook changes, or a change is stimulated through an intervention measure or a change in farming practice, then the import and inclusion of cheaper GM soya bean protein in livestock feed will prevail. All for inefficiently produced products we do not need to eat.

Availability of arable land to grow additional sources of protein and energy for livestock is also an issue against meeting the demand for other crops (including crops for energy production). Improving the yield and quality of home-grown protein and energy crops would also be an effective option for both conventional and alternative crop types. However, this would require significant investment in plant breeding and agronomy, which is currently not supported within the market.[12]

Citations:

1. Lapegna, P. (2016), *Soybeans and Power: Genetically Modified Crops, Environmental Politics, and Social Movements in Argentina*. Oxford University Press, London, UK

2. Marks, H. F. (1989), *A Hundred Years of British Food & Farming: A Statistical Survey*, edited by D. K. Britton. Taylor & Francis, London, UK

3. Lapegna, P. (2013), 'Notes from the Field: The Expansion of Transgenic Soybeans and the Killing of Indigenous Peasants in Argentina.' *Societies Without Borders* 8(2), pp. 291-308

4. Otero, G. and Lapegna, P. (2016), 'Transgenic Crops in Latin America: Expropriation, Negative Value and the State.' *Journal of Agrarian Change* 16(4), pp. 665-674

5. Lapegna, P. (2013), 'Notes from the Field: The Expansion of Transgenic Soybeans and the Killing of Indigenous Peasants in Argentina.' *Societies Without Borders* 8(2), pp. 291-308

6. Binimelis R et al. (2009). '"Transgenic Treadmill": Responses to the Emergence and Spread of Glyphosate-Resistant Johnsongrass in Argentina.' *Geoforum* 40(4), pp. 623–33

7.   Altieri, M. and Pengue, W. (2006), 'GM soybean Latin America's new coloniser.' *Seedling* 13-17

8.   Lapegna, P. (2016), *Soybeans and Power: Genetically Modified Crops, Environmental Politics, and Social Movements in Argentina.* Oxford University Press, London, UK

9.   Lapegna, P. (2013), 'Notes from the Field: The Expansion of Transgenic Soybeans and the Killing of Indigenous Peasants in Argentina.' *Societies Without Borders* 8(2), pp. 291-308

10.   Lapegna, P. (2016), *Soybeans and Power: Genetically Modified Crops, Environmental Politics, and Social Movements in Argentina.* Oxford University Press, London, UK

11.   Jones, P. J. et al. (2014), Replacing soya in livestock feeds with UK-grown protein crops: Prospects and implications. CAS Report 19

12.   Ibid.

CHAPTER VI

Are We Kind to Farmed Animals?

'[…]if one person is unkind to an animal it is considered to be cruelty, but where a lot of people are unkind to a lot of animals, especially in the name of commerce, the cruelty is condoned and, once large sums of money are at stake, will be defended to the last by otherwise intelligent people.'

Ruth Harrison, *Animal Machines* (1964)

34. Animal Welfare

Collectively, UK livestock farmers are responsible for the health and well-being of around 1.2 billion farmed animals a year. The industry claims to uphold the highest animal welfare standards within the global livestock economy. From factory farming to a system of a life worth living, I have experienced them all, chicken (meat and eggs), pigs (indoor and outdoor), beef, dairy, sheep and goats. I have taken in deep breaths of fresh country air as I have watched animals free-roam. I have breathed in hot, stuffy, dusty, ammonia-laden air in chicken broiler and egg-laying battery units.

Industrialised farming is not a choice for a lot of farmers. Farmers are compelled to satisfy market forces to stay in business, coerced to adopt less kind, increasingly intensive animal production systems, to lower costs and maintain a living.[1] If you farm in a place which is inconveniently located to the market, as with many dairy farms, and you cannot produce enough product to make it worth your while economically, you may also be in trouble.

During my work I have also spent time in abattoirs, seeing most types and stages of animal slaughter, except poultry and pig asphyxiation. My hosts did not spare me any experience. Slaughterhouses were simply a necessary part of the human food chain, however unpleasant these places were. Without passing judgement, and with little thought, I did my job and moved on to the next.

People I speak to in the industry acknowledge that animal husbandry involves unkindness. Even the best husbandry ends with an act of slaughter. Unfortunately, with demand for cheap animal-based foods, traditional mixed farming in the UK has given way to even less kind, more intensive forms of production – and it continues to give way.

The UK's Welfare of Farmed Animals legislation allows 'necessary suffering or injury', for instance in respect to their freedom of movement, provided it is in accordance with 'good practice' and 'scientific knowledge'. Breeding procedures likely to cause suffering or injury are permitted provided they are minimal or cause momentary suffering or injury or might necessitate interventions which would not cause lasting injury. Electric currents can be used on animals, other than for the purpose of immobilisation.[2]

Animals are sentient. They experience pleasure and suffering, and a range of feelings (joy, frustrations, fear, hunger, playfulness, etc.) that matter to them.[3] To be proud of livestock farming means, by default, that we must be proud to do all the unkind things we do to farm animals, which I will explain in this chapter. If we are not proud to cause such unkindness, then doing it would be nonsensical – wouldn't it?

A few people think we share a mutually beneficial relationship or 'contract' with the animals we farm. We provide for them, and in return they provide for us. This cannot be further from the truth. They do not choose the way we keep and use them; we do. No animal chooses death over life or to die for what we have provided them in their shortened lives. The animals we eat are born into servitude.

UK animal welfare ranks better than many countries, but causing a little less pain, suffering or injury does not make the way we farm and kill livestock kind or justified. Animal welfare standards do not give farmed animals the life that the livestock industry, and those who sell us their products, want us to believe that they have. The images of 'happy' farm animals in advertisements and on food packaging do not reflect the truth. As long as we farm in the name of commerce, where money is at stake, we can never be kind.

After fifty years, the unnecessary way we treat farm animals finally dawned on me. What once echoed in my well of silence began to resonate loudly. We all need to listen and rid ourselves of this cancer, untangle animal production from our food system and adopt a peaceful whole-food plant-based lifestyle.

The way we treat farm animals in the UK is guided by the following 'Five Freedoms':

- Freedom from thirst, hunger and malnutrition (by ready access to fresh water and a diet to maintain full health and vigour).
- Freedom from discomfort (by providing an appropriate environment including shelter and a comfortable resting area).
- Freedom from pain, injury and disease (by prevention through rapid diagnosis and treatment).
- Freedom to express normal behaviour (by providing sufficient space, proper facilities and company of the animal's own kind).
- Freedom from fear and distress (at least until transport and slaughter) (by ensuring conditions and treatment which avoid mental suffering).

When you read through this chapter, please be mindful of these freedoms. Question whether they always happen on our farms, and in particular factory farms.

One of the best examples of animal welfare in the UK happens at Kite's Nest Farm, Worcestershire. The farm's owner, Rosamund Young, who is also the author of *The Secret Life of Cows*, produces organic beef in a system where the cows are given a large amount of freedom. Her cows can come in or stay out as they please; calves stay with their mothers beyond suckling; the cows dine on carpets of herb-rich pastures, hedgerows and trees; and each one is allowed to exhibit natural behaviours, form complex relationships and fully interact with each other.[4]

Cow life may be better at Kite's Nest Farm, but their meat is expensive when compared to meat produced on industrialised farms. High animal welfare is defined by the price people are able, or prepared, to pay, not by an animal's best interests. In reality, high animal welfare is simply a misnomer.

Animal welfare activist Ruth Harrison, in her book *Animal Machines* (1964), revealed the indignity and suffering that industrialised production systems imposed on farm animals, and the disconnect in public understanding between rapid turnover, high-density stocking, mechanisation, low labour requirements, efficient conversion of food into saleable products and animal welfare.

George Monbiot, in his Channel 4 television documentary *Apocalypse Cow: How Meat Killed the Planet*, remarked on the absurdity of happy, mixed livestock farms portrayed in children's picture books, in stark contrast to the harsh reality of industrialised farming.

In *The Secret Life of Cows*, Rosamund Young sets out the 'clear moral obligation [farmers have] towards their animals', the 'misplaced conceit to believe that any [human-made] environment can equal or better the natural one', and how 'wherever the pursuit of maximum profit has led to intensification it is the animals that have suffered the most'.

Mark Fisher, in his book *Animal Welfare Science, Husbandry and Ethics*, discusses the harms and benefits of animal husbandry practices, the acute pain and 'compromises resulting in animals experiencing suffering – the work that they do, the social disruptions they experience, the confinement and boredom they are exposed to and the resources that they are deprived of'.

The unkindness we show to livestock denies their sentience, and despite Ruth Harrison's intervention in the mid-1960s we continue to give precedence to our selfish needs. There is no mutual contract with the animals we farm, just pain, suffering and death. All livestock farms treat their animals as machines/commodities; some are just more mechanised than others.

Whilst these observations are most pertinent to factory farms, they apply to all livestock production systems. We need to stop doing all the unkind things we do to animals. This does not mean abandoning farms, farmers, the rural economy and all other people and businesses dependent on livestock farming, but transitioning to a more secure and sustainable – plant-based – food system.

Renowned British journalist, author and pasture-based farm proponent Graham Harvey champions the need to wrestle our existing food production system away from ruthless corporate interests to one more in tune with the foods evolution prepared for us.[5] Harvey's vision for farming and the British countryside is a step in the right direction and one realistically within our grasp. Though we must go further, much further, and in fact, all the way!

We must stop filling our bellies with animal pain and suffering. We need to disturb the sounds of silence and stop all unkindness to farm animals.

Citations:

1. Carruthers, J. (1991), *It pays to be humane – a review of the evidence*. Centre for Agriculture Strategy, Paper 22
2. The Welfare of Farmed Animals (England) Regulations 2007. *Schedule 1: General conditions under which farmed animals must be kept*. (The regulations are similar across the devolved administrations.)
3. Fisher, M. (2018), *Animal Welfare Science, Husbandry and Ethics: The evolving story of our relationship with farm animals*. 5M Publishing, Sheffield, UK
4. Young, R. (2003, 2017), *The Secret Life of Cows*. Faber & Faber, London, UK
5. Harvey, G. (2016), *Grass-Fed Nation: Getting Back the Food We Deserve*. Icon Books, London, UK

a. Beef, Dairy Cows & Their Slaughter

i. Beef & Dairy Cows. In England around 51 per cent of prime carcass beef is derived from the dairy herd, 30 per cent from hill and upland suckler herds and 19 per cent from lowland suckler herds.[1] There is a drive to increase the dairy-beef herd, driven in part by bad publicity surrounding the slaughter of unwanted dairy calves soon after birth.

Unwanted dairy calves are slaughtered because not all female dairy calves are needed to replace their mothers – this on top of the 50 per cent of unwanted male calves. By artificially inseminating (AI) dairy cows with sexed semen from elite dairy cows, it has become easier to produce higher-performing replacements for the milking herd. This in turn has allowed farmers to impregnate more dairy cows with sexed semen from elite beef bulls to produce more desirable beef calves for meat production (male and female).

AI is not used for the benefit of the cow, or in the interests of kindness. It is about improving a herd's economic return by enhancing its genetic merits. There is also a cost and management benefit from not having to keep and service dairy cows with a bull.

An AI procedure starts with the physical constraint of the cow in a crush. This is necessary to protect the cow and farmer from physical harm. Cattle may be handled in this way between one and five times over a three-week period. The need to restrain a cow speaks volumes of how unpleasant this act can be to a cow. AI is not a passive procedure for the cow and farmer.

AI is not an advancement over Mother Nature for animal welfare. It is a regressive procedure which suits modern livestock husbandry and profitability, condoned and defended by otherwise intelligent people.[2]

With fine margins and pressure on the farm-gate price of beef and dairy, farmers are under pressure to use AI. The consumer has little knowledge and choice when it comes to AI.

AI starts by relaxing the cow after its initial restraint in a crush. A paper towel is used to wipe clean the cow's vulva, before the inseminator inserts a fully gloved and lubricated arm into the cow's rectum. Entry into the rectum is eased by the inseminator forming a cone with his/her fingers, whilst keeping the cow's tail to one side with the other hand. Before performing insemination, the inserted arm is manoeuvred to eject any dung present. Next the inseminator locates the cow's cervix with his/her fingers, and then applies downwards pressure onto the cow's vagina with an elbow. This is done to part the lips of the vulva. The whole procedure has to be performed carefully to avoid damaging the cow's sensitive rectum wall.[3] The AI gun, loaded with a semen-filled straw, is then inserted and guided into and through the cervical canal. Once through, the semen is squeezed onto the cow's uterine horns and the gun withdrawn.

If not done carefully, other than not achieving successful insemination the procedure can cause bleeding and trauma in the uterine horns and break the cervical seal if the cow is already pregnant, resulting in the loss of the calf.[4]

Before any of this, semen has to be collected from a bull. Several methods can be used, including an artificial vagina, transrectal massage and electro-ejaculation.

An artificial vagina is a collection vessel inserted into a cow's vagina to collect the semen from a bull allowed to mount her. Alternatively, a vessel may not be used and the semen simply collected from the cow's vagina.

Electro-ejaculation involves inserting an electrical stimulator probe into the bull's rectum. The bull must first be restrained in a cattle crush. An AC charge, 12–24 amps, is delivered to stimulate the bull's nerves and accessory sex glands to induce an erection and illicit ejaculation, which is collected in a pot for processing and use in subsequent AI. Transrectal massage can be used to prepare the bull for the process. This involves pushing an arm up to the shoulder into the bull's rectum and manually stimulating the nerves surrounding its prostate and pelvis.

The method is not without pain. However, in animals it is difficult to understand the amount of pain caused without carefully observing the animal's response to an unwanted stimulus. Signs are: irregular movement of hind limbs, kicking, moving back and forth, muscle spasm at the thighs and abdominal region, arced back, lying down or becoming recumbent.[5] Whilst pain can be controlled to some extent, there can be no certainty that the procedure is pain-free. Safety precautions include being prepared for shock and movement/jerking from the bull. The method is promoted, and authorised for use in the UK under the Veterinary Surgeon's Act 1966, Schedule 3 Amendment Order 1982. The same procedure may be used to test a bull's fertility.

How these procedures, and how a cow must suffer during them, are considered an act of high animal welfare is hard to fathom. In the UK such suffering is allowed provided it is brief and unlikely to cause lasting injury. Being brief and not lasting does not make this act kind or acceptable.

What happens to a dairy calf soon after birth cannot be considered kind either, to neither mother nor her baby. Soon after birth, typically within twenty-four hours, the calf is dragged, carried, or placed in a wheelbarrow and wheeled away. A distressing act, and the reaction often shown by the mother is heartbreaking.

In all my time on dairy farms I have only witnessed this act once, and the memory will last my lifetime. Watching a YouTube clip of a calf being removed from its mother, and I recommend everyone reading this does, is hard and impactful enough. Seeing it first-hand is even worse. I would imagine the first time for a farmer or a farm helper would be a difficult experience too, but the impact would quickly wane once repeatedly performed – 1.9 million times a year in the UK.

In my situation the calf was removed from its mother quickly after birth, wheeled away and placed in a strawed pen. I watched the calf try and fail to stand up at least ten times, before I had to move on. The mother bellowed in constant distress and was still bellowing as I left the farm forty-five minutes later.

A cow's milk is money, which is not allowed to be wasted on her a calf. However unkind, the act is another defended to the very last by otherwise intelligent people. The practice is considered reasonable and necessary for human health. It has to be done, in part to cover the low farm-gate milk price buyers are willing to pay. Do it enough times and the act, however unkind, becomes normalised/customary.

Calves in turn may be placed in separated pens for up to eight weeks, where they are unable to socialise and frolic with each other. Unwanted females and dairy-bred beef calves may be moved on after two weeks to a rearing unit, where they continue to be weaned for another six weeks on a milk and concentrate ration. At three months they may be moved on again to a finisher unit. This may all happen on the dairy farm or on separate farms.

Dairy-bred calves are most often associated with high-input systems. They are eventually slaughtered at 12–18 months of age. Dairy-bred cows, and beef-bred cows from suckler herds, where calves are kept with their mother for at least five months, tend to be reared in moderate-input systems. They are slaughtered at 18–24 months of age. Beef cows from suckler herds may also be kept in low-input systems, where they live for 24–30 months on a grass-based system.

In all systems the removal of calves from their mothers is a harrowing experience for both.

Most male calves also endure castration, which can cause stress and pain. A 'benefit' of the procedure is a reduction in the animal's testosterone levels, resulting in a lessening of sexual activity, lower aggression and an improvement in meat quality for human taste.

Calf testicles can be removed surgically or by placing an external constricting rubber band/ring or clamp at the base of the scrotum. Constriction compresses the arteries and veins, stopping arterial flow to the testicles. Responses to stress and pain caused by the procedure include struggling, kicking, tail-swishing, agitation and head-turning.

In the UK an anaesthetic does not need to be used if a rubber band/ring or clamp is applied within the first week of the calf's life. Calves over two months old are not allowed to be castrated without anaesthetic, and after that age only by a vet.

Calves may also be disbudded or dehorned. Buds are the stage before horn material shows. Either method mustn't be done without anaesthetic, unless chemically cauterised within the first week of life. Not all beef herds are dehorned.

Thanks largely to Ruth Harrison (1964), the rearing of white veal calf in the UK, a very cruel practice carried out for many years, is a thing of the past. However, low-animal-welfare veal farming is still regularly practised outside of the UK.

Instead, the UK produces rosé veal from calves kept on bedding. Here, calves are given space to lie down and stand up, groom, move, explore and interact socially to an extent. They receive twice the amount of fibrous food compared to continental veal calves.[6]

Continental veal calves are given up to 60 per cent of the legal UK space allowance, are often kept on slatted floors, which contribute to their discomfort, injury and lameness, and are likely to be kept in cramped and sometimes lightless conditions. They may experience ill health, lethargy and weakness linked to an EU minimum iron supplement requirement.[7] Calves can also be transported great distances on the continent and elsewhere, which is detrimental to their health and well-being, suffering cramped conditions, bruising, extremes of heat and cold, hunger and thirst, illness, and even death.

Narrow veal crates once used in the UK and the rest of the EU are banned, but versions are still permitted and used in the US and other countries. Here, incarcerated calves are normally denied solid food and fed only milk or milk replacer right up to slaughter.

Ruth Harrison's 1964 account of veal farming and other shameful livestock practices in the UK's past are truly shocking. Unkindness is still performed today, with new methods of unkindness invented in the name of commerce to serve convenience and profit.

Another problem with farming cattle is the diseases they may suffer, often associated with the production system we subject them to. One of several ailments is mastitis. Mastitis is a particularly painful condition that causes the mammary gland tissues to become inflamed and infected. The disease is endemic within dairy herds and is most commonly transmitted through contact with the milking machine, contaminated hands, bedding and other equipment.[8] Mastitis comes with a hefty bill to the industry, with a national average incident rate of about 30+ cases per 100 cows.[9] The condition may even result in premature culling. Medications such as antibiotics are required to treat cows. A dairy cow's teats may also become infected with warts, which are highly infectious, detrimental to her health and cause difficulty during milking as the warts make it difficult to form an effective seal between the cluster and teats.

Lameness is another common condition, causing pain, discomfort and impaired walking. Endemic diseases, such as the widespread bovine viral diarrhea, infectious bovine rhinotracheitis and Johne's disease (paratubrerculposis), exist within UK herds – bovine tuberculosis too. Exotic diseases occasionally break out, including blue tongue and foot-and-mouth.

Farmers do not want their animals to be sick, but sickness suffered within herds is an inevitable consequence of the way they are farmed.

Hooves, knees and joints can suffer from standing on hard concrete floors for considerable lengths of time. In particular, factory-farmed animals may never get to stand on the soft, herb-rich earth as nature intended. All forms of confinement induce stiffness in their joints and muscles. Like us, cows evolved to move around freely. They suffer the same pains of discomfort that we do when overly inactive. However, whilst we can get up and move about, cows kept indoors are much more restricted. Some comfort is given in the individual cubicles dairy cows are provided with, but nevertheless cows spend much of their day between milking standing on hard-floored passageways. The more a cow is confined, the more often they are likely to develop a variety of health problems and their life ended early at the slaughterhouse.

Housed and over-crowded dairy cows spend each and every day in direct or close contact with their excrement, exposing them to noxious gases like ammonia, which volatilises from their excreta. Standing in excrement can also cause bacterial infections in their hooves.

In our silence we see cows as all the same, when in reality they are as individual as our dogs and cats. They have their own personalities, wants and needs. Given compassion, they show their intelligence and affection. They are inquisitive animals, which like to venture and explore. They are sociable, form relationships and even fall out with each other. Just like we do. At times, like us, they can also be silly and fatuous.

We feed cows what we want them to eat, to serve our needs, not what they actually need for their best health. We treat dairy cows like machines and milk them like robots. The life of a dairy cow is to serve us 305 days of a year, with a short period between pregnancy to recover. We suck milk from them up to three times a day, causing sores and irritations which can lead to infections. We breed and take more milk than a cow would produce to nourish her calf, which stimulates her to produce more milk to meet the demand we expect of her body. If we were to stop milking, she would begin to stop producing milk. And so the cycle goes on, and well beyond the time taken to wean a calf. A dairy cow that does not fulfil her quota is a dead cow.

Beef cattle fair better, and at least for now are much less likely to be housed in confinement. Things are changing though, with a few US-like concentrated animal feeding operations (or CAFOs) springing up to maximise profit, each confining 1,000 cows or more.

Unlike the US, though, one cattle problem we are not currently concerned about is the use of hormonal growth promoters (HGPs). HPGs are banned in the UK to protect human health rather than out of any concern for animal welfare. The UK has also been active in phasing out the routine use of antibiotics, including restrictions on certain types like chloramphenicol and ampicillin. Antibiotics can also act as a growth promoter, but their use contributes to antibiotic resistance.

To compete with a future trading deal with the US, it is likely more beef cattle will be confined this way to lower costs. Should this come to fruition, and before rushing out to fill shopping carts with bargain beef, shoppers should consider David Simon's economic assessment of the rigged American meat and dairy sector.[10] Each US state has a statute prohibiting cruelty to animals, '[b]ut in response to industry lobbying, most states have adopted an exception to their anticruelty statutes for farm animals'. These exceptions are termed 'customary farming exemptions', or CFEs. CFEs essentially allow states to exempt all farm animals from the laws protecting them from cruelty, 'provided the act is done while following generally accepted agricultural practices'. According to Simon's research three-quarters of US states have some form of CFE, many of which have been adopted over the last few decades. Simon explains how CFEs work:

Customary farming exemptions essentially remove from lawmakers the authority to decide what constitutes cruelty to farm animals, instead turning that decision over to farmers themselves. For example, if one farmer decides it would be expedient to chop off all or part of an unanaesthetised animal's body part, like an ear, tail, beak, or genitals, and others in the industry follow suit, that procedure becomes customary and protected as a CFE. Thereafter, those who engage in what would otherwise be criminal animal cruelty are exempt from prosecution.

Simon gives some examples of CFEs:

- *Crushing or severing the testicles of unanaesthetised animals.*
- *Slaughtering chickens while they are awake and alert.*
- *Killing unwanted male chicks or spent laying hens by suffocation, starvation, or disposal in a garbage can or wood chipper.*

The driver for CFEs is the need to minimise costs, as 'from a pure economic perspective, it doesn't pay to worry about an animal's pain or suffering'.

Consumers must not expect individual farmers to undertake practices that will make them uncompetitive in the marketplace. Livestock producers will do what is necessary to compete, or else they will not be livestock producers for very long.[11]

Whilst the UK does not have CFEs, we do have laws that exempt farm animals from procedures and treatments that would not be allowed on our dogs and cats. Procedures like AI are becoming customary to be competitive and slotted into the definition of high animal welfare.

The gap between the law and what is allowed to happen to farm animals is slightly closed by animal welfare assurance schemes. However, these still do not provide anywhere near enough protection.

In the name of economics, to feed us food we do not need and food that can damage our health and our planet, the livestock farming industry carries out many forms of unkindness which would not be countenanced otherwise. The whole process has become normalised, in a similar way to how CFE works, which we hide under the disguise of high animal welfare. Normalisation is further bolstered by consumer ignorance, indifference and our hypocrisy towards animal welfare.

Farm animals are no less sentient, nor do they feel any less pain and discomfort than pets.

ii. Clarabelle. In 2015 a story was published about a rescued Australian dairy cow named Clarabelle. During her time as a milking cow, all her babies were removed. When her milk production waned, Clarabelle became uneconomical and destined for slaughter. On saving her, the rescue centre, Edgar's Mission, discovered she was pregnant. Close to her expected due date, Clarabelle began acting oddly. To the mission's surprise she had secretly given birth and hidden her calf in an area of tall grass.[12]

This is not uncommon. There are other similar stories of ex-dairy cows giving birth and hiding their babies to prevent them being found and taken away. Dairy cows are highly maternal and do not forget the experience, but each year we take their babies away and make them relive it.

iii. Their Slaughter. UK beef and dairy cows are slaughtered in a restraining pen. Most are stunned with a shot from a penetrating captive-bolt pistol to the head, aimed between the eyes and ears, which targets the brain and renders the animal unconscious. A shackle is then attached to a hind leg, the cow hoisted upside down onto a conveyor belt and carried along to the bleed area. Here the major blood vessels in the neck are severed with a sharp knife, a procedure called sticking, depriving the brain of oxygen. Bleeding-out has to be done quickly before the cow recovers consciousness.

Stunning is not infallible; a repeat shot may be required, and some cows may become conscious before being bled.

Electric stunning, a discharge to the brain via a large pair of tongs, may also be used on calves. The brain and heart can also be targeted simultaneously with an electric current to kill the cow outright. Bleeding-out is not required where the animal is electrocuted, though a successful kill cannot always be guaranteed.

An estimated 150,000 pregnant cows are slaughtered each year.[13] When hung and bled, the calves are cut out from their mother's womb, fall out onto the floor and if still alive dispatched. Unborn calves do not exist within the legal system, and as such there is no guarantee to their treatment. For instance, they may be clubbed to death or their throat slit.

Citations:

1. AHDB (2009), *Change in the Air: The English Beef and Sheep Production Roadmap – Phase 1.*
2. Harrison, R. (1964), *Animal Machines,* Vincent Stuart, London, UK
3. Short, W. (2015), '8-step guide to artificially inseminating a dairy cow.' *Farmers Weekly,* 5 November 2015.
    https://www.fwi.co.uk/livestock/livestock-breeding/8-step-guide-artificiall y-inseminating-dairy-cow
4. Ibid.
5. Baiee, F. et al. (2018), 'Modification of Electro-Ejaculation Technique to Minimise Discomfort during Semen Collection in Bulls.' *Pakistan J. Zool.,* 50(1), pp. 83-89
6. Compassion in World Farming, *Farm Animals, Calves Reared for Veal.* https://www.ciwf.org.uk/farm-animals/cows/veal-calves/.
7. Ibid.
8. AHDB, Dairy. *Technical guidance for your farm: Mastitis.* https://dairy.ahdb.org.uk/technical-information/animal-health-welfare/mast itis/#.Xjk7Whf7RR4
9. Ibid.
10. David Simon (2013), *Meatonomics: How the Rigged Economics of Meat and Dairy Make You Consume Too Much—and How to Eat Better, Live Longer, and Spend Smarter.* Conari Press, Newburyport, MA
11. National Pork Board (2003), *Swine Care Handbook.* National Pork Bord, De Moines, IA
12. Schelling, A. 'Mother Cow Hides Newborn Baby to Protect Her from Farmer.' *The Dodo,* 25 February 2015.
    https://www.thedodo.com/dairy-cow-calf-baby-rescue-1010627123.html
13. Singleton, G. H. and Dobson, H. (1995), 'A survey of the reasons for culling pregnant cows.' *Vet Rec.* 136(7), pp. 162-5

Pigs & Their Slaughter

i. Pigs. Around 5 million pigs are farmed in the UK at any one time. 60 per cent of the breeding herd are kept indoors and 40 per cent outdoors. Of the piglets destined for meat, only a very small percentage spend their whole lives outdoors. The majority of indoor pigs are factory-farmed, with the remainder typically reared in straw-bedded units. Those kept on straw may be RSPCA Assured pigs, but despite better conditions many still suffer tail-docking. Concerns of unkindness to pigs are mostly associated with intensive factory farm practices.

In the UK, before 1999, it was still legal to keep and tether sows in stalls. The practice was banned under the Welfare of Livestock Regulations 1994, with transitional time to become compliant. Tethers in the rest of the EU were not banned until 2006 and stalls from 2013, but only after the first thirty-five days of gestation. They remain in use in many European countries. Fortunately, the UK has moved on from these particular dark days, but the light still remains very dim in many other areas.

A modern indoor factory farm requires high capital investment. Space for pigs is at a premium to maximise a decent return in a market which at times has been difficult for farmers. Enough space is allowed for each pig to lie down, stretch out, defecate and for rudimentary activity. As an idea, about half of the floor is exposed when all pigs are standing; pigs at the standard liveweight for slaughter (110 kg) are provided 0.65 m^2 for all activities.

Dry sows are given enough space to enable them to feed without competitive bullying. When lactating, the sows are placed in confining crates, essentially a railed metal cage with a separate creep area for her piglets. The crates are only large enough for the sow to lie down and stand up, and with a little wriggle room to move forward or backwards. Enough headroom is provided below the lowest rail to allow the piglets access to both rows of the sow's teats. A very few farms use 'enriched' crates, which allow the sow to turn around. The floor may be plastic, concrete or sometimes metal, and slatted, situated above a pit or channel to collect the sow and piglets' excrement.

Concrete-floored pens with straw may be used to keep sows between farrowing and for post-weaning pig growth cycles; however, fully or part-slatted floors may be used. Whenever pigs are housed on hard concrete floors and slats, it can cause considerable discomfort and lead to lameness, injury and infection but is nevertheless allowed. The main reason cited for confining sows in a crate is the potential danger she presents to her piglets (or rather to a farmer's profits). Constant confinement makes her weak and less stable when she stands and lies down, putting the piglets at risk of getting crushed. Sows suffering from confinement stress may also show unusual aggression towards the stockperson, and towards other sow's piglets if not kept separated. Aggression towards other piglets wouldn't normally happen in the wild, as the sow moves far away from the group to farrow.

The unkind way sows are confined is driven by commercial interests, and for ease of stockmanship. It has nothing to do with the interests of sows, who suffer to allow farmers to make enough money in a market unwilling to pay for more humane, and with it more costly, animal welfare practices. Repeated crating and pregnancy take their toll, and the sow eventually becomes economically unviable and often too large for the crate. Her reward for all the hard work, the removal of her babies and other suffering she endures is a one-way trip to the slaughterhouse.

Whilst confined, industry guidance suggests that a sow's strong maternal, natural nesting behaviours can be adequately expressed by giving her some straw/paper to lay on, or a piece of rope/toy attached to the crate. This is a situation that cannot be taken seriously by otherwise rational people but is sadly defended to the very last.

The crate becomes a sow's home about a week before giving birth and until the end of farrowing. A stay of up to five weeks. Shortly after weaning, the sow is re-impregnated and gestates for around four months before being confined again for the next nursing period. A sow averages around 2.3 litters per year, with around 13.6 live pigs born indoors per litter (12.1 outdoor). On average around one indoor piglet is born dead/mummified, with an average pre-weaning mortality rate of 11.6 per cent (12.7 per cent outdoor).[1] Sows are repeatedly denied bonding with their young and fulfilling their maternal instincts.

Piglets (<4 weeks or no more than 5 kg in weight) found in poor health and considered untreatable are commonly killed on farms by a 'trained' stockperson or vet by lifting the piglets by their back legs above the stockperson's head and then throwing them down forcefully onto a hard surface to smash their heads. A sharp blow to the head over the brain with a suitable tool, sufficient to break and depress the cranium, may also be administered.[2] This act is permitted in law (Animal Welfare Act 2006), provided it is considered an 'emergency' (to avoid suffering) and that the act is carried out swiftly. It is allowed under farm assurance schemes, including Red Tractor and even RSPCA Assured. Violent methods of euthanasia like this are considered acceptable and humane for farm animals but would be considered unacceptable and inhumane by most people for pets. Older pigs may be restrained and stunned with a captive-bolt gun, then dispatched (e.g. throat slit and bled out, or pithed using a rigid rod – physical destruction of the brain to ensure rapid death following captive-bolt stunning) before regaining consciousness – or shot.

After farrowing and weaning stages the piglets are moved on to grower and then finisher stages. Thereafter, they are slaughtered for their meat.

The main commercial breed used is the Large White. Most commercial pigs in the UK are a hybrid of predominantly Large White and Landrace genetics. These sows produce a large litter of piglets, which grow quickly, efficiently and have a lean body. Farms do not always manage each stage, with some stocking weaned pigs only to grow and finish. Pig units operate on continuous batch cycles of farrowing, weaning, growing and finishing to provide sales throughout the year. Accommodation is generally cleaned after each stage is complete and the pigs move on. The large units may mate, farrow and wean on a one, two or five weekly basis. Around 92 per cent of the meat they yield is produced on 1,600 UK 'assured' farms.

In factory farms tail-biting is a big problem, and can lead to severe pain, lameness, infections, abscesses in the spine, joints and organs, sometimes resulting in total carcass condemnation. While farmers may look for alternative methods to tail-docking, including the provision of toys, the practice is not common and performed without pain relief. Tail-docking is performed to prevent more painful conditions resulting from tail-biting, though the biting still occurs frequently despite docking, because the cause is the system rather than the piglets having a natural tail. Pigs' teeth may also be clipped or ground down and ears notched without pain relief.

Boredom, lack of space to escape other pigs, disease, inappropriate environmental conditions or inadequate provision of food are the main root causes of tail-biting. Mouthing at bars that imprison them, which can cause further sores, is another sign of a pig's psychological distress. Housed pigs are also largely unable to express their natural rooting behaviour, though they try in vain.

Surgical castration of piglets in the UK is less common than in the majority of pig-producing countries, as most assurance schemes discourage the practice, but it still does occur, and without anaesthetic and pain relief. However, the practice is common on the continent and in other countries, often without anaesthetic/pain relief, so meat eaters need to think long and hard about what pig meat they are buying as we currently import over 60 per cent of pig meat consumed here. The purpose is to reduce the risk of boar taint in the meat. UK law prohibits castration without an anaesthetic on pigs aged seven days or older, which can encourage earlier boar castration on commercial farms that are not part of an assurance scheme.

Castration involves an incision through the scrotum and vaginal tunic. The testicle is grasped, traction applied and twisted and pulled away along with a portion of the spermatic blood vessels and cord. The wound is left undressed to heal and an antiseptic applied in some instances.[3] We have the means not to tail dock and castrate pigs by adjusting their environment and groupings. Unfortunately, this is another example where economics drives animal welfare and not kindness.

Artificial insemination (AI) is commonplace within the pig industry and has become indispensable on factory farms. It involves an insemination rod, lubricated to smooth the rod's passage, and towels to clean the vulva before insertion. The rod is pushed up and twisted into the sow's cervix. When in place, a semen bag is attached to the rod and the semen squeezed with a little pressure into the sow's uterus.

Boar semen needs to be sourced first, either from a commercial stud enterprise or self-collected on-farm. Semen is commonly gathered from a trained boar using a mounting dummy. The boar is encouraged to make a few thrusts, before the handler extends the penis to display the erection. The handler then grasps the end of the boar's penis and holds it until the pig stops thrusting and ejaculates into a thermos. In traditional pig-farming businesses with a smaller number of pigs, sows are more likely to be covered by a boar. However, as costs of AI lower, more and more pigs are being serviced in this way.

Pigs are clean animals, but the conditions they are kept in on factory farms result in them becoming coated in each other's excrement. In their natural environment, pigs will seek out specific dunging zones and rest on clean areas. Pigs are also noted for their high intelligence and are able to outcompete dogs in many cognitive tests. Not that intelligence should be the defining factor in animal welfare, because it is a subjective human term. They are wonderful social creatures, with inherent, strongly-embodied behaviours, which they are not allowed to exhibit in factory farms.

Sows in factory farms are not treated humanely. The level of unkindness shown is not acceptable. The cramped conditions they and other pigs are kept in are stress-promoting and render them open to numerous viral and bacterial diseases. Antibiotics are regularly needed, amounting to nearly 50 per cent of all UK farm antibiotics used. Before moving on to their slaughter it is worth reflecting on packaging terminology used in relation to the way pigs are kept outdoors.

Outdoor pigs can mean one of three things: 'outdoor bred', where animals are born outdoors and then reared indoors; 'free-range', where pigs live all their life outdoors; and 'outdoor-reared', a halfway house system where pigs are grown to around 30 kg outdoors and then 'finished' indoors. Packaging in supermarkets can be very confusing, with all three descriptions and pictures used giving the impression of equal treatment.

ii. Stella. Stella was a sow believed to have lived on a factory farm, who escaped when in transit and luckily found her way to a friendly family. Her tail had been cut off, and she suffered paralysis on one side of her face – possibly a consequence of the fall in her bid for freedom. She was taken in by the Rooterville Animal Sanctuary in Florida, who described her as 'an amazing animal who is affectionate and joyful', and she lived out her days at the sanctuary.[4]

Pigs are beautiful, gentle, friendly and highly intelligent animals. Whilst they can be dangerous if hungry or fearful, in particular a sow with piglets, these are normally characteristics of captivity on factory farms, not when kept in a loving environment where they can fulfil their natural behaviours.

iii. Their Slaughter. Pigs are slaughtered at around five to six months. The main method used, in around 86 per cent of cases, is gassing. In 2014 Animals Australia, in a world-first, filmed inside a gas chamber where pigs spend their last moments of life. The short film leaves it up to the viewer to judge whether gassing is humane.[5] This was followed up by an investigation by Animal Liberation Victoria in 2015, with a more graphic and higher-quality view of the process.[6] Both videos are harrowing and look to be out of a horror movie. Watch for yourself, but if watching the videos would be too distressing, let me describe.

The process involves stressed pigs being coerced in small groups into a railed cradle, which is then lowered into a gas chamber pre-filled with a minimum 70 per cent carbon dioxide air concentration (UK minimum) to kill the pigs.[7] Pigs find the carbon dioxide gas aversive and show signs of extreme distress for 30–60 seconds, including attempts to escape, gasping,[8] squealing and head-throwing. Some of the movement may occur after the animal has lost consciousness.[9] From a Humane Slaughter Association (HAS) account, the movement that occurs before unconsciousness is 'not' a welfare concern![10]

Also, according to the HAS:

This system has many welfare benefits including: reduced risk of potential human error compared with, for example, electric stunning [induced unconsciousness with electric tongs] in which there is a risk of incorrect placement of electrodes.

Not a particularly good advert for electrical stunning, which is the method used for most of the remaining 14 per cent. The HAS conclude that a non-aversive gas mixture would be 'preferable', but this is not yet 'commercially available' (i.e. an alternative, less aversive gas system would cost too much money!). You can't help but be concerned when you read words like 'preferable' in relation to humane slaughter. Research published in 1997 concluded that 'during killing with a high concentration of carbon dioxide, pigs would have to endure a moderate to severe respiratory distress induced with this gas for a considerable period of time prior to the loss of brain responsiveness'.[11]

Imagine your last moments, deprived of oxygen and in pain before falling unconscious. We are no different to pigs in that respect. Close to 11 million pigs were slaughtered in the UK during 2019,[12] with 86 per cent treated in this way.

Citations:

[1.] AHDB (2019), *UK pig facts and figures.*
https://pork.ahdb.org.uk/media/277092/uk-pig-facts-and-figures_2506_190507_web.pdf

[2.] Pig Veterinary Society (2013), *The Casualty Pig: Interim Update April 2013*

[3.] The British Kunekune Pig Society. *Castrating Boars. Castration of Kunekune Boars and the anatomy of the boar.*
https://www.britishkunekunesociety.org.uk/articles/Castration.aspx

[4.] Mahaffey, A. (2014), *Stories of Rescued Pigs That Will Make You Squeal with Joy!* https://www.onegreenplanet.org/animalsandnature/three-stories-of-rescued-pigs-that-will-make-you-squeal-with-joy/

[5.] Animals Australia (2014), *If this is the 'best', what is the worst?*
https://animalsaustralia.org/features/not-so-humane-slaughter/

[6.] Quinn, L., 'Horrific video shows pigs being "forced into small cages with Tasers" before they are "lowered into gas chambers" and killed.' *Daily Mail Australia*, 7 December 2015.
https://www.dailymail.co.uk/news/article-3348745/

Horrific-video-shows-pigs-forced-small-cages-Tasers-lowered-gas-chambers-k
illed.html

7. Humane Slaughter Association (HSA), *Frequently Asked Questions*.
    https://www.hsa.org.uk/faqs/general#n3

8. Verhoeven, M. et al. (2016), 'Time to Loss of Consciousness and Its Relation
    to Behaviour in Slaughter Pigs during Stunning with 80 or 95% Carbon
    Dioxide.' *Front. Vet. Sci* 3(38).
    https://www.ncbi.nlm.nih.gov/pubmed/9232122

9. Humane Slaughter Association (May 2007), *No. 19: Carbon Dioxide
    Stunning and Killing of Pigs*.
    https://www.hsa.org.uk/downloads/technical-notes/TN19-carbon-dioxide-
    pigs-HSA.pdf

10. Ibid.

11. Raj, A. et al. (1997), 'Welfare implications of gas stunning pigs: 3 The time
    loss of somatosensory evoked potentials and spontaneous
    electrocorticogram of pigs exposed to gases.' *Vet J.* 153(3), pp. 329-39

12. Defra (2020), *United Kingdom Slaughter Statistics – August 2020*.
    Accessed at https://assets.publishing.service.gov.uk/government/
    uploads/system/uploads/attachment_data/file/858176/slaughter-statsnotic
    e-16jan20.pdf and https://pork.ahdb.org.uk/prices-stats/production/uk-
    slaughtering-forecasts/

c. Broiler Chickens, Laying Hens & Their Slaughter

i. Broiler Chickens. Broilers are chickens grown for meat. Life begins
in a hatchery, where parent birds are bred to produce viable eggs. Upon
laying, the eggs are removed and placed in incubators which provide
optimum conditions for the chicks to develop. The chicks start to hatch
from around twenty-one days, and at one day old are placed in a
mixed-sex flock broiler unit. The first stage of their life is crucial to their
development, with ambient temperatures of 32 °C to 35 °C and relative
humidity of 60 to 70 per cent.[1] Conditions are modified thereafter as
the chicks mature, until reaching a slaughter weight within five to seven
weeks of stocking; depending on the system. The average broiler weight
is 2.2 kg, requiring a substantial growth rate a day.

The growth rate is too fast for the broiler to develop a fully
functioning immune system, necessitating the need for vaccines to
avoid the risk of viral infection like Newcastle disease, infectious
bronchitis, avian pneumovirus, and so on.[2]

The broiler unit typically accommodates 10,000 to 40,000 birds per
shed, in which they never enjoy natural light and fresh air. The birds
remain in the sheds until slaughter weight. The cramped conditions they

endure are stress-promoting and render them open to respiratory diseases, including the viral infections I mention above.

Under EU law the maximum stocking density is approximately nineteen birds per square metre, which if you do the maths equates to a floor space smaller than an A4 sheet of paper supporting a combined weight of around 47.5 kg. Under the RSPCA Freedom Food label, the maximum stocking density is 30 kg per m^2 or around twelve birds per square metre.[3] Growth rate is limited to 45 g per day, requiring a slower-growing bird breed.

With such restrictive space the birds become increasingly unfit, which benefits the business as exercise increases the feed cost to reach slaughter weight. Fast and unnatural growth results in a juvenile bird of obese adult proportion. A bird's bones grow at a slower rate than their meat, and at point of slaughter their skeleton is hardly able to support their weight. Some collapse before reaching slaughter weight.

Lameness is a common problem within housed flocks, which makes it hard for the birds to walk. The most afflicted birds may not get to food and water, causing premature death by starvation and dehydration. Difficulty walking is shown by a bird's legs and feet splaying under the weight of their obese bodies. The extent of leg disorders in UK broiler farms was estimated in a study by Knowles et al. (2008).[4] The study assessed the walking ability of 51,000 birds, representing 4.8 million birds within 176 flocks. At a mean age of forty days, over 27.6 per cent of birds studied showed poor locomotion and 3.3 per cent were almost unable to walk. Scaled up, nearly 300 million chickens may be suffering from poor locomotion every year, with 30 million almost unable to walk.[5] These horrific figures exclude another 40 million birds which will die of various afflictions per year before even being sent to slaughter (based on an average Red Tractor mortality rate of 4 per cent).[6] One can only imagine how much more lame birds suffer, in particular during crating and transport to the slaughterhouse.

As the birds struggle to stay upright, they may drag their abdominal region through contact with bedding contaminated by urine and excreta. The birds may suffer lesions on their breasts, hocks and feet. The growth rate can also lead to premature death through lung and heart failure. Total mortality in broiler units is around 4 per cent, meaning around 1,600 birds out of a flock of 40,000 birds flounder and suffer a particularly horrific death in any forty-day period. Dead and even dying birds may be cannibalised. Mortality rates are much higher for most other poultry, at up to 8 per cent for indoor turkey and duck.

Before harvesting, food is normally withdrawn, resulting in hunger and stress. The most modern systems may use a catching machine, otherwise birds are caught and crated by hand. Bruising, bone and joint dislocation and breakages are not uncommon during catching, placement in crates and subsequent transport. The risk of injury is greater when birds are caught and carried by one leg, a method which is quicker to perform. Stressed poultry also defecate more, which, when transported and slaughtered, can increase the risk of infection and meat contamination.

Ducks, geese and turkeys are handled differently. Ducks are generally grasped by the neck with, but not always, a supportive hand underneath. Bigger poultry are caught and held around their middle with two hands, whilst covering their wings. Turkeys are held by coupling a closed wing and holding a leg. At least that is what's recommended. Less care may occur with repetitive catching as the catcher tires or through boredom, risking a reduction in animal welfare over time.

During transport to the slaughterhouse, poultry can suffer from hyperthermia and hypothermia, particularly when wet. Especially at risk are end-of-lay hens with reduced plumage.[7] With the closure of many local slaughterhouses, the journey is often longer, increasing this risk and other stressors. The industry is highly critical of any suggested measures to restrict the transport of chickens in freezing or very hot conditions.

Broiler chicken welfare is better under RSPCA Freedom Food but still involves pain, suffering and unkindness. Unfortunately, RSPCA involvement helps normalise the practice to consumers.

The minimum expectation from a free-range system should be one that maximises daytime access, has a well-designed range and offers the opportunity for hens to run, investigate and express their natural behaviour. The more crowded and restricted a free-range system is, the more likely a significant percentage of birds will not venture outside. This is not a chicken's choice; it can come down to a number of reasons, including a lack of grass outside and flock hierarchy. In large flocks those at the bottom of the pecking order can become very intimidated and too frightened to go outside.[8] This applies to free-range layer hen systems too. Slow-growth, organic systems with small flocks maximise well-being and provide a better life, in this case lasting around seventy days.

ii. **Laying Hens.** The current egg laying potential of a modern commercial hen is around 300 eggs a year. For context, the red junglefowl, the ancestor of laying hens, lays a dozen or so eggs a year. Industrialised hens are farmed well beyond their natural capacity – so much so that the productive life of a factory-farmed laying hen is only around twelve months. When she is sent to be slaughtered, another hen is incarcerated in her place.

Across the EU 'barren cages' were outlawed and replaced with 'enriched cages', giving each hen a slightly larger space allowance and a limited opportunity to perch, dust and nest. In my time I have experienced both systems, with enriched caged birds only marginally less wretched than their predecessors.

Space allowance is a little more than an A4 sheet of paper, with no opportunity to exercise or experience natural sunlight and fresh air. The main cause of mortality is associated with the large number of eggs laid, which depletes the hen's calcium store and causes brittle bones (osteoporosis). This is compounded by a lack of exercise.[9]

The sheer scale of enriched caged hen houses, home to thousands of soulless, crowded and bored birds, trapped within a life not worth living, stacked into several tiers, makes it extremely difficult to inspect injuries and mortality. Many go unnoticed. Enriched cages are unkind to hens on many more levels than the hens are stacked. The only solution is to phase out this abhorrent practice.

Chicks have the tip of their beaks cut off using an infra-red beam without anaesthetic, to reduce future feather pecking and cannibalism. This unkind act is to prevent them hurting each other during laying. The chairman of the British Free-Range Eggs Production Association (BFREPA), James Baxter, has been quoted as taking issue with beak-trimming being referred to as mutilation.[10] The defence of BFREPA is that it stops hens causing harm to each other, such as pecking and cannibalism, in later life. However, feather-pecking and cannibalisation are a consequence of farming and keeping the birds in unnatural conditions. It is mutilation; it can't be anything other. The beak is extremely sensitive, and research is very clear that beak trimming causes pain and suffering. Given we do not need to eat eggs for our health, then it is unnecessary. Therefore, whether we like it or not, farming eggs to eat is an unkind act carried out simply for our pleasure. Beak-trimming is also only one form of unkindness hens will suffer up to and including their slaughter.

Barn and aviary systems, whilst kinder to hens, still restrict their freedom and ability to express the full spectrum and level of natural behaviours. Small free-range flocks, and ideally organic systems, are preferable.

Care should be taken on how free-range eggs are chosen in the supermarket. 'Value' free-range eggs are derived from houses which accommodate thousands of birds. As with broiler chickens, although daytime access is granted, many hens never venture outside as those at the bottom of the pecking order can become very intimidated and frightened to go outside.[11] Some may also become habituated, a learned condition that suppresses their natural inquisitive nature, during the first weeks of captivity when they are not let out. Free-range birds may also suffer feather loss and injury from pecking and ammonia gas, which can build up to high levels in sheds, and other injuries, with the more seriously injured culled on site. Bone breakage is another significant issue, given chickens are bred to produce eggs in cages, not for healthy bones and outdoor living. Dead and dying chickens may also be cannibalised. Factory-farmed free-range chickens do not live the happy lives consumers are led to believe. I would expect many genuine, small-scale, free-range chicken farmers would agree with me that the term 'free-range' has been hijacked by Big Agriculture and has lost its true sense and meaning.

End-of-lay birds have to be caught and crated for transport. As with broilers, in all systems, and particularly intensive in ones, the level of osteoporosis and poor general fitness can result in a high risk of bone fracture or joint dislocation during catching. More so when caught by one leg.

In an egg layer's hatchery male chicks are not allowed to live longer than a day. They have no commercial value and have to be disposed of. At a day old they are either sent down awake on a conveyor belt into a macerator to be ground up alive, or gassed to death in closed cabinets. Chicks who die by gassing, called controlled atmosphere killing (CAK), may be sold on as feed for reptiles and other animals. The fact that their death may have a purpose does not make farming and gassing them any kinder.

iii. Little Miss Sunshine. Little Miss Sunshine is a rescued Australian layer hen. She was kept in a barren cage for eighteen months, unable to express her natural behaviours and stretch her wings. She was saved from slaughter in 2013 by the Edgar's Mission animal sanctuary, where she was allowed to feel joy, flourish and interact with her human friends.[12]

She was not rescued alone, and this happy story stretched to the other 1,080 hen-layers kept with her – four to a cage. The farmer could no longer face caging and then sending 'spent' hens to slaughter, and quit the industry.[13]

iv. Their Slaughter. In the UK gas slaughter is the main method used to kill chickens, hens and turkeys. They are placed, still in their transport crates, into either closed cabinet, tunnel or pit gas systems and left until dead. Tunnel and pits are pre-filled with CO_2. Cabinets are only filled with CO_2 once the birds have been placed. CO_2 causes the painful formation of acids in nasal and mouth fluids.[14]

In the Meyn cabinet system, birds are gassed in five-stages. They remain conscious in the first two stages for a total of two minutes.[15]

In the Marel Stork tunnel system, birds are kept in the first of two phases until the birds are rendered unconscious, or in a multi-phase system where the birds are transported through five sections and exposed to an increasing level of CO_2 (28–70 per cent).[16]

An example of a pit type is the Linco system, where birds are lowered into the pit in several stages and exposed to an increasing level of CO_2 (5–70 per cent). They are held in each stage for a short period, for at least six minutes in total.[17]

According to the Farm Animal Welfare Council (FAWC), none of these methods produce immediate insensibility in a way similar to electric stunning.[18]

Poultry may also be electrically stunned. On arrival the birds are shackled upside down by their legs onto a conveyor belt, which moves along to an electrically charged water-bath. Shackling is a painful process, in particular for birds with broken legs from prior handling. FAS also states that 'live shackling may cause considerable pain and distress, which are likely to be exacerbated when heavy birds or fracture-prone, end-of-lay hens are shackled'.[19] As their heads are dipped into the water it creates an electrical circuit between the water and the shackle, which stuns the bird. Unconscious, though that is not always guaranteed, they are moved along the conveyor belt to where their necks pass through a mechanical cutter. The cutter severs the major blood vessels to complete the kill. Stunning may be ineffective if the water-bath is not adequately charged, signs of which include rhythmic breathing, tension in the neck and the presence of a third eye reflection.[20] The solution is not to start again; with a throughput as high as 10,000 birds per hour there is only grace to rectify the fault for followers.

Some systems may apply stun-and-kill to induce cardiac arrest, avoiding the use of a high current, which may affect the meat quality.[21]

Slaughter without pre-stunning is permitted under an exemption for religious purposes, in this case for Jewish (Shechita) and Muslim (Halal) communities, whose beliefs require full animal consciousness. In both situations the animal's throat is cut with a knife. However, some within the faiths have a more relaxed interpretation of their holy laws than others. In 2017, FSA data shows that 81.5 per cent of animals (including sheep, goats and cattle) slaughtered for religious purposes were stunned first.[22]

When chicks and adult poultry are gassed, they are not rendered unconscious/killed instantaneously and experience pain and suffering.

Citations:

1.   Compassion in World Farming (2019), *The Life of: Broiler Chickens*. https://www.ciwf.org.uk/media/5235306/The-life-of-Broiler-chickens.pdf
2.   Ibid.
3.   RSPCA (2017), *RSPCA welfare standards for meat chickens*. https://science.rspca. org.uk/sciencegroup/farmanimals/standards/chickens.
4.   Knowles, T. G. et al. (2008), 'Leg Disorders in Broiler Chickens: Prevalence, Risk Factors and Prevention.' *PLoS ONE* 3(2): e1545 doi:10.1371/journal.pone.0001545
5.   Knight, A. and Wiebers, D. (2020), *A British Pandemic: The Cruelty and Danger of Supermarket Chicken*. Open Cages
6.   Ibid.
7.   Farm Animal Welfare Council (2009), *Report on the Welfare of Farmed Animals at Slaughter or Killing. Part 2: White Meat Animals*
8.   Young, R. (2003, 2017), *The Secret Life of Cows*. Faber & Faber, London, UK
9.   Compassion in World Farming, *Farm Animals, Egg Laying Hens*. https://www.ciwf.org.uk/farm-animals/chickens/egg-laying-hens/.
10.   Ryan, C., 'Analysis: Increasing pressure to ban infra-red beak treatment.' *Poultry News*, 13 February 2019. http://www.poultrynews.co.uk/ health-welfare/welfare/analysis-increasing-pressure-to-ban-infra-red-beak-treatment.html
11.   Young, R. (2003, 2017), *The Secret Life of Cows*. Faber & Faber, London, UK
12.   Animals Australia, *Chickens: How a hen named Little Miss Sunshine became a TV star*. https://www.animalsaustralia.org/features/ little-miss-sunshine.php.
13.   Ibid.
14.   Farm Animal Welfare Council (2009), *Report on the Welfare of Farmed Animals at Slaughter or Killing. Part 2: White Meat Animals*
15.   RSPCA (2018), *Killing Systems for Farmed Chickens and Turkeys*.
16.   Ibid.
17.   Ibid.
18.   Ibid.
19.   Farm Animal Welfare Council (2009), *Report on the Welfare of Farmed Animals at Slaughter or Killing. Part 2: White Meat Animals*

20. Ibid.
21. Ibid.
22. Food Standards Agency Board Meeting (20 September 2017), *Animal Welfare Update. Annex 2*

d. Breeding Chickens, Turkeys & Ducks

i. Breeding Chickens, Turkeys and Ducks. Egg-laying hens and broiler chickens start life in a breeder unit, which may be part of a closed, self-supply production system or supplied by a specialist unit. In the UK hens are typically mated by a cock rather than by artificial insemination (AI). Each cock mates around ten hens a day. After the hens have been fertilised and eggs laid, the cycle begins again. There is no let-up for the hen and the cock, at least not until she/he is physically spent and no longer economically viable.

In nature hens choose their breeding partner from competing males and lay a dozen or so eggs a year. In breeder units they are not given a choice and are required to produce a constant supply of eggs throughout the year. Repeated mounting results in feather loss and the formation of wounds on the hen's back, which are not only unsightly but can become very sore. The damage is caused by the cock grabbing hold of the hen's neck feathers and 'treading' to steady itself, which happens over and over again without sufficient recovery between mating.

Turkey breeding has created a bird too big to procreate naturally, or at least effectively. This has led to two commercial opportunities. One, it allows for an even bigger bird to be created, and secondly, more turkeys can be covered by a single male's semen.

Turkey semen is collected by securing the bird upside down on a person's lap or in a specialised cradle. The area either side of the bird's copulatory organ is sexually stimulated by his 'handler' to stimulate ejaculation. Finger pressure is then applied to expose the organ from the folds of flabby skin that surround it to ejaculate semen from its now bulbous ducts. The semen is then collected in a pipette. The process may be repeated until the bird's semen supply is fully depleted.

In other parts of the world semen from chickens is collected in a similar way, for subsequent AI.

AI is administered by upturning the female turkey to expose her abdomen, which is squeezed to open her cloacal vent and cause her oviduct to protrude. Semen is then injected into the vent. An insemination machine, loaded with semen, may be used for rapid application in large flocks.

Another concern with breeders is the need to restrict their diet. Bred to grow exorbitantly fast, it is necessary to limit what they can eat or else they would end up at table bird proportions. A restricted diet leads to stress, frustration and chronic hunger,[1] a situation that is hard to reconcile with 'freedom from hunger'.[2]

Ducks are beautiful waterfowl, much loved by adults and children when they live amongst us in our parks, canals, rivers and lakes. They are among the most visible of the animals we eat, and we are witness to the multifaceted behaviours they exhibit – their social interactions, bathing and copious preening to waterproof their feathers and to keep them warm to name a few. We show great delight at the sight of newborn ducklings. Yet in a farmed situation there is no legal necessity to give ducks access to the open water they need to fulfil these powerfully embodied behaviours. When it comes to bathing, most commercially reared ducks are given little more than a head bath. Full-body bathing is essential for a duck's good health. It is a welfare must, and should not just be something given by a farmer able to gain a premium from its provision.

Being able to gain a premium is problematic to all forms of livestock husbandry. I have met many farmers who wish they had access to local markets and customers prepared to pay a premium price. Sadly, this is not the case in many instances and farmers find themselves farming in concert with the wants of the customers available to them, post-farm-gate producers, retailers and through decades of government failures to protect lives worth living.

The life for many ducks is a rearing shed similar to those that broiler hens are kept in with equally inadequate space allowance, which can be as little as eight ducks per m^2, contributing to lower health outcomes through a lack of exercise, feather damage, poor leg health and lameness, and ammonia gas. Ducks can also suffer eye problems when kept under artificial light, and have a nervous tendency. Slaughter time comes between forty-two and fifty-six days of age, when they reach 3.1 and 3.5 kg respectively.

The UK does have a very small percentage of free-range duck farms, with daytime access to a range but not necessarily access to open water. The problem with supplying open water to so many birds is heavy faecal contamination and the health and welfare consequences that go with it unless regularly cleaned. This adds a cost few farmers are prepared, or can afford, to provide, rather than a reason not to. Open water is normally provided in organic production systems. An interim step would be making the RSPCA Freedom Food standard the minimum legal requirement, as this includes full-body bathing. Full-body access to open water is not a luxury; it is an absolute biological necessity for all waterfowl to ensure good eye, nostril, leg, plumage and other health essentials. The best solution, though, is simply not to farm ducks.

In 2010 about half of farmed ducks were reared to the RSPCA standard, which tailed off dramatically when the duck industry took animal welfare into their own hands (via Red Tractor, which operates to a significantly lower level of duck welfare). In fact, according to the RSPCA, by the end of 2014 no ducks were being reared to their standard,[3] as a result of Red Tractor. Not the best advert for an industry more motivated by profit, or a need to remain viable, over animal welfare. The practice is also another example of a breach of the Five Freedoms, given the ducks' inability to express normal behaviour.

Ducks are biologically adapted to a life around water. They need water to perform their natural water-related behaviours, such as head-dipping, wet preening, wing-rubbing, bathing and different types of shaking movement.[4]

The contrast between the affection we show to ducks in the park and the ones we want to eat is disturbing.

One forbidden practice in the UK is foie gras (fatty liver) production, from ducks and geese. To produce a fatty duck (or goose) liver, which is enlarged by six to ten times its usual size during the process, the birds are force-fed copious amounts of food (e.g. boiled maize and fats) through a tube inserted into their throats. The feeding process starts when the ducks are around twelve weeks old and lasts for around two weeks (about seventeen days for geese).

The ducks may be confined in pens or kept individually in confined cages during feeding. To ease feeding, ducks kept in cages, may have their heads put through an opening in the front of the cage. This makes it easier to grip and restrain the bird's neck and limit its resistance. Others are repeatedly caught and grasped.

Hand-feeding can last up to a minute, or just a few seconds if an automatic pump is used. Foie gras production is extremely cruel and should have no place in our society. France is the largest producer, with other European countries, such as Hungary, Bulgaria, Spain and Belgium, to a lesser extent.

Whilst forbidden here, we do import foie gras from these countries instead!

Production of duck eggs, which are becoming more popular in the UK, also has its welfare issues in not fulfilling the birds' biological needs. A typical layer duck will start producing about an egg a day from 20–28 weeks, depending on her breed, commonly laying around 160 eggs during her economic cycle, which ends when egg production drops below 50 per cent.

Like duck meat, duck egg production does not enjoy many legislative safeguards or guidelines (e.g. from the Foods Standards Agency), leaving it up to the industry and/or to an individual businesses' welfare ethics.

Citations:

1. Compassion in World Farming, *Farm Animals. Chickens Farmed for Meat.* https://www.ciwf.org.uk/farm-animals/chickens/meat-chickens/
2. RSPCA, *What are the Five Freedoms of animal welfare?* https://kb.rspca.org.au/knowledge-base/what-are-the-five-freedoms-of-animal-welfare/.
3. RSPCA, *Watertight: the case for providing farmed ducks with full-body access to water.*
4. Ibid.

e. Sheep & Their Slaughter

i. Sheep. Sheep mainly graze outdoors on grass, with conserved silage and hay used to supplement feeding when necessary. Whilst there are three broadly defined systems – lowland, hill and upland farms – it is much more complicated than that, with some sheep moving from one system to another. For example, lambs may be brought down from the hills to be finished on lowland pasture and/or indoors, and not always on the same farm. Lambing may also take place indoors and outdoors.

Lambs bought in and finished as store lambs can be intensively finished on a concentrate diet. A high-protein feed like rapeseed meal may be fed as part of an intensive system to achieve the desired growth rates, with lambs achieving their slaughter weight within ten to fourteen weeks. Other farmers may feed their lambs a silage-based diet and achieve a heavier slaughter weight over a longer period. Lambs may also be fed brassica crops, such as stubble turnips and kale, and other foods. A flock of store lambs can also be used by dairy farmers to clean off grass from autumn and into early winter.

The weaning age on sheep farms is variable. In intensive systems lambs may be weaned within a few weeks. Other lambs may be kept with their mothers for up to twenty weeks but generally somewhere between ten and sixteen weeks.

A lamb's slaughtered age is highly variable. The average age is believed to be between six and seven months, within a range of ten to fifty-two weeks.[1]

Weaning is not always completed naturally. In intensive systems mother and baby can be abruptly separated at an early age and lambs reared artificially. Lambs and ewes have been shown to form a strong, lasting mutual bond and the experience of separation can cause marked emotional stress.

When artificially reared, lambs, unable to express themselves, may develop behaviours such as non-nutritive suckling on the navel or scrotum of other lambs.[2] This behaviour can lead to sores and infection.

Ewes show their stress through vocalisation, standing head raised and ears erect, orientated to where their lambs are being kept; and a period of decreased resting and rumination.[3]

Natural weaning needs to be completed to reduce separation stress to a minimal level. Where lambs are removed prematurely the experience is particularly noxious.[4]

Whilst many ewes in the UK are still naturally bred to a ram, an increasing number are being artificially inseminated. This involves the collection of semen from stud rams, either using an artificial vagina (AV) or the recovery of semen from a 'teaser' ewe's vagina. The process can be very quick. An electro-ejaculator may also be used.

The technique for electro-ejaculation involves inserting a lubricated probe into the ram's rectum. Once settled, the handler sexually stimulates the ram's accessory sex glands and sympathetic nerves, found close to the brim of the pelvis, with a gentle massage. This primes the ram for ejaculation. If a manual pulse ejaculator is used, bursts of stimulation lasting four to six seconds are administered. If the ram fails to ejaculate within five attempts, it is recommended to rest the ram for around ten minutes before repeating. Other types of ejaculators allow a gradual increase of charge to ejaculation. Over-charging can be traumatic.[5]

Vaginal insemination involves the insertion of a pipette with an attached syringe to administer the semen. After cleaning the vulva, the tail is held up and out of the way and the pipette pushed into the cervical canal located deep inside the vagina. Semen is then syringed onto the posterior folds of the cervix.[6]

A more concerning form of AI is laparoscopic artificial insemination. Here, the ewe is turned on her back onto a sloping rack (about 45 degrees), head down and feet shackled above to keep her still. The area around her udders is shaved. A rigid laparoscope (fibre-optic telescope) is then inserted through her abdominal wall a few centimetres in front of the udder on the one side, to enable the operator to pinpoint the ewe's reproductive tract. An AI gun is then inserted in a similar position on the other side. The operator uses the telescope to visualise the womb. Gas may be administered via an attachment to the telescope to push the stomach, intestines and bladder out of the way to create space inside the womb and make it easier to locate the ewe's uterine horns. Semen is than placed surgically onto both of her uterine horns. The wounds are sprayed with an antiseptic, and an oral antibiotic is given. The same procedure may be used on goats.[7] Farmers talk about high animal welfare one day, and the next will say, 'Where's my sheep rack, laparoscope and AI needle?'

Another concerning issue with sheep is castration and tail-docking, practices that the British Veterinary Association (BVA) and the Sheep Veterinary Society (SVS) have described as a procedure that causes pain and suffering. BVA commented:

> *Castration and tail-docking are sometimes routine husbandry procedures and, because they are painful, efforts must be made to reduce their use, especially where there are alternative management techniques.*

The reason for tail-docking is to reduce the risk of fly-strike, which itself has negative consequences for an animal's welfare. Where absolutely necessary (i.e. where alternative preventive measures have failed), the BVA recommend the use of local anaesthesia and appropriate analgesia, a practice they also acknowledge can be impeded by a lack of licensed products for use in sheep. The same advice applies to castration, and they strongly recommend that either practice is only carried out in consultation with the farm's veterinary practice.

Both the BVA and SVS call for suitable products to be granted UK licences, calling for manufacturers to begin the approval process.

This action was criticised by the National Sheep Association (NSA) for taking place in the public arena and without prior discussion with them and other interested groups, though they did support the recommendation in principle, where practical and affordable forms of pain relief to raise welfare conditions were available. NSA chief executive Phil Stocker said:[8]

> *[...] until pain relief was readily available, farmers should be trusted to make the best management decisions and should not*

be 'ostracised' or put at a competitive disadvantage when there were no assurances lower-welfare produce wouldn't be allowed in the country under new trade deals.

This raises a number of questions, such as why:

- the NSA do not want issues like this raised in the public arena, as the pain and suffering caused by the practice of tail-docking and castration is likely to damage reputation and perhaps inappropriately lay the blame on the sheep farmer's doorstep.
- pain-relieving licensed products are not available already, when animal welfare should be the top priority, as well as why the sheep industry is not leading the call for licensed products.
- competitive disadvantage and a lack of assurances on lower-welfare imported sheep meat should be the driver for animal welfare over relief from pain and suffering.

Farmers can get caught in the crossfire, driven to use what means are available at a cost they can afford, given the farm-gate price they are expected to take and the competition they face. Some farmers look to avoid tail-docking and only do it as a last resort where they consider the alternative, fly-strike, to make it an economic necessity. However, there should be no last resort. Human beings are capable of remarkable things but seemingly can't offer sheep or other livestock pain relief. This is a shocking reality of livestock's longer shadow. The defence is that it is necessary to remain in business as doing the right thing, even if we may do things better than most other countries, would put us at a competitive disadvantage.

Consumers need to be openly educated on this and other welfare issues and be prepared to pay the price for kinder practices. Trade agreements must not allow UK farmers to be undermined by the import of more animal cruelty at any stage of production, including slaughter. Better still, stop eating sheep; and sleeping in, standing on or wearing their clothes.

ii. Amari. Amari is a Dorper sheep rescued by the Farm Sanctuary, South California, along with her twin lambs. She was originally rescued by a local man, concerned that if not protected she would be slaughtered for meat. Without the means to look after her properly, he contacted the sanctuary. Her pregnancy was detected on arrival, and she gave birth soon after. She was provided with a warm, safe and secure home, and nourishing food, but despite her care remained frightened around humans.

Farm Sanctuary reported that 'she likely lost lambs to slaughter in the past, and has no way of knowing this will not happen again'. However, she enjoys the company of the sanctuary's other sheep.

iii. Their Slaughter. Sheep are normally slaughtered using electric tongs positioned to deliver an electric current to the brain, which renders the animal temporarily unconscious. The major blood vessels in the ewe's neck are then severed with a sharp knife.

Religious slaughter without pre-stunning is permitted under an exemption. The area around the neck may be shorn before sticking. The procedure is normally carried out with the restrained sheep (in a cradle rack) laid on its back. The sheep is held, the neck extended and a clean cut delivered across the neck to sever the major blood vessels. As with any sticking, an outburst of blood follows, and the animal must not be shackled and hung until it has bled out.

The Jewish (Shechita) and Muslim (Halal) methods are similar. In either case, pre-stunning may be used, depending on how strict these religious practices are interpreted by each faith. Some believe the animal must be fully conscious, whilst others believe that pre-stunning is acceptable as long as the heart is still beating before sticking.

A Food Standards Agency (FSA) survey found that between 29 January and 4 February 2018, 46 per cent and 25 per cent of sheep slaughtered in England and Wales were by Halal stun and Halal non-stun respectively.[9]

The Farm Animal Welfare Council (FAWC) reviewed neck-cutting without pre-stunning and came to the following conclusions:[10]

It is difficult to measure pain and distress during the slaughter process in an objective scientific manner and subjective indicators, such as behavioural responses and vocalisation, are prevented from being displayed because of the degree of restraint and the severance of the trachea respectively. By the same token, it is impossible to state with objectivity that an animal would not feel pain and distress following such a procedure.

When a very large transverse incision is made across the neck a number of vital tissues are transected including: skin, muscle, trachea, oesophagus, carotid arteries, jugular veins, major nerve trunks (e.g. vagus and phrenic nerves) plus numerous minor nerves. Such a drastic cut will inevitably trigger a barrage of sensory information to the brain in a sensible (conscious) animal. We are persuaded that such a massive injury would result in very significant pain and distress in the period before insensibility supervenes.

Additionally, on one visit, we observed the slaughterman place his hand into the neck wound of cattle immediately after the cut had been made, presumably to try to ensure the free flow of blood from the severed carotid arteries. [...] This procedure in itself is, in our view, likely to cause further unnecessary pain and distress and is also unlikely to achieve its objective.

The FAWC also commented on the scientific evidence on the time to loss of brain responsiveness, which shows it may take between five and seven seconds for sheep, and twenty-two to forty seconds in adult cattle, to become insensible. The time 'may be extended should occlusion of the carotid arteries take place'.[11]

Occlusion refers to a phenomenon observed in a proportion of cattle, particularly in calves, when the carotid arteries have been severed transversely. Very rapidly after the cut, the carotids may, by virtue of their elasticity, retract into their own external connective tissue coat. The connective tissue becomes filled with blood, which then clots, thereby occluding the flow of blood from the severed arteries by sealing the cut ends. Because the heart is still beating, the blood pressure in the anterior aorta is maintained and hence also in the vertebral artery. This latter vessel supplies the brain and is not severed during the neck cut. Occlusion therefore has the effect of delaying insensibility for a considerable period and therefore increases the time during which an animal may be experiencing severe pain and distress.[12]

Citations:

1. Texel Sheep Society, *Survey of age of English lambs at slaughter*. https://texel.uk/survey-of-age-of-english-lambs-at-slaughter/
2. Napolitano, F. et al. (2008), 'Welfare implications of artificial rearing and early weaning in sheep.' *Applied Animal Behaviour Science* 110, pp. 58-72
3. Ibid.
4. Ibid.

5. Sheep Vet Society, *Guidelines on the Examination of Rams for Breeding.*
    https://www.sheepvetsoc.org.uk/sites/default/files/SVS%20RAM%20GUID
    ELINES.pdf

6. Menzies, P. (2015), 'Artificial Insemination in Sheep.' in *MSD Veterinary
    Manual*, https://www.msdvetmanual.com/management-and-nutrition/
    management-of-reproduction-sheep/artificial-insemination-in-sheep

7. American Boer Gat Association, *Laparoscopic Insemination of Sheep and
    Goats.* http://abga.org/wp-content/uploads/2016/01/
    Laparoscopic-Insemination-of-Sheep-and-Goats.pdf

8. Price, R. 'Castrating and tail-docking lambs should be "last resort." *Farmers
    Weekly*, 26 February 2020. https://www.fwi.co.uk/livestock/health-
    welfare/ xcastrating- xand-tail-docking-lambs-should-be-last-resort

9. DEFRA/LCWM (2019), *Results of the 2018 FSA Survey into Slaughter
    Methods in England and Wales*

10. Farm Animal Welfare Council (2003), *Report on the Welfare of Farmed
    Animals at Slaughter or Killing, Part 1: Red meat*

11. Ibid.

12. Ibid.

f. Fish & Their Slaughter

i. Fish & Their Slaughter. It is commonly cited that over two trillion
fish are slaughtered each year to feed the world directly or indirectly,
with little thought given to the pain, often prolonged, and fear, we
inflict. Most are caught from the sea, with an ever-increasing number
coming from fish-farming enterprises.

Electrical and percussive stunning in a water-bath is increasingly used
on farmed fish, but traditional killing through asphyxiation is still widely
practised. Fish may also be subjected to live chilling with CO_2 in holding
water, chilling in an ice-slurry bath and bled without stunning. All methods
take several minutes to bring about insensibility and are not humane.[1]

The science is very clear. Fish are just as sensitive to pain as any other
animal. They are sentient, but unable to communicate the pain we make
them suffer. Their silence must be listened to and their welfare
protected.

Best practice, essentially being the least unkind to fish, but still
unkind, involves stunning and killing before exsanguination. All other
forms of killing cause avoidable suffering before death.[2]

Like any method of slaughter, there isn't a humane method; all
slaughter is unkind.

According to the World Wildlife Fund[3] each year bycatch, the
incidental capture of non-targeted species, kills:

- 300,000+ small whales, dolphins and porpoises;
- 100 million sharks;
- 250,000+ loggerhead and leatherback turtles;
- Thousands of seabirds;
- Billions of unwanted fish and invertebrates.

This is simply not acceptable, with incidental captures discarded like trash. Bycatch can be reduced, and there are some success stories, but far too few. There is also the long legacy of fishing gear entanglements, which will continue to take their toll on non-target species. The incident rate will continue to grow as more fishing gear is lost over time. However, whilst we are concerned about things like plastic straws, little attention is thrown towards the mountains of discarded/lost fishing gear which swirl endlessly around our oceans, inflicting continuous misery on sea life, large and small. Perhaps because people do not eat straws.

One option to reduce the burden on ocean resources is to feed salmon and other farmed fish more plant-based foods, like maize, rapeseed and soya. However, this would take up cropland needed to grow food directly for ourselves, and in the case of soya contribute to further deforestation. The obvious answer is to stop eating the flesh of the animals we so inefficiently consume, but that would dare disturb the sound of silence around that and the errant fishing gear floating around in the oceans.

Many of us, including myself a few years back, take pleasure from being unkind to fish for fun. At around twelve I took up coarse fishing and in 1993 game fishing for trout. In between I indulged in sea fishing too. Even as I took up fishing, I remember questioning adults about whether fish could feel pain and suffer. I was told an outright no, and from that point was touched by the sound of silence. I dared not disturb the enjoyment and mental relief I found in hooking, playing with and later eating the fish I caught.

Each coarse fish I lifted out of the water I handled with the greatest of care. Each trout I caught I dispatched clinically with a hefty priest (a bat-like tool made for the purpose). Each sea fish I caught was dropped in a crate to die of asphyxiation over a long period, without a second thought. Fish don't feel pain and suffer, I kept convincing myself, as adults once told me and I continued to pretend in my own adulthood.

There are around 900,000 freshwater anglers in England and Wales and about 750,000 people who fish in the sea each year[4] who want to, or have to, say 'Oh no they don't', but the uncomfortable and inconvenient truth is 'Oh yes they do'. Fish are sentient, and when we fish for them we are not being kind. They feel pain and suffer, however hard that may be to reel in and digest. Unfortunately, the amount of money that can be made from fishing is more important to people and organisations than the pain and suffering the practice causes.

A significant number of my fellow fishmen also failed to take their damaged tackle, beer/soft drink cans and plastic rings, luncheon meat and sweetcorn tins, plastic bottles, sandwich wrap, crisp packets, etc., etc., back home with them. I would often spend part of my time walking the banks picking up and bagging their discards, which would otherwise entangle and harm wildlife.

The sound of silence is a funny thing. In the back of my mind there was always a nagging doubt. I was never really 100 per cent convinced, but that's how the sound of silence works. I had taken in only what I wanted to hear as I loved fishing, like I once loved eating meat. I wanted to keep on fishing and as a result dared not listen. I also liked to think, like many anglers do, that fisher-people are river conservationists, keeping a beady eye out for things like livestock pollution (from farms we then illogically buy cheap, externalised, and heavily subsidised pollution from, which we then complain about). We need to stop and think: 'Are my food choices, and the price I'm willing to pay for them, a part of the problem?' The answer is a resounding yes. Furthermore, being a conservationist is not a justification for being unkind to fish or any other animal in any language.

We listen in the same vein to the animals we farm, slaughter and eat. We justify what we do to animals for the short pleasure we get each time we catch and/or eat them, because we have been conditioned to believe it is necessary. There is no let-up in that conditioning. Livestock's longer shadow stretches across the rural landscape, towns, cities, rivers and oceans before ending up down our throats.

Livestock farming is only necessary because we have made it that way, through the decisions we have made and those that have been made for us. In the UK we are so reliant on livestock farming that taking it away would have a dramatic impact on our food security. It is that reliance that makes livestock farming a necessity and why what we do in the name of animal welfare to feed us animal-based foods is considered reasonable and necessary. And what would we eat in the winter, once we have eaten all the crop harvests? We can change that, though it would take time to dismantle, not as quickly as others and I would like to see, but it is a process we must do with urgency.

We are not kind to the animals we eat, and we are arguably the unkindest to fish, a species we are least interested in when it comes to animal welfare. To satisfy our ferocity to eat fish, we come up with sustainable fishing standards where sustainably means little more than to sustain human interests over other species. We do not fish sustainably in the interest of marine life and other land species that rely on marine life. We fish to satisfy greed, and like 'sustainable' deforestation we do it by cloaking these practices in silence and with false narratives. In fact, deep trawlers constantly 'deforest' the ocean floor with their enormous nets, at a scale vastly greater than land-based deforestation. Astronomical areas of our oceans have been turned into deserts, devoid of flora and fauna.

If you were to capture a carp at a fishery and fail to place it on a wetted unhooking mat, you would quite rightly be vilified by your fellow anglers – the same anglers who would drive a hook into a fish's mouth, drag it around the water, each way and every way, before netting and temporarily depriving the fish of oxygen. But that's okay, because 'fish do not feel pain', as otherwise intelligent adults pretend, as long as it is unhooked on a wet mat and despite what the science says about the rest.

Fish possess nociceptors, respond to noxious/painful events, react positively to analgesics, and show avoidance learning. All indicative of a conscious avoidance of suffering.[5]

And for all fish netted or hooked from the sea and left to die by asphyxiation, what would be the difference in suffering between that and throwing a weighted hessian bag full of kittens into a river to drown? Nothing, other than kittens are more highly valued by society than fish, and if caught in the act we would end up in court. We do the equivalent to perhaps over two trillion fish a year, without a second thought and without repercussion.

The deep-dwelling fish we commercially or recreationally drag up from the bottom of the ocean suffer immensely from decompression too, as their gas-filled swim bladders expand and crush neighbouring organs.[6]

Symptoms of decompression, as described by Jonathan Balcombe (2016),[7] include:

- esophgeal eversion, where the esophagus turns inside out and exits the mouth;
- exophtalmia, bulging of the eye from its orbit;

- arterial embolism, a sudden interruption of blood flow due to blockages by gas bubbles;
- kidney embolism;
- haemorrhage;
- organ torsion, damaged or displaced organs surrounding the swim bladder;
- cloacal prolapse, where the fish's rectum turns inside out and exits its body.

Fish are also crushed and asphyxiated under the weight of each other as the huge nets they are caught in are winched in, tightened and hauled onto the fishing vessels. Longlined hooked fish can languish, impaled for hours or even days before being dragged on board.[8] There is no kind way to catch and kill fish, with multi-millions of non-targeted species also casualties of these barbaric practices.

The slaughter which takes place doesn't stop there, with millions more marine animals caught up in the millions of tonnes of deadly broken, lost and discarded lines and nets which litter our oceans. Nets which eventually settle to the bottom under the weight of their victims, including fish, birds and marine mammals.

We are desperately unkind to these fish, which we also feed to billions of livestock (e.g. farmed fish, pigs and chickens), which we are also unkind to in an endless cycle of misery.

I fished for over forty years. I hid in silence amongst the UK's 1.65 million coarse and sea anglers. Whilst I can't take back the suffering I caused, I cut out the myths, let in the light and extinguished the shadows I accrued for my selfish interests and ended my silence.

Citations:

1. Humane Slaughter Association, *Frequently Asked Questions*.
   https://www.hsa.org.uk/faqs/general#n3
2. European Food Safety Authority (2004), 'Opinion of the Scientific Panel on Animal Health and Welfare (AHAW) on from the Commission related to the Welfare Aspects of the Main Systems of Stunning and Killing the Main Commercial Species of Animals.' *The EFSA Journal* 45, pp. 1-29
3. World Wildlife Fund, *Bycatch*.
   https://www.worldwildlife.org/threats/bycatch
4. Environment Agency (2019), *Watersports Participation Survey 2017*
5. Fish Welfare, Eurogroup for Animals (2016),
   http://www.eurogroupforanimals.org/
   wp-content/uploads/E4A-Briefing-Fish-Final-Oct-2016.pdf.

6.   Balcombe, J. (2016), *What a Fish Knows: The Inner Lives of Our Underwater Cousins*. Oneworld Publications, London, England

7.   Ibid.

8.   Ibid.

g. Rabbits & Their Slaughter

i. Rabbits & Their Slaughter. Another small meat production system still in existence in the UK, perhaps surprising to some, is rabbit farming, using breeds such as the New Zealand White, Californian and Carolina. Rearing can take place in free-range systems, open-floor pens, cage and pen systems, or the bunnies may be cramped together, eight or more in a cage with a small-gauge wire floor to stand on, leading to painful abscesses, ulceration and other foot infections. Respiratory infection/distress is also common, with farmed rabbits receiving above-average amounts of antibiotics compared to other factory-farmed livestock.

Space allowance for caged rabbits can be little more than an area of an A4-sized piece of paper each, with a headroom of 45cm.[1] If you kept pet rabbits in such conditions, you can quite rightly expect a visit from the RSPCA, but it is not a problem for rabbits bred for meat. They are simply left to suffer 'necessary pain and injury', along with extreme stress and boredom, like other livestock, through the denial of their basic natural instincts, to feed us meat we do not need to eat.

Whilst rabbit farming is on a small scale in the UK (there were around 400 rabbit farmers by the late 1990s housing between 25–30 and 300–400 does[2]), you can pop over to other European countries (mainly Italy, France and Spain) and take your pick from around 180 million farmed rabbits, of which around 119 million are reared in commercial, cramped, wire mesh systems.[3] The remaining 61 million are reared on backyard farms, which may also include wire mesh systems, particularly for the does. Figures for slaughtered rabbits in the UK are not available, which needs to be corrected for awareness and transparency. The caged method is also common in the US, China and other countries, with China the largest producer.

Rabbit farming applications are still made in the UK, with a 2021 application made in Cornwall for a 250-doe rabbit breeding and rearing farm wanting to produce up to 10,000 battery-bred rabbits a year.[4] The application created local outrage, and rightly so. The application was turned down, but others by the same applicant have been approved in other parts of the UK. In this instance does were to be caged throughout their lives, with their kittens housed on a barn floor with access to a small outdoor run.

Rabbits bred for meat endure these horrific conditions for around eight to twelve weeks, when they are then harvested at around two kilograms in weight, with breeding rabbits kept until between eighteen and thirty-six months of age.[5] Each doe can have up to eight litters per year, with up to eighty kittens in total (separated from her at three to four weeks old) over her shortened life expectancy.

Farming rabbits in this way is sickening but no less sickening than farming chickens in so-called 'enriched cages' and sows in crates. If you would happily eat livestock other than rabbits without names, then you may wish to explore the ethics behind speciesism. Most people have had the opportunity to see the natural behaviours of these beautiful, sentient animals close-up and would never dream of eating them. Other livestock are animals too, equally beautiful and sentient. When we farm these animals, like farmed rabbits, we stop calling them animals and call them livestock instead to make eating them palatable.

Free-range bunnies are available too, kept in larger enclosures, but like their caged compatriots they still go on to be slaughtered, skinned, hanged, butchered and processed.

To get a good appreciation of what farming rabbits in cages involves, Compassion in World Farming did an investigation in 2014. Undercover investigators visited sixteen rabbit factory farms across five countries (Italy, Greece, Czech Republic, Poland and Cyprus), and what they uncovered is a very hard watch.[6]

Like other livestock, rabbits are commonly subjected to artificial insemination. For AI they are held on their backs, which anyone who has ever handled rabbits understands is a stressful and frightening position for any rabbit.

Slaughter generally involves dispatching with a blow to the head or by a penetrative/non-penetrative captive bolt, followed by bleeding. Bleeding may also happen when a rabbit is still conscious due to a dispatch failure, with some extremely unfortunate bunnies bled without stunning.

Citations:

1. RSPCA, *Farming Rabbits*.
   https://www.rspca.org.uk/adviceandwelfare/farm/
   rabbits/farming.
2. *Rabbits – a pest to some, but a source of profit to others. Farmers Weekly*, January 1998. https://www.fwi.co.uk/news/rabbits-a-pest-to-some-but-a-source-of-profit-to-others
3. European Commission (2017), *Overview Report: Commercial Rabbit Farming in the European Union*
4. Animal Aid, (26 February 2021). *No appetite for rabbit farming in the UK.* Accessed at: https://www.animalaid.org.uk/no-appetite-for-rabbit-farming-in-the-uk/.
5. RSPCA, *Farming Rabbits*. https://www.rspca.org.uk/adviceandwelfare/farm/rabbits/farming
6. Compassion in World Farming, *Exposing the biggest secret of the Cage Age*. https://www.youtube.com/watch?v=we1xeip5P6I

h. Welcome to the Slaughterhouse

A slaughterhouse is a place where animals are taken and killed, to provide us with food we are told we need to eat, and where animal welfare is supposed to be centred on minimising animal cruelty, but more realistically is governed by economics.

Slaughterhouses are soundproofed rooms where the noises we do not want to hear are made. Places where we cut out the facts and only let out the brisket, pork loin, lamb chops, chicken breast, and turkeys for Christmas. Places where we turn down our lights for fear of what we may see. There is no kind way to slaughter an animal; we simply hide away from and normalise slaughter to make us feel better about ourselves when we eat them.

Approximate number of animals slaughtered per year in the UK:[1]

Sheep (ewes, rams, other sheep and lambs)	15 million
Pigs (clean pigs, sows and boars)	11 million
Cattle (heifers, steers, bulls, cows & calves)	2.7 million
Broiler chickens	1.1 billion
Broiling fowl (incl. spent hens and breeders)	57 million
Turkeys	15 million
Ducks	14 million
Geese	1 million

In all, the number of farm animals slaughtered in the UK is just shy of 1.2 billion. 1.2 billion animals we have not been kind to. As consumers move away from beef, concerned about price, cow burps and red meat's association with colorectal cancer,[2] they increasingly turn towards chicken, pork and farmed Atlantic salmon. Whilst they're right to be concerned, the increased demand for these meats exposes a higher number of animals to unkindness, in their life and in their death. The animals we slaughter are not just a number. They are distinct, sentient individuals.

The coastline of Great Britain, excluding islands, is 17,819 km long. By the time it is fully grown each broiler chicken has a space allowance of around an A4 sheet of paper (0.21 x 0.297 m). 1.1 billion broilers lined up lengthways occupy a smoothed length of 326,700 km, enough to circle the coastline over eighteen times, or take you 85 per cent of the way to the moon (384,400 km). A frightening statistic.

Citations:

1. Calculated from Defra Slaughter Statistics
2. World Health Organisation (2015), International Agency for Research on Cancer

i. Animal Transport

In 2019 the Farm Animal Welfare Committee (FAWC), an expert committee of the Department for Environment, Food, and Rural Affairs (Defra) and the Devolved Administrations in Northern Ireland, Scotland and Wales, published *Opinion on the Welfare of Animals during Transport*.[1]

The committee acknowledges that 'some pain and distress are unavoidable in all animal sectors even with current knowledge, husbandry and farming practices' and that 'difficult ethical and practical decisions have to be made when dealing with suffering, sometimes imposing a lesser act that may still cause short-term pain or distress...'

This clearly demonstrates that if you want to eat animal-based foods, and because the provision of animal welfare has a commercial cost which beyond a certain point, it is argued, cannot be afforded, we have to accept causing each animal we eat a degree of pain, suffering and distress. Or put more bluntly, if you eat animal-based foods you are complicit in causing animal pain and distress for no reason at all. This is justified on the basis that animal-based foods are a need, not a want, and what we allow to happen to them is based on the price we want to pay. The reverse is true: animal-based foods are a want not a need and there should be no price set on animal welfare.

On transport the publication reported:

All transport movements are stressful for animals, with a number of contributing factors that influence this, including catching, moving/herding, loading, unloading (either individually directly or in modules), the actual journey, driver quality and access to food, water and rest.

During transportation animals will be exposed to a number of potential risk factors for poor welfare including thermal loads, motion, vibration, acceleration, impact, fasting and the withdrawal of food and water, behavioural restrictions, social disruption and mixing with unfamiliar animals, noise and air contaminants.

The longer the journey the more harm is potentially caused due to longer exposure to the various stressors the animals suffer, including both mental and physical health.

Transport of animals is a continuation of the 'unavoidable' pain and distress endured by farmed animals during all stages of their life from conception to slaughter. Some species fair better than others and some farming practices are less unkind than others, but all cause an animal pain and distress in the end.

In the UK there is a legal duty to transport animals in a way that does not cause injury or unnecessary suffering. Animals must also be fit to travel. That is a misnomer, based on what you have just read.

Unfortunately, some animals transported are unfit to travel and some will inevitably suffer and/or be injured. Some animals have their food withdrawn before transport, causing hunger and stress, which would also be a breach of the Five Freedoms.

For example, some dairy cows sent to slaughter will be lame. Lameness is a painful condition, causing discomfort and impaired walking.

Furthermore, according to Knight and Wiebers (2020),[2] nearly 300 million chickens may be suffering some form of lameness – and 30 million almost unable to walk – on top of other pain and stress they suffer even before they are sent to be slaughtered. In addition, bruising and bone and joint dislocation and breakages are not uncommon during catching, placement in crates and during transport. When transported poultry can also suffer from hyperthermia and hypothermia, particularly when wet. Especially at risk are end-of-lay hens with reduced plumage. This is over 1 in 3 birds, so when tucking into chicken each year you can be sure of being complicit in this animal cruelty.

With the closure of many local slaughterhouses in the UK, overall journey times have increased, causing prolonged suffering.

Calves can also be transported great distances on to the continent and elsewhere, suffering cramped conditions, bruising, extremes of heat and cold, hunger and thirst, illness, and even death.

Scotland is the only nation in Britain that still exports young calves to Europe, a journey that can last as long as six days en route to places like Spain and Italy. This is not because of a legal ban, but because English and Welsh ports have refused to carry them.

The government expresses a desire to ban all exports of live animals for slaughter or for further fattening abroad, a position not tested, due to European Court judgements which have ruled against trade restrictions within the European Economic Community. Scottish ministers have been against such a move, in order to protect Scottish beef farmer profits rather than calves.

Animals transported over long distances abroad experience prolonged overcrowding, exhaustion, dehydration, pain and stress, and many die as a result.[3] When animals are moved beyond the European Community, their legal protection can no longer be guaranteed either.

Animal abuse in transport, which would include many animals unfit to travel before loading, was uncovered in 2016 by the Bureau of Investigative Journalism.[4] Between 2014 and 2016, 5,333 animal welfare breaches were recorded, 4,005 falling within the most serious category (Category 4). A further 11,176 were Category 3 incidents. The source data was provided by the Foods Standards Agency (FSA), who are responsible for regulating and reporting these kinds of animal welfare abuses.[5]

Category 3 incidents are 'major' non-compliances which are likely to compromise animal welfare, but there is no immediate risk to the animal – although the non-compliances may lead to a situation that poses a risk to animals.

Category 4 incidents are critical non-compliances considered to pose a serious and imminent risk to animal welfare or are ones where avoidable pain, distress or suffering has been caused.

600 instances involved the arrival of dead animals, including sheep,

chickens and pigs. In one incident 574 out of 6,072 birds were found dead after suffering hyperthermia. In a consignment of 220 pigs, 33 were dead on arrival. One cow sent from a farm had a face lesion completely full of maggots.[6]

These incidents involve many animals but is recorded as a single incident.

One of the most commonly recorded welfare incidents in transport are birds trapped between the crate and the module they are loaded into on the transport lorry. Basically, imagine being pushed into a chest of draws with a part of your body hanging out. The following are actual reported incidents by meat hygiene inspectors, involving broilers:[7]

[Broiler] 2 trapped birds with head crushed by the crate on top-severe welfare issue; catching system should be reviewed as current transport system (S31) & (S38) is causing pain, injury and unnecessary suffering; design of system should be modified to prevent birds getting trapped; 1 bird with wing trapped between module and a crate weighing more than 80 kg; severe bruising/fractures; upside down birds – birds suffocated under other birds and/or feet crushed by the crate on to; Reported to TS [Trading Standards]: (S40).

[Broiler] 3 birds with wing trapped between module and a crate weighing more than 80 kg upside down birds – birds suffocated under other birds and/or feet crushed by the crate on top; several severe bruising/fractures; Reported to TS: (S40)

[Broiler] 4 birds with wing trapped between module and a crate weighing more than 80 kg upside down birds – birds suffocated under other birds and/or feet crushed by the crate on top; Reported to TS: (S40)

[Broiler] 2 lorries with wet crates and birds arrived on site; Yellow curtain on top of module not properly placed – birds exposed to accidents/injury (risk of birds to enter alive in the steam module wash); transport department not having any robust procedure to prevent this from happening although catchers, foremen and lorry driver are responsible for checking the lorry to ensure welfare of birds is not compromised; Reported to TS: (S40)

Other common sets of incidents involve: cattle and sheep being transported in the last stages of pregnancy; cattle, pigs and sheep with bruising, lameness, injured legs, foot rot and lesions; sheep with ingrowing horns; broilers and turkeys with bruising and fractures; overstocking; livestock found dead on arrival; and cattle inaccurately stunned or stunning not effective at the slaughterhouse. A few particularly horrifying accounts have been mentioned above, and here are some more:

[Pig] One lorry with few lame pigs with abscesses, one with deformed legs. One pig found [dead on arrival]. All bad pigs placed in one pen with large adult sows and boar. For over 12 hour journey.

[Pig] Severe lesions on the joints, bursitis and abscesses on front and hind legs. More than 95% of animals affected.

[Pig] 1 cull sow DOA, cause of death unknown. Plus 4 very lame pigs on the same lorry. (There was one very lame boar, hardly could bear weight on any leg, and was screaming in pain when was forced to walk. When boar went down one of drivers was forcing that boar to move by lifting its tail. This action was stopped immediately by Slaughterhouse Animal Welfare officer, and pig was humanely slaughtered on the lorry. There was one very lame pig on front leg, could not bear weight on that leg at all.)

[Sheep] Massive tumour ulcerated and bleeding in sheep's mouth.

[Cattle] Cattle with badly injured left forefoot. It was a chronic condition with visible necrotic changes. Whole foot was enlarged and without a hoof. Part of the leg was also clearly swollen.

[Cattle] Joint lesions in both front limbs, animal unable to place front hooves on ground when walking, bearing weight on forearms instead.

[Cattle] A bovine animal was brought today to (S31) & (S38). The animal had clearly suffered an old injury which resulted in losing her eye. As the owner confirmed upon enquiry, it

happened two months ago. At the point of delivery to the abattoir the animal's face lesion was completely full of worms due to an injury which had not healed properly which – in my professional opinion – causing unnecessary and avoidable suffering to the animal.

[Cattle] Eye malign tumour.

[Lairageman] Lairageman, were observed hitting 3 bulls, with a wooden stick and using an electric goad on the animal's body, more than necessary and on one animal, which refused to move in the race, hitting his head and neck and also hind quarter between metal pole and iron gate close to the race before entering the lane communicating with all pens.

[Hen] During ante-mortem and post-mortem inspection bruising/fractures were found in the hens. Pathological-anatomical autopsy pointed to the damage incurred prior to arrival at the slaughterhouse.

In one weather related incident 250 broilers were found dead on arrival, with many other similar incidents occurring involving high numbers of dead on arrival, generally due to weather/poor health conditions. The industry is opposed to restrictions on transporting chickens on freezing or excessively hot days.

Citations:

1. FAWC (2019), *Opinion on the Welfare of Animals during Transport*
2. Knight, A. and Wiebers, D. (2020), *A British Pandemic: The Cruelty and Danger of Supermarket Chicken*. Open Cages
3. Compassion in World Farming, *Our Campaigns: Ban Live Exports*. https://www.ciwf.org.uk/our-campaigns/ban-live-exports/
4. Wasley, A. and Robbins, J. 'Severe Welfare Breaches Recorded Six Times a Day in British Slaughterhouses.' *The Bureau of Investigative Journalism* [online], 28 August 2016. https://www.thebureauinvestigates.com/stories/2016-08-28/severe-welfare-breaches-recorded-six-times-a-day-in-british-slaughterhouses
5. Ibid.
6. Ibid.
7. Food Standards Agency, Animal Welfare database

j. Animal Welfare Abuses

Unfortunately, in any walk of life not everybody follows the rules. Some farmers/farmhands and slaughterhouse workers are no different, and some are involved in animal welfare abuse.

Livestock does not always do what is desired by handlers. The resulting frustration can spill over and result in unfortunate acts to coerce animals into compliance. These may be a one-off moment or in some cases it can be systemic.

In 2019, Surge, a UK-based animal rights organisation, conducted an undercover investigation of dairy farms across the UK. The investigation, part of Surge's Dismantle Dairy campaign, exposed various animal welfare abuses, including punching, kicking, beating, tail-twisting and force-feeding calves. One calf was aggressively hit. A farmworker was also caught touching a cow in what appeared to be an intimate way. Footage also showed cows reacting emotionally and defensively to their calves being removed shortly after giving birth. However, this particular practice is allowed and is not considered an animal welfare abuse in law or in assurance schemes.

Surge was widely criticised for 'trespassing', 'secretly filming' and for being 'disrespectful' to dairy farmers.

Further investigations by Surge and others have shown abuses on UK goat, turkey and duck farms, which are extremely upsetting.

At slaughterhouses a lot of undercover footage has been taken showing shocking abuse of animals, which along with other investigations led to the introduction of mandatory CCTV in slaughterhouses, in May 2018, to prevent future incidents of this nature.

The level of abuse at slaughterhouses themselves was, as already mentioned for transport, brought to the public's attention in 2016, by the Bureau of Investigative Journalism.[1] Abuse was systemic.

In a two-year period from July 2014, vets and meat hygiene inspectors working for the FSA reported a total of 9,511 animal welfare abuses, 3,375 of which occurred during stunning/killing. Half of all incidents were Category 4 breaches – the most serious, involving avoidable pain, distress or suffering. Over half of the reported incidents happened during transport, before the animals arrived, which cannot be picked up by CCTV.[2]

For the five-year period of 2011 to 2016, the bureau uncovered 16,370 breaches, 6,241 of which were Category 4.

Examples of animal welfare abuse cited, included: a cow being violently slammed against a wall, a slaughterhouse worker beating bulls with a wooden stick and an electric prod, a haulier kicking cattle during unloading, sheep being grabbed and dragged by their wool and ears, pigs being lifted by their ears and tail, and pigs and chickens being immersed in scalding water while still alive.[3] The actual reported accounts by meat inspectors for some of these incidents are set out in the preceding section.

At the time of the investigation the journalists reported a shortage of meat inspectors, meaning the number of breaches was likely to be underreported. They quote Paul Bell, a Unison Officer: 'Simply there are not enough staff to monitor animal welfare in areas like the killing rooms.' Bullying and harassment of staff was also reported by Unison.[4]

The FSA pointed out that the number of incidents was tiny in comparison to the huge number of animals slaughtered.[5] One incident is too many, and in reality, more animals are abused because each separate incident may involve multiple animals. Add in the transport of 330 million already suffering, sick, lame broilers and the figures are as poorly as the animals themselves.

In more recent FSA data, 2019 inclusive, there were 218 Category 3 incidents and 210 Category 4 incidents. This gave a combined total of 428 incidents, which compares to 541 incidents between 2014 and 2016 inclusive (of which 130 were the most serious).[6]

In 2019 there were eighty more Category 4 incidents than happened over the two-year period between 2014 and 2016. The substantial increase in reported incidents may be related to the presence of CCTV.

Citations:

1. Wasley, A. and Robbins, J. 'Severe Welfare Breaches Recorded Six Times a Day in British Slaughterhouses.' *The Bureau of Investigative Journalism* [online], 28 August 2016. https://www.thebureauinvestigates.com/stories/2016-08-28/severe-welfare-breaches-recorded-six-times-a-day-in-british-slaughterhouses

2. Ibid.

3. Ibid.

4. Ibid.

5. Ibid.

6. Food Standards Agency, *Animal Welfare Non-Compliances in Approved Slaughterhouses*. https://data.gov.uk/dataset/7d0fe757-cfc4-44f4-b023-edf0afe478be/animal-welfare-non-compliances-in-approved-slaughterhouses.

k. Final Thoughts

Livestock farming exists because it is considered necessary. A number of reasons are given, including:

- We need to eat animals to meet our dietary requirements.
- If we do not farm animals, then we would not have enough food to feed ourselves.
- Eating animals is part of our circle of life.
- We evolved to eat animals.
- If we do not eat animals we will shrink and crumble to dust.
- If we do not farm animals they would no longer exist.
- If we do not farm animals most of our agricultural land would have little to no economic value.
- Livestock farming provides the landscape people love to enjoy.
- Livestock farming provides and preserves biodiversity and is good for the environment.

None of this has any merit, as I've shown throughout *Livestock's Longer Shadow*. We have merely made livestock farming a necessity through the choices we have made.

We *only* farm animals because we *want* to eat meat, fish, eggs and dairy, *not* because we have to; and as a consequence, we are okay with being unkind to animals. We do not have to eat animals, wear animals, stand or sit on animals or sleep in or under their clothes. We can be kind to animals instead, and by doing so, be kind to ourselves and our planet.

Hope lives in kindness.

Tim

CHAPTER VII

Contributing Authors

We must fight against the spirit of unconscious cruelty with which we treat the animals. Animals suffer as much as we do. True humanity does not allow us to impose such sufferings on them. It is our duty to make the whole world recognize it. Until we extend our circle of compassion to all living things, humanity will not find peace.'

Dr. Albert Scheitzer (1875–1965)

Pig Farming – Facing the Truth

Dr. Alice Brough, BVM&S MRCVS

From as far back as I can remember, I never wavered in my determination to become a veterinarian. I have always felt a deep connection with nature; my grandfather was a farmer, and I enjoyed a rural upbringing with plenty of mud, trees and animals. My older sister decided to become vegetarian at a young age, and I used to think she was silly to deprive herself of such culinary delights. I loved and had a tremendous affinity with animals, but like most people I believed what I was told about farming and taking the lives of animals for our own benefit; it was necessary and humane. I went through my whole childhood with this belief, and even when working on farms during school holidays (I worked in all sectors to gain experience across each), I managed to dismiss many of the issues in the same way that the vets and farmers seemed to be doing. I assumed that they were perhaps unique to that particular farm or an unusual occurrence that surely wouldn't be affecting many animals. Through vet school, I fell prey to the conditioning of the farm animal education provided, even though I was learning about everything that could go wrong and how to treat it; welfare in Britain is second to none, we look after our animals. All of this began to change when I first took employment on a pig farm, and the illusion continued to fall away piece by piece after I graduated, as I visited farm after farm encountering the same hideous conditions, disease and suffering.

We all have defining moments in our lives, and so many of mine revolve around pigs. I have experienced some extremely traumatic personal losses and challenging periods both in childhood and as an adult, but everything somehow pales into insignificance alongside the scale of suffering I am so aware of in animals, happening right under our noses. It's like a switch was flicked in me at some point in the first couple of years as a practising vet, and I now feel there is little else worth my time than fighting for these animals that we so callously use without thought.

On my first day at the intensive breed-to-finish pig farm I took a two-week placement on at age fifteen, as I was being shown around by one of the staff we came across a piglet that had been badly hurt by his mother accidentally stepping on him. On this farm, and as is the case for the majority of Britain's breeding pigs, mothers were confined to farrowing crates; she was trapped in a cage unable to turn around to see where her piglets were in relation to her feet and 250 kg body. The worker proceeded to pick the piglet up by his hind legs and deftly swing him in an arc over his head to bring his skull crashing down onto the concrete floor.

I had to momentarily take a seat on the edge of the crate; I thought he was joking when he said that they 'knocked' piglets that were not viable. This is legal, and the most commonly used method of euthanasia for piglets under four weeks old. It's something that I went on to do myself frequently, employed at this same farm throughout my 'gap year', and indeed throughout my career as a vet. This always, quite rightly, horrifies people when shown in undercover investigations, but as long as we continue to farm pigs, it will and must remain in practice to ensure that suffering is not prolonged.

A heart-wrenching nuance to this is that it is often performed right in front of the piglet's mother, or even using her cage bars as the hard surface on which the piglet is killed. I couldn't bear the thought of not killing them outright the first time so always used the floor (as there was no chance of missing) and as much force as I could muster to ensure the skull shattered beyond repair. I would gather up the piglet, giving them a quick moment of being held into my chest, and then say I was so sorry over and over as I took a deep breath and carried out the swing. Once it was done I would heave a sigh, gently place them down on the floor, checking that there was no sign of life remaining, and sit for a moment on the edge of the crate to apologise to the mother sow, giving her a stroke if she would allow it. I did this every time, and it never got easier.

I found it hard to reconcile killing every piglet that was deemed too small to survive, so one of the workers and I cooked up a plan – using the lid of a spray can, I would hand-milk the sow and syringe feed the 'poor doers'. If they were really struggling I would put them in my jumper for warmth and feed them regularly. I managed to bring quite a few piglets through the vulnerable stage in this way, but then had to wonder whether they would have been better off dying than enduring what was yet to come. I suppose it was so that I didn't have to kill so many myself and could feel that I had tried my best.

The abject misery is palpable walking through a crate-based farrowing house, connecting with the desperate eyes of mothers peering out through the bars of the cage scarcely larger than their own bodies. Almost worse was being unable to connect with those who had become so depressed that they would barely move from lying on their sides, staring blankly at the wall a few inches from their face. Occasionally a sow would just stop eating; I would fill up her feed with the rest and come back to find it still there at the end of the day, and I would watch her waste away whilst nursing her piglets for four weeks, before being shot or loaded onto the cull lorry for slaughter.

Sometimes I would come in to find piglets strewn about the front end of a pen, ripped in half or mangled and dying. This happens in farrowing crates, usually in gilts (first litter) when the stress of the environment causes them to 'savage' their own babies. When it came to feeding time, a cacophony of screams would erupt, and sows would bang their cage bars and feeders with their nose until I got to them, sometimes drawing blood in anguish; not surprising considering they cannot move and have no other stimulation.

Most days I was tasked with 'litter clipping': cutting off the tails and teeth, tattooing through the ears with thick pins, and administering injections of iron and antibiotic to each piglet around twenty-four hours after birth. I wore ear defenders because they squeal throughout and receive no anaesthetic or pain relief.

One of the workers used to take pleasure in killing finisher (50–110kg) pigs with a crowbar, and another would throw piglets into the weighing trolley from a great distance like it was a game. The rest were kind and did their best, but the reality is that there are lots of people out there like the first two I have described. I have met many over the course of my career, and this type of behaviour is often picked up in undercover investigations. Of course, these people display immaculate behaviour during official inspections, so it is very difficult to prevent as most of the time nobody knows what is happening behind closed doors.

The two weeks that I spent at that first farm, aged fifteen, were undoubtedly shocking, and there was no end to the novel horrors I encountered during the five months working there again at eighteen. The sadness that it made me feel never dissipated over the course of the next fifteen years (and still hasn't), but in order to do my job well, I had to learn new ways to deal with the suffering and death I witnessed so frequently over my four years in practice, and it was not something I could ever become desensitised to. This early farm work set me on a career path within this industry, hoping I could get myself to a position where I could directly influence the welfare standards for these animals. The reality sadly fell short.

On graduating from the University of Edinburgh in 2015, I went straight into 'mixed' practice (companion, equine and farm animal), with an emphasis on training up in pig medicine. We were taught very little on pigs at university, and it is not a popular career choice; approximately 150 vets look after all of the commercial pig holdings in the UK. I was the only person in my year with an interest in pigs, and that was only because I simply could not get on with my life knowing how bad it was for them, not because I thought it was a job I would love.

I was assigned around fifty farms as their primary vet, conducting routine quarterly inspections for 'Red Tractor assured' farms and dealing with any emergencies or requesting additional visits for those or my colleagues' clients in between. The farms ranged from small independent to large corporate pyramids. I saw the odd pet pig or sanctuary, much to my delight, but this was overshadowed by days spent at slaughterhouses on a fairly regular basis to observe disease or vaccination efficacy within my clients' herds.

Routine visits were scheduled up to three months in advance, and where I dealt with corporates, I was expected to get round multiple farms in a day - up to six. Emergency calls were picked up by any vet not scheduled for a visit or already in the area, as long as it didn't compromise health status and biosecurity (i.e. not contaminating one farm with the diseases of another).

I was regularly called out to deal with outbreaks of disease that required diagnostics or treatments, but often I would stumble across horrifying scenes at a routine inspection that had either been missed or deemed non-important by the farmer, so commonplace were the issues. The level of disease, vice (tail, ear, flank and vulva biting) and death on farms is staggering, and there was rarely a day that I did not find something to make my stomach churn. The conditions that we force animals into - the unnatural groups, with the deleterious genetic traits that we have created; the cramped, squalid and barren environments that are not set up for these animals to thrive - all result in having to provide support in the form of pharmaceuticals (usually antibiotics), and accepting that there will be a lot of 'deadstock' along with the livestock. From the industry's own figures, over two million pigs will die before reaching slaughter each year.

A prominent part of my job was performing post-mortem examinations; it was important to establish the possible reasons for clinical signs presenting in the herd by exploring the carcasses of pigs that had died unexpectedly, or euthanising those that were displaying said signs. I carried a captive-bolt gun and a long screwdriver which was used to 'pith' the pig – scrambling the brain after stunning with the bolt gun – and a blade to sever the blood vessels in the neck if I was in need of sterile brain samples. I did an awful lot of killing; the numbers would stretch into four figures over the years. I knew that euthanasia was something that I would have to perform a lot as a vet, but with the pigs it felt very different. This was somebody's fault; the care was inadequate, or the sheer volume of demand for the flesh of these animals meant that we have forced them into ever more industrialised situations that make them sick and chronically stressed.

There are certain memories that stick more than others when I think back, and they still feel fresh no matter the time elapsed. It's almost impossible to convey the scale of what I was seeing in Britain's pig farms, so I shall relate only a few of the moments that remain so prominently in my mind. Be assured that what I describe does not, and cannot, fully capture the inhumanity of our treatment of the animals we rear to eat – the scenes that I talk about are never unique, and this is captured perfectly in the alarmingly similar farm exposés that we see frequently in the news.

At one routine, at first glance unremarkable, visit to a Red Tractor assured finishing unit (the pigs were around five months old and close to slaughter), I discovered an outbreak of tail-biting. Pigs default to chewing when stressed, and often this manifests as chewing the tails of other pigs; it is a common occurrence despite the tail mutilations we perform on around three-quarters of pigs in an attempt to prevent it. At this farm I found approximately twenty pigs 'off their legs' – dragging their back ends because the infection from the tail bite had spread to the joints. I had to shoot many of these pigs, with the assistance of a vet student because the farmer was unavailable, despite my announced visit. When I went back the next day after registering my serious concerns, the farmer began by delightedly explaining that he had hand-planted the several thousand daffodils lining the drive himself. My heart dropped – he had found the time to make his driveway attractive but not to provide basic care to the sentient beings he was being paid to look after. I didn't want to jeopardise the relationship so that I could retain some level of oversight on what happened to these pigs, so I identified all the problems and indicated what must be addressed diplomatically. Unfortunately on this occasion, and indeed several others, doing my job correctly got me banned from the farm and I never worked for that client again. As far as I am aware, the farm is still in operation.

In a similar vein, on visiting another finishing unit (this time a 'high welfare' straw-based system), the farmer was visibly frustrated when I arrived, and it was clear he wished I hadn't turned up that day. I found one poor soul at the back of a pen, lying in her own excrement and unable to get up and walk. I soon realised she had her whole back end eaten out and infected from a severe tail bite which had gone well into the spine and some of the flesh around where the tail should have been. The stench of her rotting flesh was sickening. I told the farmer we must kill her immediately, there was no chance of recovery. I ran to fetch my bolt gun from the car, and returned to see him dragging her towards the passageway by her ears. I screeched for him to stop and he reluctantly let go and began shooing her forward. She couldn't use her back legs so she just became more distressed and the farmer more frustrated, so before I could put my gun down to lift her he started kicking her – directly into the rotting wound at the base of her spine. It must've been excruciating; the thought still makes me wince. I screeched at him again and ran to shoot her. Leaving that farm in the hands of this person at the end of the visit left me with this gripping feeling that I was failing; as a vet whose sole purpose is to protect the welfare of animals, I realised I couldn't actually do anything to protect them because I couldn't be on all farms at all times and there are no real consequences for abusing animals marked out as food. Undercover investigations have unveiled this sort of disregard for basic animal sentience time and time again, and I have seen it in so many people charged with the care of thousands of thinking, feeling animals. If I saw it at pre-arranged visits and it's caught on sporadic investigations, it must be fairly prevalent.

I never quite understood why so many farmers failed to even provide a humane and timely end to a suffering pig's life; it is a surprisingly common phenomenon. Repeatedly telling the same people that a secondary action (pithing or bleeding) is required after using a bolt gun, or that once a pig has been deemed irretrievable they must be euthanised at the first opportunity, was soul-destroying. One farmer was so careless with this that his stockperson told me he had found pigs still alive in the dead bin in amongst decomposing corpses two hours after attempted euthanasia. Some put it off if the dead bin was full, some were too squeamish, and some perhaps simply not concerned by the drawn-out slow death they were inevitably inflicting. I couldn't understand it; a total and worrying lack of empathy.

I find pigs easy to read; their eyes are almost human, their facial expressions and body language plain, and they have many different vocalisations to communicate with us if we take the time to listen. Their suffering always seemed so blindingly obvious to me. I have also had some beautiful moments on farms when I was left to carry out inspections alone; singing to small pigs as they followed me round a shed gazing up inquisitively, scratching and playing with big pigs while they chattered and jumped about like excitable dogs. These moments always ended in a dull ache as I realised I would never see the same pigs twice – they would be dead and replaced with a new batch before my next visit – and when I left they would have no distractions from the barren concrete and nauseating build-up of their own waste.

The first time seeing some new way for an animal to suffer was always a shock. I recall clearly the day I found a pig that was still alive and conscious with a ruptured umbilical hernia; I had seen the fatal aftermath plenty of times and always remarked upon what a horrible way to go it must be. Umbilical hernias are common in modern pigs due to a combination of genetic and management factors, and when they grow large enough that the protrusion grazes on the floor, they have the potential to rupture and the intestines spill out of the body. I spotted this pig at the back of a pen with perhaps forty penmates – the farm itself held 9,000. I wasn't sure at first what the problem was. I assumed lameness as this is so commonplace, so I rolled him over to see if there were any obvious limb abnormalities. It physically hurt to see what was uncovered as I rolled him. I felt the blood drain from my face, and the air escaped my lungs in a strange groan. I shouted at the farmer to get the gun as quickly as he could while I batted away the rest of the pigs, who were diving in relentlessly from all sides. I'll never forget it as long as I live; his wide eyes, the way he'd had to lie down onto his own shredded intestines to prevent the others from continuing to eat them. He was cold to the touch but very much aware of everything that was happening. Pigs should be euthanised at the first sign of abrasion on a hernia, but I recently reviewed some undercover footage (May 2021) that showed tens of pigs with ulcerated hernias becoming infected and on the verge of bursting open. Those farmers know how it ends, yet they do not provide the swift death readily available to them. It seems to be a complete failure to empathise, which frightens me – no wonder we have violence, division, oppression and war when seemingly normal people have the capacity to look the other way when confronted with overt horror that they could so easily prevent.

Things like these hernias are missed often until the pig has died, because the nature of our modern farming systems mean that one person can be charged with ensuring the well-being of thousands of animals. Systemic disease presenting as meningitis – convulsions, rolling eyes, fever – is common in pigs and awful to see. I went along to one of my largest farms for a routine check one day, and the stockperson had clocked off at lunchtime, having been through the pigs in the morning. Towards the end of my inspection I spotted a pig looking odd on his own. He was fitting and hot, with his eyes flickering erratically from side to side. He had urinated and defecated where he was and become covered through the convulsions. He was around 70 kg so I propped his front up on my legs towards the drinker. He slurped desperately as I directed a stream towards his mouth. He was still 'there' and fighting to survive, and I was left with a difficult decision to treat him and entrust the farmer to continue his care or to put him out of his misery. He couldn't drink or eat for himself so I held him to the drinker until he was finished, then I decided to administer a steroid to bring down the inflammation and fever, and an antibiotic to treat the infection. I left instructions for the farmer, but to this day I wish I'd just shot him. He had no life to enjoy and was in severe pain and distress for that period, but I knew my treatments could work relatively quickly, and I hoped that when I left he wouldn't just pass away alone in the dark in his own filth. I didn't sleep and could only relax a little once I had spoken with the stockperson in the morning; he had survived and was up and eating. What life did I give him though, and for such a short period before his life was taken by another person in a far more stressful manner than I could have done?

As a vet who had sworn to put the welfare of animals first and foremost, and above all commercial considerations, every day I had to think 'In this moment, what is best?' Weighing up the options and realising they are all terrible was hard, but invariably the best option for a sick or injured animal in the food industry is death, delivered as swiftly as possible. I made an effort to treat individual animals where I could (which is unusual as this is usually left to the farmer), but it was often such a dilemma once I was experienced enough to know how a disease or ailment might progress.

I felt that I provided a humane ending for so many animals over those four years; some weeks it was relentless, every day another score to kill, looking them in the eyes as I aimed my weapon. But it became frighteningly obvious to me that what is being sold as a humane ending (slaughter) to consumers was far from it. 'Humane' to me means ending suffering. Nobody would send their old decrepit dog to a slaughterhouse to be put down. The animals that go to slaughter are, in theory, relatively healthy and so young – at only a fraction of their natural lifespans.

As described earlier in this book, the majority of pigs in the UK are killed in gas chambers. For the remainder, 'electronarcosis' is the norm – an electrical current passed through the brain to cause loss of consciousness. This method is open to human error and equipment failure, which is starkly obvious when one stops to observe this process. At smaller abattoirs, it is possible to flit between watching the pigs be killed and appraising the organs hung on hooks. The last time I visited an abattoir, I took some time to watch, to be a presence for the pigs in their last moments. What I saw was awful. Two out of three were 'botched', with dire consequences. The electrical stunner is supposed to be applied to either side of the head to pass a current through the brain. Once the pig drops, shackles are attached around the back leg and the pig is slowly lifted off the ground. Following an electrical stun, pigs ordinarily drop into 'tonic-clonic' seizures and their legs paddle. Sometimes this causes the shackles to loosen and the pig to fall four or five feet face-first onto the slippery blood-soaked floor below. There they scrabble and panic, trying to get up, while some poor slaughterhouse worker has to try to accurately thrust a blade into their throat without being hurt in the process.

If this bleeding fails to render the pig unconscious quickly enough, they may be lowered into the scalding tank whilst still alive and breathing; 68 degrees Celsius and filthy. There is footage online showing just this – a pig swimming, screaming and bleeding in the tank trying to escape. The smell of hot blood, gut contents and fear is always brought back to my memory when I smell flesh being cooked or burned. Another reason I could no longer bring myself to eat animals; it all ends in this. An undignified, terrifying and painful death for these sentient creatures.

That same visit, I had plodded back over to the organ hooks to do my job and started chatting with one of the workers. We discovered that it was both of our birthdays that day, so had a laugh and a totally normal conversation as we covered our hands and aprons in blood, examining and slicing at hearts and lungs. Except that is not normal, is it? The workers there were all friendly, kind and helpful, but what they were having to do, and how they were having to compartmentalise what was actually going on in front of them, would surely have some long-lasting effects that they took home at the end of the day? It certainly did for me. I still have recurring nightmares where I am having to slaughter pigs on an endless loop.

I have described here a small selection of moments, all of which still sit very much at the forefront of my consciousness at all times, but it is reflective of the whole. The stories that tend to stick in people's minds are those that involve overt cruelty by human carers, and it is the same for me; it pains me to see how extraordinarily badly some can fail in their duty. Undercover investigations show perfectly what I saw in practice, and the industry is well aware of what is happening; the marketing used to detract from this reality is criminally misleading. And to stress, this abuse and heinous disregard for living beings is not unique to the pig industry. Welfare in Britain is not what it's made out to be, and a large part of that is what is deemed legal and appropriate – the standard conditions and practices that so clearly do not meet an animal's needs. It may or may not come as a surprise, but all of the problems I have described here were found on Red Tractor and RSPCA Assured farms, at visits arranged well in advance.

I've described individuals here, and what their individual experience meant for me, but that is not how they are seen by the industry. When we are fed this blanket line of 'farmers love their animals', I cringe at my memories. I find it jarring when they refer to those sentient beings, bred only to be killed as babies, as their 'property', or 'stock'. I have to say here it is of course not all farmers who behave so abhorrently, but there is no way for a consumer to tell from a packet on a shelf which slab of flesh has come from which farm and whether they have been abused or neglected. This was the driving reason that I decided I could no longer justify funding this cruelty by purchasing animal products; no label can guarantee welfare.

I'll admit that I found socialising quite hard towards the end of the four years, and I felt distinctly 'broken'. Being with family or friends eating dead animals made me extremely uncomfortable, and I couldn't really find joy in anything, knowing what is happening to billions of animals that nobody cares about. The longer it went on, the more frequently I wept on the drive home. Some visits were so harrowing that I would clench the wheel and just scream as I drove off. Bouts of depression left me staring into space, driving back on autopilot without phone calls or music. In moments of desperation, I considered swerving my car into the path of an oncoming lorry. In moments of clarity I recorded what I had seen. None of this is right, and all of it is for a completely unnecessary part of the human diet; that is what is so very hard to reconcile. I held great resentment that I was having to do this because of other people's thoughtless choices.

Eventually, I had to accept that my daily witnessing of unnecessary pain, suffering and death for the farmed animals in my care was destroying me.

In the end it was a choice between my own mental health and being there for the pigs when I could. Every day I was conflicted, taking life in order to end suffering, administering drugs to prolong a life not worth living, and being powerless to act in a way I wanted to because quite simply the farmer is the customer, the customer is king, and the assurance schemes are paid by the farmer to assure them. I only lasted the four years I did because I couldn't bear to leave the pigs. It felt horrendous walking away in the end, but I feel that what I'm doing now will make more of a difference for them in the long run and I have more power to effect change and bring about awareness than I ever did on farms.

As a consequence of those four years as a pig vet, I became vegan in the last two, which was challenging enough in an industry where vegans are lovingly referred to as 'terrorists'. I sat through presentations to farmers and vets on how to avoid being caught out by activists, I witnessed the spitting fury unleashed when an exposé dropped. I heard farmers discussing the £50,000 they had spent on security and CCTV systems to prevent undercover filming whilst simultaneously spending nothing on improving the conditions and problems that activists would seek to film.

Enough was enough, and when I finally left the industry I determined a new path as an activist. I remember the first 'vigil' I went to with the Animal Save Movement, an organisation which bears witness to the lives passing in slaughterhouses. It was held at the slaughterhouse that I had frequented most as a vet, and when the smell from the first lorry hit me it all came flooding back like some hideous flashing montage from a film. Since then I have endeavoured to share as much as I can about my experiences and use this knowledge to fight for better. As I write, I have just launched judicial review proceedings against the UK government, calling for an end to factory farming. I attend protests, talk to people in the streets, give speeches, assist in farm investigation reviews, and just share whatever knowledge I have where I think it can be useful. Tim and I have both contributed to the documentary *The End of Medicine*, which explores the nature of pandemics and antimicrobial resistance in relation to our use of animals. I know I will never falter now. The public need to know what an animal goes through to end up on their plate.

But we are slowly getting there. I believe the younger generations especially are more cognisant of the truths of animal agriculture. The industry puzzles over why more young vets don't want to go into pig work, and there are active drives to engage students and younger members of the profession. I am astonished that the industry does not understand why the students don't want to be party to any of it.

I cannot see how a nation of so-called animal lovers, a civilised society, an industry that claims 'world-leading welfare', can possibly still operate with this level of horror in 2021. Every farmed animal endures unnecessary and avoidable suffering. I am thriving as a vegan, and so are many elite athletes – out-performing their contemporaries in many cases. There is no justification for what we do to animals, on an industrial scale, by the trillions each year.

I do not regret any of my decisions, because I would not possess the knowledge needed to change things if I hadn't forced myself through it. I just wish the progress was quicker. I wish everybody could see what I've seen.

Confronting Cognitive Dissonance and Speciesism

Eloise Bailey, BA (Hons) MRes (Anthrozoology)

Have you ever wondered why you eat certain animals while welcoming others into your home and onto your sofa? Why you may be disgusted by the thought of consuming crocodile or cat, but won't bat an eyelid whilst devouring a bacon sandwich? It's a good question, but for most people is likely one that hasn't crossed their minds.

It crossed my mind when I was eight years old. My dad was driving my brother and I home from school, and we passed a small field containing lambs just off the motorway. I'd seen these lambs before, of course, but this time the animals were grazing underneath a large billboard promoting 'British Lamb'. I adored animals, as do most children, and was confused by what I was seeing, so I asked my dad if we ate lambs. He told me that yes, we do. Pointing to the field, I asked if we ate those particular lambs, and my dad told me about animal farming as candidly as he could while being sensitive to my age. At this point in our lives my dad ate meat with no moral qualms, and so I will always be thankful for his honesty in that moment – most children receive a different response if they question the system of animal farming; their concerns put to bed with white lies from well-intentioned parents. After a few tears and curious questions regarding other animals, I declared at eight years old that I didn't want to eat animals ever again. I never did.

For me personally it seemed a very simple decision to make, and so for a long time I struggled to comprehend why it wasn't as simple for everyone else. Children are naturally empathetic and are drawn to animals, so it's not a reach to think that far more people would be vegetarian had they had the opportunity to learn where meat comes from while they were young. Most of us aren't given a chance to question our eating habits at that age, and it becomes much harder for us to change once we are older and fixed in our habits. You may think that to eat meat is your own personal choice, but the truth is that what is and isn't acceptable to eat is a decision that has been made for us. Our regard for animals is largely a societal attitude that most of us abide by and believe in. This is shown by how different societies vary massively in what they do and don't eat - cats and dogs are regularly consumed as food in South Korea; spiders in Cambodia; guinea pigs in Peru and horses in Italy; to name a few examples that we in Britain would turn our noses up at or even be disgusted by - around the world, vegans and meat eaters alike feel so strongly about the consumption of cats and dogs, who they regard as beloved pets that they actively campaign to have the practice banned in other countries. There were no set categories for which animals were food, pets, experiments or pests, until somebody decided where they each belong for us. We become aware as we grow up that animals are connected to meat, of course, and understand that certain animals are killed for the food we eat, but by then we have already established that it's 'the way things are'. Most people don't like to question 'the way things are' when it comes to meat-eating because it results in a moral conflict known as cognitive dissonance.

Cognitive dissonance is experienced when we encounter an inconsistency between our behaviour (in this case, eating animal products) and our values (that we love animals, or at least are against animal cruelty). Cognitive dissonance can be resolved in one of two ways - by changing our values and convincing ourselves that our current behaviour is morally justified, or by changing our behaviour to fit our true values. The former is usually the preferred option for meat eaters, because it is easy to separate the pig from the pork - after all, all of the leg work has already been done for us. Thanks to the way that meat is marketed, we can simply bury our heads in the sand and not think of the animals involved at all. The use of language when marketing products is powerful; words such as 'sausage', 'bacon', 'ham', 'veal' and 'game' work to distance the meat from the animals that each product comes from. To label a packet of bacon as what it actually is - 'pig flesh' - wouldn't be in the interests of the animal industry, and they know this.

If we can't avoid the distancing of meat from animals, we can instead rationalise our behaviour in other ways - convincing ourselves that we need animal protein to be healthy; believing that animals don't feel pain in the same way that we do (particularly in the case of fish); or by deciding that certain animals are unworthy of our moral consideration because they are deemed less intelligent than us. But how do we define intelligence, and why would we have the right to treat those less intelligent than us in any way we wish? Humans believe that they are the epitome of intelligence and as such judge other species by comparing them to ourselves. The truth is that we are all animals and are all intelligent in species-specific ways. If for argument's sake we are to compare their intelligence against each other, we could use the comparative example of dogs and pigs, two species that have been proven to have a very similar level of intelligence. Both are efficient problem-solvers, can discriminate between others of their own species, harbour long-term memories, have their own individual personalities and, most importantly, are sentient beings.[1, 2, 3] Here in the UK we consider dogs to be members of our families and have enacted laws to condemn those who neglect or abuse them, while we pay people to gas pigs by the million each year just so that we can eat their bacon. The difference we see between these two animals is purely a societal perception – one that is not truly our own. We have given these animals different values based on our own human preferences, which can be described as 'speciesism'. Speciesism is the act and attitude of considering members of one species to be morally more important than members of another species,[4] be it human superiority over animals or the consideration of one animal's life as more deserving of moral consideration than another's, such as favouring a dog over a pig. We know that both animals experience pleasure, pain, fear and much more[2, 5] but have decided that one should be a pampered pet and the other not much more than a commodity. Speciesism encourages the discrimination of animals, facilitating the belief that our human desire to unnecessarily exploit animals is more important than an animal's life.

It takes a considerable effort to confront our own cognitive dissonance and our speciesism. It requires us to question our very identities and to confront the unfavourable idea that we may not always be the ethical, kind people that we like to believe ourselves to be. We can choose the easy option and fool ourselves into believing the advertisements of happy hens and laughing cows frolicking in spacious fields, or we can choose to see the difficult reality – the endless sheds packed full of chickens buckling over their own unnatural weight, the bins full of dead and rotting piglets who couldn't even stay alive long enough to be crammed onto the lorries headed for the slaughterhouse, and the cries of cows as their newborn calves are taken from them. We can choose to hear the distressing calls of animals as they fight not to die, and we can choose to watch their final moments as they are stunned, cut and gassed. The reality of animal farming is hidden from public view for a reason, to allow us to carry on ignoring our consciences and go against our natural empathy. What would your eight-year-old self have thought?

Citations:

1. Colvin, C. M. and Lori, M. (2015), 'Thinking Pigs: A Comparative Review of Cognition, Emotion, and Personality in Sus domesticus.' *International Journal of Comparative Psychology*, 28(1). https://escholarship.org/uc/item/8sx4s79c
2. Broom, D. M. (2015), 'Sentience and pain in relation to animal welfare.' *Proceedings of XVII International Congress on Animal Hygiene*, pp. 3-7. Košice, Slovakia: International Society for Animal Hygiene. https://www.researchgate.net/publication/289790582_Sentience_and_pain_in_relation_to_animal_welfare
3. Horowitz, A. (2014), *Domestic Dog Cognition and Behaviour: The Scientific Study of Canis familaris.* Springer, Heidelberg, Germany
4. Horta, O. (2010), 'What Is Speciesism?' *The Journal of Agricultural and Environmental Ethics* 23, pp. 243-266. https://www.academia.edu/531921/What_Is_Speciesism.
5. Balcombe, J. (2009), 'Animal Pleasure and Its Moral Significance.' *Applied Animal Behaviour Science* 118, pp. 208-16. https://www.wellbeingintlstudiesrepository.org/cgi/viewcontent.cgi?article=1004&context=acwp_asie

Plant-based Health

Dr. Shireen Kassam MBBS FRCPATH PhD DipIBLM

It took me thirteen years of clinical practice as a doctor in the UK to come across the overwhelming scientific literature connecting our food choices to our health. Once discovered, this knowledge cannot be forgotten or ignored. Although the science may initially appear complicated, the take-home message from decades of nutrition research comes down to a few fundamental truths. The only foods that have the ability to prevent the vast majority of chronic illnesses and maintain optimal physical and mental well-being are derived from plants. Fruits, vegetables, whole grains, beans, nuts and seeds in their natural forms are the only foods that are essential for human health. All other foods are optional, unnecessary and in some cases harmful. Meat, eggs and dairy act merely to crowd out the healthy foods and distract us from the essential foods and nutrients that most of us are severely lacking in. In conjunction with the overconsumption of processed foods, diets high in animal-derived foods are contributing to the rising burden of non-communicable diseases and driving the development of common illnesses such as heart disease, hypertension, type 2 diabetes, certain cancers and dementia, to name a few. Unhealthy diets are now contributing to one in five deaths globally and in the UK and these deaths are due to cardiovascular diseases, type 2 diabetes and cancer. This means that doctors like me are treating patients with diseases that need never have happened. Having discovered this truth, I had to take action.

Health professionals have also been called to action by their professional bodies to join the fight against the climate and ecological crises. These existential threats are directly impacting human health and unless we act now, the world as we know it will never be the same. One of the major contributors to both these crises is animal agriculture and the related loss of animal habitats. Animal agriculture results in significant greenhouse gas emissions, biodiversity loss, land-use change, ocean and water pollution and deforestation. Whilst animal agriculture takes up more than 80 per cent of farmland it only supplies 18 per cent of calories globally. At the same time, we are just not eating enough of the healthy plant foods. As I write, we are over a year into the Covid-19 pandemic, which is undoubtably a result of the way humans treat other animals and their habitats. When you consider that three out of four new and emerging infections in humans arise in animals, the only way to prevent the next pandemic is to stop using animals for food. Health professionals have been asked to lead by example and help their patients transition to a more sustainable way of living, which includes a predominately plant-based diet.

With this in mind, in 2017 I founded Plant-Based Health Professionals UK, a non-profit organisation whose mission is to provide evidence-based education on healthy whole-food plant-based nutrition for the prevention, management and, in some instances, reversal of chronic illness. Since our launch the community of health professionals using plant-based nutrition in their clinical practice has grown exponentially.

So many clinicians are tired of firefighting and are now using nutrition to address the root cause of our common diseases. This has led me to develop the first UK University-based course on plant-based nutrition at Winchester University and the response to this has been overwhelmingly positive. In 2021, Dr. Laura Freeman and I launched the first CQC-regulated online lifestyle medicine healthcare service, Plant Based Health Online. I feel the tide is changing from the grassroots level and the emphasis of healthcare is shifting towards the promotion of healthy, sustainable lifestyles.

Our food system is broken. It is harming human health, the planet and of course our fellow animals. We need to demand a fundamental change to farming, food production, nutrition education, food policy and pricing and country-based dietary guidelines. We need to be able to support communities to build a better future by making healthy foods accessible and affordable to all. A doctor's code of practice demands that we first do no harm. For me this now includes an ethical and moral duty to not harm humans, animals or the planet.

Giving up Livestock Farming – Sheep

Sivalingam Vasanthakumar BSc(Agri)MSc(Sust.Agri)

Farming has been a part of my life since I was a child. My family are originally from Tamil Nadu in Southern India and were brought to the tea plantations of Sri Lanka as indentured workers during the British Empire. In India's ancient and hugely controversial caste system, we were part of the Vaishya caste, made up of tradesmen and farmers. I grew up on a mixed vegetable and dairy farm on the outskirts of Colombo, the capital city of Sri Lanka. Our values in terms of animal welfare were very much wrapped up in religion, as cows are seen as sacred and therefore not slaughtered for meat. My mother always said that animals can't talk or express their feelings, so it is our duty to give them the best care possible. For us, farming was never an industry or business; it was our way of life. Therefore, the finances in terms of keeping animals was not discussed; if a cow fell ill, we would treat it without thinking about the cost; we saw them all as living beings and our responsibility to look after.

As a child, I always had a particular love for cattle. Our cows were kept with their calves until weaning; each morning we would tie up the calves and as the cows groomed them, we would milk them by hand, leaving enough for the calf to drink throughout the day. They were never dehorned or disbudded, and during the festival for cattle in January, we would decorate their horns with paint and tie garlands of flowers around their necks. Due to the heat and humidity, we washed our cows twice a day to keep them cool. This was my job as soon as I returned home from school, I would draw water from the well with our cowman and together each cow was tended to. This was the best part of my day, racing to try and keep up with the cowman.

In the 1970s Sri Lanka endured a critical food shortage due to mismanagement of farmland, limited incentives to growers and the government's form of 'welfarism' consisting of free food rations. In response to this, the party in power stopped cash crop imports and had a slogan of 'Produce or Perish' in an attempt to kick-start agricultural production again. In response, on our farm we began planting vegetables on a large scale, particularly intercropping chilli with groundnuts. I still remember our first bumper harvest of chilli, which sold for a good price in the market, buying us a knap-sack sprayer for chemical pesticides. Up until that point we had always used permaculture principles without knowing what that even was; it was just the way we were able to efficiently utilise all the space we had. However, we were increasingly influenced by western farming models and it was so exciting to consider being able to use what was advertised to us by chemical company reps. It is only looking back now that I realise how detrimental this was to our traditional ways of farming.

My parents dreamt that I would one day study medicine, but my heart was always in farming. I studied agriculture in India from 1978 to 1983, and my plan was to graduate and return home to continue farming with my mother. Unfortunately, at that point the civil war broke out in Sri Lanka and we were forced to sell our cows and my family moved to India. It was a very sad day to watch our cows go. I then came to the UK in 1985 to study a postgraduate diploma and then a master's in Sustainable Agriculture at Wye College, University of London. I was so excited at the idea of the big farms, tractors and combine harvesters that I saw in the UK.

After finishing my master's degree, I worked on the university dairy farm as a stockperson. Part of this job was calf-rearing; we separated calves from their mothers at forty-eight hours and moved them into single rearing cubicles, where they were trained to drink from buckets. This upset me a lot. The stress on the cow and calf was huge, but I felt that it was my job and therefore something I had to do and not question.

I also was responsible for taking weaned bull calves to be sold at market, something that also troubled me. These practices were just not common in Sri Lanka. However, in retrospect, I realise that even at home when we sold bulls at one year old, most of them would go for slaughter – in Sri Lanka there are no rules or regulations to monitor the conditions of slaughterhouses and they are often gruesome. Whilst Western commercial livestock systems bring a set of unique animal welfare problems, I have begun to realise over time that there is no system of livestock production that is entirely free of pain and suffering.

After many years of working on commercial livestock farms, and then alternative inner-city educational farming projects, I decided to rent 20 acres in Kent and rear goats for meat. Our first six nanny goats were crossbreeds and looked just like the goats we kept at home. It was a great feeling to start the business, and I had big plans to develop and keep expanding the herd. I was very keen that, as browsing animals, my goats should not have to live off grass like they often do in the UK. Every day I would drive around and cut fodder from trees and tie them into huge bundles for the goats. They loved it, and it made me feel like I was back home again. There was one particular nanny goat, Clover, who looked just like my childhood goats. I still remember going to check them one morning and finding her with triplet female kids, all standing and drinking, a very special memory.

However, amidst the enjoyment of farming once again and looking after my own animals, I began to increasingly struggle to take the male kids for meat. We used to leave the trailer in the field so that they would get used to going in and out for their feed, making it easier to load them on the day. In the morning they would all be sleeping in the trailer and I would just shut the trailer gate, hitch the trailer and drive to the slaughterhouse. Whilst driving I would think about how I had betrayed them, and when we arrived they would refuse to come down the ramp.

In the end I closed the goat farm and we emigrated to the US, but I returned to continue farming again. In 2014 I bought 20 acres of land in Devon; and soon after purchased the first twenty ewes, with plans to expand to seventy. At the same time I started a food business, cooking and selling masala dosa, a South Indian savoury pancake filled with potato curry and eaten with coconut or tomato chutney and dhal. The plan was that we would also make lamb curry to be sold at the markets to take away. However, when it came to taking the first group of ram lambs to slaughter I couldn't bring myself to do it, and sold them to another farmer, who would continue to fatten them and then sell them for slaughter himself.

Dissociating myself from their slaughter made me feel slightly more comfortable about what I was doing, although I knew that it defeated the point of having the sheep in the first place! The next year we had twenty-one ram lambs that were fat and ready to go for slaughter, I didn't feel that I could sell them on as I had done the year before, and wasn't sure how I would be able to take them to slaughter myself, so I kept delaying making any decisions about them. My youngest daughter was studying Veterinary Medicine in Edinburgh at the time and phoned me one evening to tell me that I couldn't keep putting off the inevitable. At this point I had to tell her that I wasn't able to continue livestock farming any more, and together we decided to give the lambs to an animal sanctuary. (After lots of research and recommendation we approached Goodheart Animal Sanctuary in Worcester and they said they would take the twenty-one ram lambs and the sixty-nine ewes. Goodheart Animal Sanctuary is 95 acres and managed by a top-class, dedicated team of staff. Great sanctuary) and stop breeding sheep. It was a sad moment to decide to give up what I loved; even now I miss the routine of checking the stock each morning and having a cup of tea when I am done. But at the same time, I am just so happy to know that I will never have to make that trip to a slaughterhouse again.

Since giving up livestock farming in 2018, I have allowed myself to question even more aspects of the livestock industry. The human health issues surrounding high meat and dairy consumption, animal welfare issues across the meat and dairy industries and the devastating environmental impacts of animal agriculture. I am a huge believer that we need more small farms in the UK, with less reliance on imports and better regional distribution. Climate change will force us to rethink our food system, and I hope that this includes a decrease in livestock farming in the UK, making things better for the animals, the environment and ourselves!

I still feel very connected to farming, and would like to grow vegetables for my food business. I am in the process of buying a smallholding, with the plan of moving onto the site in spring 2021. We will grow most of the vegetables needed for the food business in polytunnels and outdoor agroforestry areas. This way I can keep my connection to farming, use the skills that I learned as a child, and hopefully inspire my customers to think more about where their food comes from and how that food is produced. I would encourage any farmer who is questioning their feelings around animal agriculture to be brave; changing your identity is hard, but it is so liberating to no longer have to justify what you know is wrong.

Resources

Selected Bibliography

Anderson, Kip and Kuhn, Keegan. *Cowspiracy: The Sustainability Secret*. Earth Aware Editions, 2016.

Anderson, Kip and Kuhn, Keegan with Wong, Eunice. *What the Health: The Startling Truth Behind the Foods We Eat*. BenBella Books, 2018.

Balcombe, Jonathan. *What a Fish Knows: The Inner Lives of Our Underwater Cousins.* Oneworld Publications, 2016.

Barnard, Neal D. *The Cheese Trap: How Breaking a Surprising Addiction Will Help You Lose Weight, Gain Energy, and Get Healthy*. Hachette, 2017.

Buettner, Dan. *The Blue Zones*. National Geographic, 2005.

Bulsiewicz, Will. *Fibre Fueled: The Plant-Based Gut Health Program for Losing Weight, Restoring Your Health, and Optimizing Your Microbiome*. Avery, 2020.

Campbell, Colin T., and Campbell, Thomas M. II. *The China Study: The Most Comprehensive Study of Nutrition Ever Conducted*. BenBella Books, 2006.

Campbell, Colin T. *Whole: Rethinking the Science of Nutrition*. BenBella Books, 2014.

Campbell, Colin T. *The Low-Carb Fraud*. BenBellla Books, 2014.

Davis, Garth. *Proteinaholic: How Our Obsession with Meat Is Killing Us and What We Can Do About It*. HarperCollins, 2015.

Fairlie, Simon. *Meat: A Benign Extravagance*. Permanent Publications, 2010.

Greger, Michael, MD. *How Not To Die. Discover the Foods Scientifically Proven to Prevent and Reverse Disease.* Pan Macmillan, 2016.

Greger, Michael, MD. *How to Survive a Pandemic.* Pan Macmillan, 2020.

Harrison, Ruth. *Animal Machines*, 1964.

Harvey, Graham. *The Killing of the Countryside.* Jonathan Cape, 1997.

Juniper, Tony. *Rainforest: Dispatches from Earth's Most Vital Frontlines.* Profile Book, 2018.

Katz, David. *The Truth About Food: Why Pandas Eat Bamboo and People Get Bamboozled.* Dystel & Goderich, 2018.

Lapegna, Pablo. *Soybeans and Power: Genetically Modified Crops, Environmental Politics, and Social Movements in Argentina.* Oxford University Press, 2016.

Lymbery, Philip with Oakeshott, Isabel. *Farmageddon: The True Cost of Cheap Meat.* Bloomsbury Publishing, 2014.

Lymbery, Philip. *Dead Zone: Where the Wild Things Were.* Bloomsbury Publishing. 2017.

Merrington, L., Winder, L., Parkinson, R. and Redman, M. *Agricultural Pollution: Environmental Problems and Practical Solutions.* Spon Press, 2002.

Monbiot, George. *Feral: Rewilding the Land, Sea and Human Life.* Penguin Books, 2014.

Oppenlander, Richard A. *Food Choice and Sustainability: Why Buying Local, Eating Less Meat, and Taking Baby Steps Won't Work.* Langdon Street Press, 2003.

Oppenlander, Richard A. *Comfortably Unaware: What We Choose to Eat Is Killing Us and Our Planet.* Beaufort Books, 2012.

Phillips, Michael. *Mycorrhizal Planet: How Symbiotic Fungi Work with Roots to Support Plant Health and Build Soil Fertility*. Chelsea Green Publishing, 2017.

Pollan, Michael. *The Ominvore's Dilemma: A Natural History of Four Meals*. Penguin Books, 2006.

Raubenheimer, David and Simpson, Stephen J. *Eat Like the Animals: What Nature Teaches Us About Healthy Eating*. William Collins Books, 2020.

Tuttle, Will. *The World Peace Diet: Eating for Spiritual Health and Social Harmony*. Lantern Books, 2005.

Simon, David R. *Meatonomics: How the Rigged Economics of Meat and Dairy Make You Consume Too Much / How to Eat Better, Live Longer, and Spend Less*. San Conari Press, 2013.

Stika, Jon. *A Soil Owner's Manual: How to Restore and Maintain Soil Health*. Independently published via Amazon, 2016.

Tree, Isabella. *Wilding: The Return of Nature to a British Farm*. Pan Macmillan, 2018.

Winter, Michael. *Rural Politics: Policies for Agriculture, Forestry & The Environment*. Routledge, 1996.

Wrangham, Richard. *Catching Fire: How Cooking Made Us Human*. Profile Books, 2010.

Young, Rosamund. *The Secret Life of Cows*. Faber & Faber, 2017.

Films and Documentaries

COWSPIRACY: The Sustainability Secret. Directed and produced by Kip Anderson and Keegan Kuhn, 2014.

Dominion. Directed and produced by Chris Delforce, 2018.

Earthlings. Directed and produced by Shaun Monson, 2005.

Forks Over Knives. Directed by Lee Fulkerson and produced by John Corry and Brian Wendel, 2011.

Land of Hope and Glory: The hidden truth behind UK animal farming. Produced by Surge, 2017.

Meat the Truth. Directed by Karen Soeters and Gertjan Zwanikken, 2007.

The Game Changers. Directed by Louie Psihoyo and executive produced by James Cameron, Arnold Schwarzenegger, Jackie Chan, Lewis Hamilton, Novak Djokovic and Chris Paul, 2018.

The End of Medicine. Directed by Alex Lockwood, produced by Keegan Kuhn and executive produced by Rooney Mara and Joaquin Phoenix, 2021.

What the Health. Directed by Kip Anderson and Keegan Kuhn and produced by Joaquin Phoenix, Kip Anderson and Keegan Kuhn, 2017.

73 Cows. Directed and produced by Alex Lockwood. Lockwood Films, 2018.

Seaspiracy. Directed by Ali Tabriza. Assistant Director, Lucy Tabrizi. Executive produced by Dale Vince and Jim Greenbaum, produced by Kip Anderson, 2021.

Plant-Based Health Professionals UK

PBHP mission

To educate health professionals, members of the public and policy makers on behalf of whole-food plant-based nutrition in preventing and treating chronic disease.

I would urge everyone to make a regular appointment with PBHP, and check out their concise and factful factsheets:

plantbasedhealthprofessionals.com

My Top Websites/YouTube Channels

Ed Winter (aka Earthling Ed)

In each generation special individuals come into our lives, and in the vegan world today we have an especially inspirational person in the UK – Ed Winters (aka Earthling Ed). Ed is a vegan, educator, public speaker and content creator based in London, England; and Co-Founder and Co-Director of Surge, an animal rights organisation. If anywhere *Hope Lives in Kindness* it is in Ed.

Earthling Ed: earthlinged.org

Surge: www.surgeactivism.org

YouTube: www.youtube.com/channel/UCVRrGAcUc7cblUzOhI1KfFg

NutritionFacts.org

NutritionFacts.org is a strictly non-commercial, science-based service provided by American physician Dr. Michael Greger, offering a wealth of free updates on the latest in nutrition research in bite-size videos. Dr. Greger is also the author of several essential books, including: *How Not to Die*, *How Not to Diet* and *How to Survive a Pandemic*. An incredible resource for your whole-food plant-based nutritional science.

NutritionFacts.org: nutritionfacts.org

Mic the Vegan

Mic the Vegan is a vegan science writer who covers a variety of topics from the health effects of a vegan diet and the environmental impact of eating animal products to the sociological phenomenon of casual animal exploitation. If you want a dose of facts verses myths then there is no better mediator than Mic, in his unique and engaging style.

Mic the Vegan: micthevegan.com/about

YouTube:
www.youtube.com/channel/UCGJq0eQZoFSwgcqgxIE9MHw

Global Vegan Crowd Funder

Global Vegan Crowd Funder (GVCF) is a not-for-profit platform that enables those who care to work together to change how things are done. Sharing funds; sharing ideas; sharing knowledge. A crowdfunding platform run by vegans, for vegan projects and initiatives. If you want to be a part of a vegan community and make a difference, check out GVCF.

GVCF: globalvegancrowdfunder.org

Other Good Sites

David Katz, davidkatzmd.com
Goji Man (Simon Hammett), www.gojimannutrition.com.
Happy Healthy Vegan (Ryan Lum/Anji Bee),
www.happyhealthyvegan.org.
Hench Herbivore (Paul Kerton), henchherbivore.com
Joey Carbstrong, www.joeycarbstrong.com
Michael Klaper, www.doctorklaper.com
Plant-Based Health Professionals, plantbasedhealthprofessionals.com
Plant Based News, plantbasednews.org
Physicians Committee for Responsible Medicine, www.pcrm.org
The Real Truth About Health,
www.youtube.com/channel/UCp_ShZAUGtFLpYkgcTrayRQ
The Vegan Society, www.vegansociety.com
Vegconomist, vegconomist.com
Viva, the UK's leading vegan campaigning charity, viva.org.uk

My Favourite UK Plant-based Eateries

Today there are numerous plant-based or 'vegan-friendly' eateries across the UK, serving a mixture of mock meat, dairy and egg alternatives and whole plant-based foods. One of the best ways to find one near you is to use the HappyCow app (www.happycow.net/mobile), available on Google Play or the Apple App Store.

Whether you are a philosophical vegan, or simply want to help meet your five or more fruit and veg a day and be healthier and kinder in tasty ways, there is a plant-based eatery for you.

Close to me are two wonderful places to eat:

Mooli Foodworks, Wellington, Somerset, who offer specialist vegan deli, ice creams and dining. You would be hard-pushed to find a better plant-based eatery. wellington.moolifoodworks.co.uk

Plantside, Wellington, Somerset, serving an array of tasty, nutritious, wholesome foods. plantsidecafe.co.uk

Further away:

The Eating Gorilla, Penrhyndeudraeth, North Wales, offering not-to-be-missed breakfasts, hot and cold lunches and treats. www.eatinggorilla.co.uk

Unity Diner, Spitalfields, London, where you can enjoy a divine tofish and chips and much more besides. www.unitydiner.co.uk/

All these wonderful places serve up more than just heathy, and *not quite* so healthy comfort foods, they have to be amongst the friendliest places you can eat.

Whatever your dietary persuasion, give them and the many other plant-based eateries a try – places where hope lives in kindness and you simply owe it to your taste buds to explore.

About the Author

Tim Bailey, BSc (Hons) MCIWM CEnv

Tim Bailey is one of the UK's foremost and prominent regulatory farm pollution experts, with a distinguished thirty-year career, from field to national advisor. He studied at the University of Plymouth and has a BSc (Hons) degree in Environmental Science, specialising in geology and hydrogeology, and is a Chartered Waste Manager (MCIWM) and Chartered Environmentalist (CEnv). Tim is also a researcher, conservationist, horticulturalist and a specialist natural history author with several published titles sold worldwide. Tim lives in Somerset, England, and is an advocate for plant-based agriculture, rewilding, animal kindness and for a whole-food, plant-based diet.

About Dr. Alice Brough

Alice graduated in 2015 from the University of Edinburgh and began her clinical veterinary work in 'mixed' practice (all species) whilst training in pig consultancy. She spent four years as a pig vet, working with hundreds of commercial pig farms across England, involving routine inspections, emergency visits, stockperson training and health monitoring in slaughterhouses. Witnessing the tremendous suffering inflicted on animals raised for food lit a fire in her, and she left the industry in 2019. Alice grew up in rural Derbyshire, with a lifetime affinity with animals and nature and today works as an advocate for animal and human rights, and environmental justice. She features in the soon-to-be-released documentary *The End of Medicine*, which explores pandemics and antimicrobial resistance with respect to our interactions with animals.

Index